U0215096

计算机技术开发与应用丛书

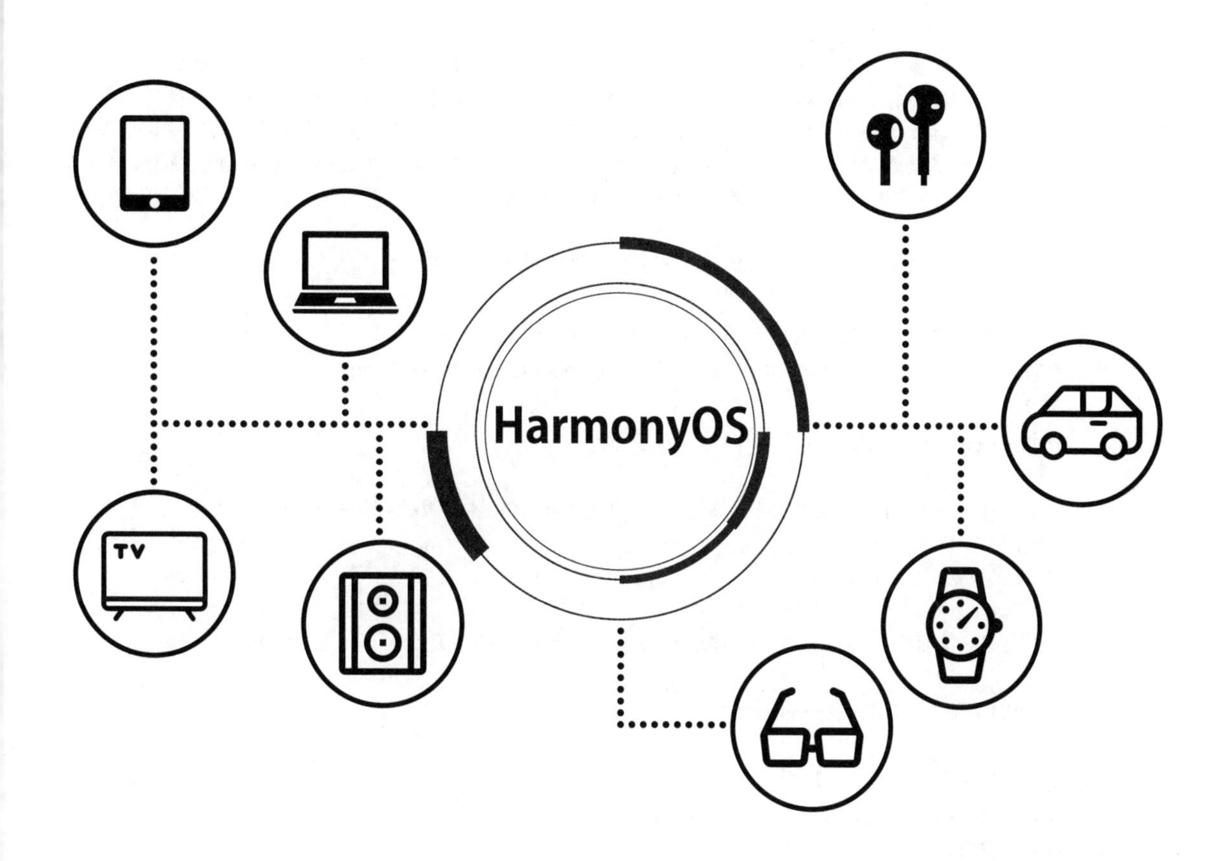

鸿蒙操作系统应用开发实践

陈美汝 郑森文 武延军 吴敬征 ◎ 著
Chen Meiru Zheng Senwen Wu Yanjun Wu Jingzheng

清华大学出版社
北京

内 容 简 介

本书系统全面地讲解了鸿蒙操作系统下应用开发所需基础知识，并通过丰富的案例实践提高应用能力。

本书共11章，第1章对鸿蒙操作系统进行概述，并搭建应用开发环境。第2～5章分别讲解Java、JavaScript语言基础及Java、JavaScript下的UI布局实现。第6章讲解鸿蒙轻量级智能穿戴设备的开发。第7章讲解应用所具备的能力(Ability)。第8章讲解本地应用数据管理及分布式服务。第9章讲解图像、相机及音视频的多媒体功能。第10、11章为应用实战，包含两个完整实践案例。本书通俗易懂，循序渐进，包含丰富的代码讲解和应用实例，操作性较强，并配套PPT和视频讲解，是鸿蒙应用开发入门者的不二之选。

本书主要面向鸿蒙应用的入门开发者，也可作为培训机构的参考用书。

图书在版编目(CIP)数据

鸿蒙操作系统应用开发实践/陈美汝等著.—北京：清华大学出版社，2021.6（2022.8重印）
（计算机技术开发与应用丛书）
ISBN 978-7-302-58201-4

Ⅰ.①鸿…　Ⅱ.①陈…　Ⅲ.①移动终端－应用程序－程序设计　Ⅳ.①TN929.53

中国版本图书馆CIP数据核字(2021)第099084号

责任编辑：赵佳霓
封面设计：吴　刚
责任校对：徐俊伟
责任印制：杨　艳

出版发行：清华大学出版社
　　网　　址：http://www.tup.com.cn，http://www.wqbook.com
　　地　　址：北京清华大学学研大厦A座　　**邮　　编：**100084
　　社 总 机：010-83470000　　**邮　　购：**010-62786544
　　投稿与读者服务：010-62776969，c-service@tup.tsinghua.edu.cn
　　质量反馈：010-62772015，zhiliang@tup.tsinghua.edu.cn
　　课件下载：http://www.tup.com.cn，010-83470236
印 装 者：三河市金元印装有限公司
经　　销：全国新华书店
开　　本：186mm×240mm　　**印　　张：**25.25　　**字　　数：**628千字
版　　次：2021年7月第1版　　**印　　次：**2022年8月第4次印刷
印　　数：3501～4500
定　　价：99.00元

产品编号：092233-01

前　言

PREFACE

HarmonyOS 自 2018 年对外流出相关的设计概念以来，就引起了广泛关注，它被认为是国产新一代操作系统的希望，是一款“面向未来”、面向全场景的分布式操作系统。2019 年 8 月，华为在开发者大会上正式发布 HarmonyOS，它提出了基于同一套系统能力、适配多种终端形态的分布式理念，能够支持多种终端设备。2020 年 9 月，华为在开发者大会上发布了 HarmonyOS 2.0，推出应用开发者 Beta 版本，并在同年 12 月推出了手机开发者 Beta 版。自此，开发者可以在鸿蒙的开发环境上开发和调试多个不同终端的应用，整体开发环境和 SDK 支持也初步成熟。

作为最早一批鸿蒙应用开发者，我们在 2020 年 6 月就投身于鸿蒙的相关工作并且参与了包括 2020 年 9 月 10 日发布会上最早展示的鸿蒙 Demo 应用的相关开发工作。之后我们有意识地对鸿蒙的应用开发知识进行梳理，希望能在鸿蒙系统到来之际为广大感兴趣的开发者提供一套较为系统且全面的鸿蒙开发讲解图书。除了本书的内容之外，针对鸿蒙的应用生态，我们还将安卓平台上二十余款非常受欢迎的组件移植到了鸿蒙平台供广大开发者使用，相关的源码已经开源到 Gitee 上。进一步地，我们也在多个社区平台上基于这些自己开源的源码进行讲解分析，让读者更好地去使用和学习。

在本书编写时，HarmonyOS 的北向应用开发能力刚刚问世，系统还未完全成熟，处于快速更新迭代的状态，因此本书内容的广度和深度有限，仅涉及 HarmonyOS 应用开发中的一些基础核心功能。且在编写过程中，由于 HarmonyOS 的多次更新迭代，本书的代码也经历了多次测试和更改，因此读者在学习过程中也难免会碰到大大小小的问题，还望读者见谅，也欢迎随时联系我们反馈问题。

读者对象

本书非常适合初学者入门，不仅涵盖了鸿蒙大部分的能力特性，还在此基础上对鸿蒙应用开发所用到的 Java 及 JavaScript 语言进行了简要的讲解，帮助对这两类语言还不熟悉的开发者能在学习鸿蒙开发之前对鸿蒙所使用的开发语言有更清晰的认知。在内容讲解上，针对代码部分也采用循序渐进的方式进行讲解，保证读者能够根据提供的代码一步步掌握书里的知识点，并且提供了直观的运行效果参考。

本书组织结构

本书针对 HarmonyOS SDK 4(Java 2.1.0.5，JavaScript 2.1.0.5)版本，对鸿蒙操作系统的应用开发基础进行了梳理和介绍。同时，也构建了一个在分布式场景非常常用的视频流直播实例作为实战的内容演练，进行了详细分析和讲解。其各章的主要内容如下：

第1章对HarmonyOS进行了综合介绍，总体涵盖了鸿蒙操作系统的系统特性、系统架构及开发环境，指导构建鸿蒙上的Hello World项目，并由此对鸿蒙项目结构、文件及日志管理工具进行讲解。

第2章在介绍HarmonyOS的Java开发内容之前，从Java语言基础、Java的类和对象及继承等多个在Java开发中必备的知识点进行简要介绍，保障读者能更好地切入和理解后续的学习内容。

第3章对鸿蒙的Java UI的常用组件、容器及动画进行了覆盖性讲解，让开发者基本了解和学习到Java UI的开发模式。

第4章在介绍HarmonyOS的JavaScript开发内容之前，从JavaScript简介、开发环境、核心语法及在HarmonyOS中针对Java和JavaScript的比较对JavaScript开发中必备的知识点进行了简要介绍，保障读者能更好地切入和理解后续的学习内容。

第5章对鸿蒙的JavaScript UI的开发框架、布局、组件、交互及动画进行了覆盖性讲解，让开发者基本了解和学习到Java UI的开发模式。

第6章从用户界面的构建及HarmonyOS所提供的基本功能、系统能力上简要讲解了鸿蒙轻量级智能穿戴设备的开发。

第7章讲解HarmonyOS中非常重要的Ability概念，详细地从Page Ability、线程及Service Ability 3个点入手介绍HarmonyOS应用所具备能力。

第8章从本地应用数据管理、分布式服务及Data Ability 3个点切入，详细介绍了HarmonyOS的数据管理能力及实现方法。

第9章从图像、音视频、相机3个多媒体常用的能力上切入，对HarmonyOS的多媒体能力和实现方法进行了详细介绍。

第10章详细讲解了HarmonyOS的组件及其使用方法，并且以实际开源的组件项目为例进行实践介绍。

第11章以分布式应用中非常重要的直播场景为目标，综合性地指导和讲解如何构建应用让两部HarmonyOS手机实现视频流直播的能力。

版本信息

HarmonyOS本身也在不断地迭代演化之中，其SDK和IDE随着版本的更新，API及应用开发特性也在不断地更新丰富。本书选取撰写时发布的HarmonyOS SDK 4(Java 2.1.0.5,JavaScript 2.1.0.5)版本进行代码梳理和讲解，IDE版本为DevEco Studio 2.0 Beta3，但是依然可能会出现本书代码与实际代码不同的情况，在这种情况下读者可以跟踪最新代码并获取最新信息。

致谢

在本书的撰写过程中，有非常多的人为我们提供了帮助，在此对诸位表达真挚的感谢。首先，感谢华为各位同事，为我们提供了一些技术上的支持和帮助。也感谢中国科学院软件研究所智能软件研究中心的罗天悦、杨牧天老师为本书提供的大力支持。特别感谢组内的小伙伴吴圣垚和马卞，大力参与本书撰写及配套资源的筹备，还有陈丛笑、戴研、刘雨琦、朱伟、熊铁翔、蒋筱斌、吕泽、邵妍洁等同学，协助完成书中内容及代码的测试验证。最后感谢清华大学出版社的赵佳霓编辑，在写作和出版过程中为我们提供的帮助。再次感谢大家！

编　者

2021年3月

本书源代码下载

教学课件(PPT)

目 录

CONTENTS

第1章

走进 Harmony

1.1 了解 HarmonyOS

当前的移动互联网创新，仍然仅局限于以手机为主的单一设备，单设备的操作体验已经不能完全满足人们在不同场景下的需求，而 HarmonyOS 正是为万物互联而生。HarmonyOS（鸿蒙操作系统）是一款“面向未来”的操作系统，一款面向全场景的分布式操作系统，它创造性地提出了基于同一套系统能力、适配多种终端形态的分布式理念，将多个物理上相互分离的设备融合成一个“超级虚拟终端”，通过按需调用和融合不同软硬件的能力，实现不同终端设备之间的极速连接、硬件互助和资源共享，为用户在移动办公、社交通信、媒体娱乐、运动健康、智能家居等多种全场景下，匹配最合适的设备，提供最佳的智慧体验。

HarmonyOS 主打“1＋8＋N”的全场景体验，其中“1”指的是主入口手机，“8”指的是智慧屏、平板、PC、音响、手表、眼镜、车机和耳机 8 种设备，“N”则指的是泛 IoT 硬件构成的华为 HiLink 生态，其中包括移动办公、智能家居、健康生活、影音娱乐、智能出行等各大场景下的智能硬件设备。

对于 HarmonyOS 的应用开发者而言，通过 HarmonyOS 所提供的多种分布式技术，使应用开发者能够聚焦于上层业务逻辑，而忽略不同终端设备的形态差异，从而极大地降低了开发难度和成本，提升了开发效率。本书将聚焦于 HarmonyOS 的应用开发，旨在带领更多开发者加入 HarmonyOS 全场景的生态建设中。下面具体介绍一下应用开发所需要的基本知识。

1.1.1 HarmonyOS 技术特性

多种设备之间通过 HarmonyOS 可以实现硬件互助和资源共享，依赖的关键技术主要包括分布式软总线、分布式数据管理、分布式任务调度和分布式设备虚拟化等。

1. 分布式软总线

分布式软总线是手机、手表、平板、智慧屏、车机等多种终端设备的统一基座，是分布式数据管理和分布式任务调度的基础，为设备之间的无缝互联提供了统一的分布式通信能力，能够快速发现并连接设备,高效地传输任务和数据。分布式软总线示意图如图 1.1 所示。

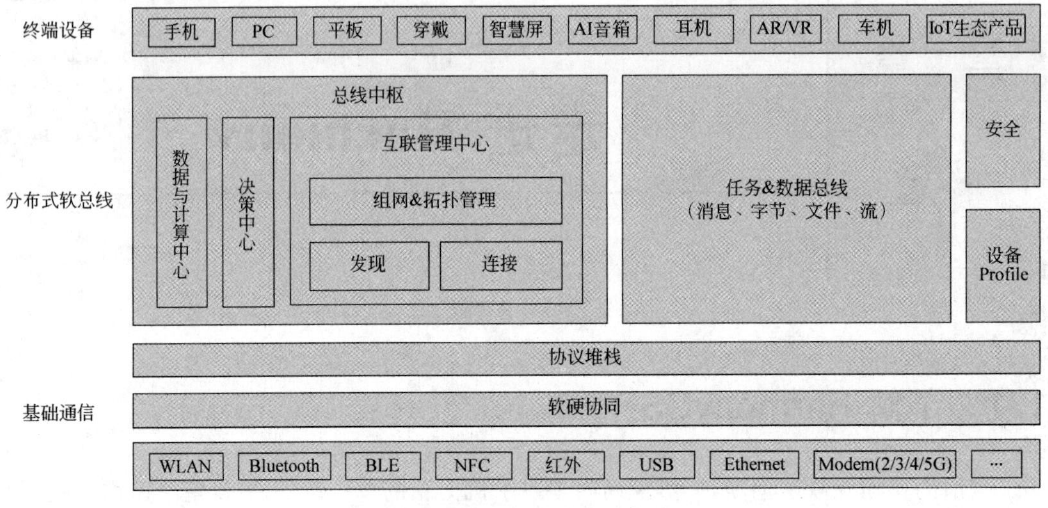

图 1.1　分布式软总线示意图

2. 分布式数据管理

分布式数据管理位于分布式软总线之上,用户数据不再与单一物理设备进行绑定,而是将多设备的应用程序数据和用户数据进行同步管理,应用跨设备运行时数据无缝衔接,让跨设备数据处理如同本地处理一样便捷。分布式数据管理示意图如图 1.2 所示。

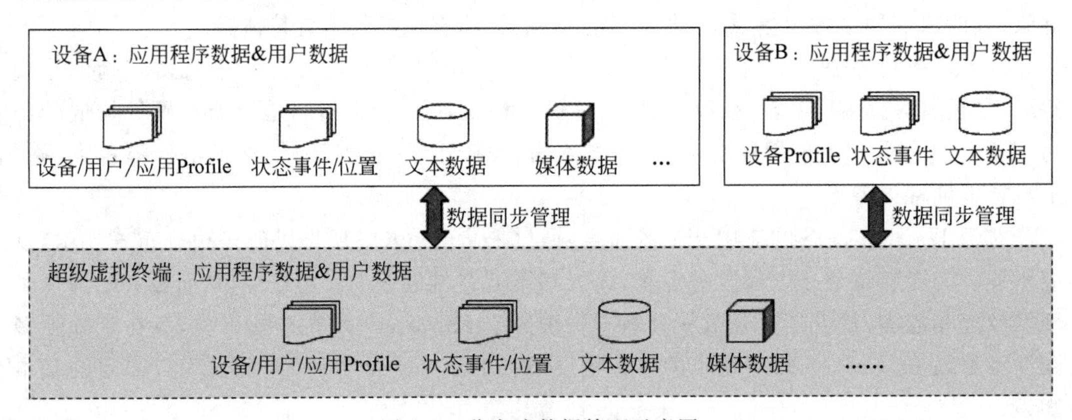

图 1.2　分布式数据管理示意图

例如基于分布式数据管理,可以通过手机访问其他设备中的照片和视频,并将其他设备中的视频转移到智慧屏进行播放,也可以将编辑在任一设备中的备忘录信息进行跨设备更新同步。

3. 分布式任务调度

分布式任务调度基于分布式软总线、分布式数据管理等技术特性,构建统一的分布式服务管理,支持对跨设备的应用进行远程启动、远程控制、绑定/解绑、迁移等操作。在具体的场景下，能够根据不同设备的能力、位置、业务运行状态、资源使用情况，并结合用户的习惯和意图，选择最合适的设备运行分布式任务。分布式任务调度示意图如图 1.3 所示。

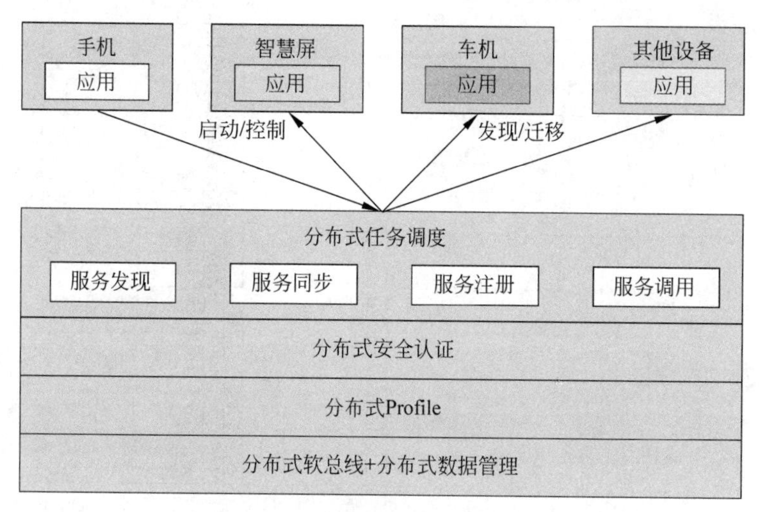

图 1.3 分布式任务调度示意图

利用分布式任务调度机制，可以实现多设备间的能力互助。例如，手机设备拍照具有美颜功能，但在家庭多人合影等场景下，手机屏幕较小，此时可以用手机控制智慧屏的摄像头，同时还能调用手机的相机美颜功能，并将最终照片传回手机。

除此之外，还可以通过分布式任务调度，实现业务的无缝迁移。例如在上车前，可以通过手机查找并规划好导航路线，待上车后，导航会自动迁移到车载大屏和车机音箱，待下车后，导航又会自动迁移回手机。

4. 分布式设备虚拟化

分布式设备虚拟化可以实现不同设备的资源融合、设备管理、数据处理，将周边设备作为手机能力的延伸，共同形成一个超级虚拟终端。针对不同类型的任务，为用户匹配并选择能力最佳的执行硬件，让业务连续地在不同设备间流转，充分发挥不同设备的资源优势。分布式设备虚拟化示意图如图 1.4 所示。

5. 一次开发，多端部署

HarmonyOS 通过提供统一的 IDE，进行多设备的应用开发，并且通过向用户提供程序框架、Ability 框架及 UI 框架，保证开发的应用在多终端运行时的一致性。通过模块化耦合，对应不同设备间的弹性部署。一次开发，多端部署的示意图如图 1.5 所示。

6. 统一 OS，弹性部署

HarmonyOS 拥有“硬件互助，资源共享”和“一次开发，多端部署”的系统能力，为各种硬件开发提供了全栈的软件解决方案，并保持了上层接口和分布式能力的统一。通过组件化和小型化等设计方法，做到硬件资源的可大可小，以及在多种终端设备间按需弹性部署。

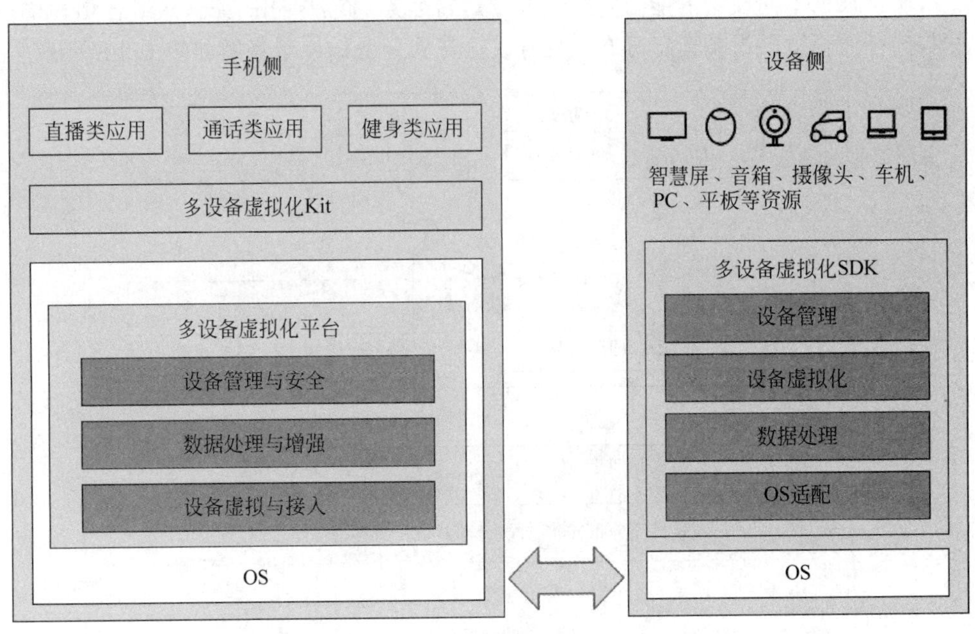

图 1.4　分布式设备虚拟化示意图

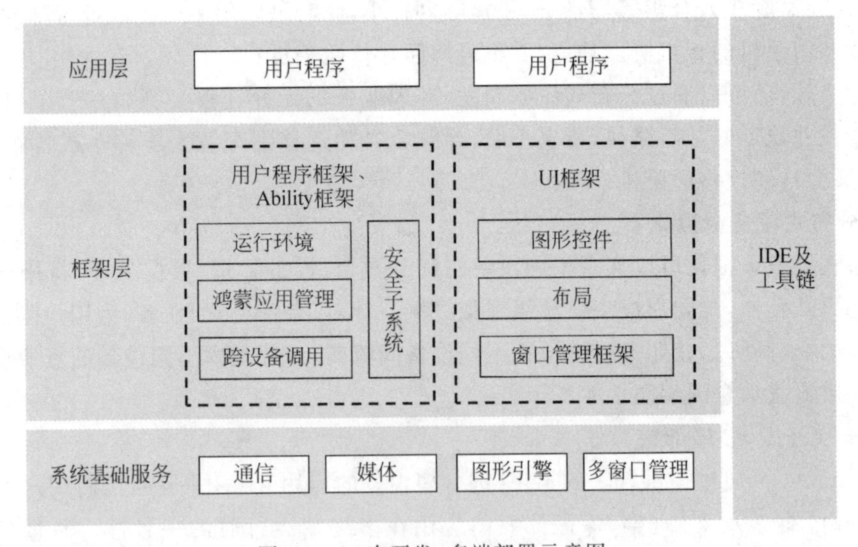

图 1.5　一次开发，多端部署示意图

1.1.2　HarmonyOS 系统架构

HarmonyOS 整体遵从分层设计，从下向上依次为内核层、系统基础服务层、框架层和应用层。HarmonyOS 系统架构如图 1.6 所示。

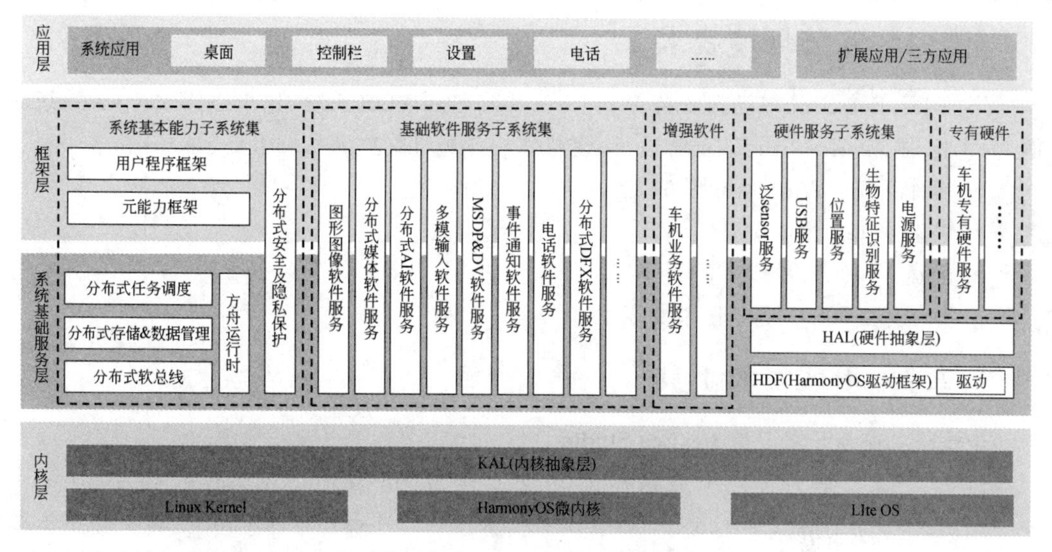

图 1.6　HarmonyOS 系统架构

1. 内核层

HarmonyOS 采用多内核设计(Linux 内核、HarmonyOS 微内核或者 Lite OS)，支持针对不同资源受限设备选用适合的 OS 内核。内核抽象层(KAL)通过屏蔽多内核差异，对上层提供基础的内核能力，包括进程/线程管理、内存管理、文件系统、网络管理和外设管等。

2. 系统基础服务层

系统基础服务层是 HarmonyOS 的核心能力集合，通过框架层对应用程序提供服务，包含系统基本能力子系统集、基础软件服务子系统集、增强软件服务子系统集、HarmonyOS 驱动框架(HDF)和硬件抽象适配层(HAL)、硬件服务子系统集和专有硬件服务子系统集。

3. 框架层

框架层为 HarmonyOS 的应用程序提供了 Java/C/C++/JavaScript 等多语言的用户程序框架和 Ability 框架，以及各种软硬件服务对外开放的多语言框架 API。

4. 应用层

应用层包括系统应用和第三方应用。HarmonyOS 的应用由一个或多个 FA(Feature Ability)或 PA(Particle Ability)组成。其中，FA 有 UI 页面，而 PA 无 UI 页面。FA/PA 均能够实现特定的业务功能，支持跨设备调度与分发，为消费者提供一致、高效的应用体验。

1.2　搭建 HarmonyOS 开发环境

随着华为 HarmonyOS 的快速发展，为适配基于 HarmonyOS 应用的快速和高效开发，华为提供了 DevEco Studio 开发工具。本书将基于 DevEco Studio IDE，手把手教你搭建 Harmony 开发环境。

1.2.1 安装环境要求

当前 DevEco Studio 已经支持 Windows 系统和 macOS 系统，这里以 Windows 系统为例，为了保证 DevEco Studio 正常运行，建议你的计算机配置满足以下要求：

- 操作系统：Windows 10 64 位。
- 内存：8GB 及以上。
- 硬盘：100GB 及以上。
- 分辨率：1280×800 像素及以上。

1.2.2 下载安装工具

步骤一：安装开发工具 DevEco Studio

首先，需要去华为官网下载最新的开发工具 DevEco Studio，下载网址为 https://developer.harmonyos.com/cn/develop/deveco-studio。（DevEco Studio 的编译构建依赖 JDK，DevEco Studio 预置了 Open JDK，版本为 1.8，安装过程中会自动安装 JDK。）

下载的安装包解压完成后，双击 deveco-studio-xxxx.exe 文件，进入 DevEco Studio 安装向导，默认系统选择的路径，或选择自己的安装路径，单击 Next 按钮，直至安装完成。注意在如下安装选项页面勾选 DevEco Studio launcher，如图 1.7 所示。

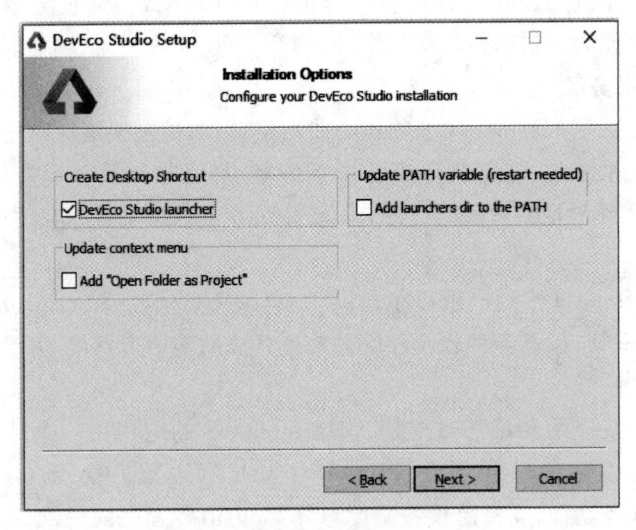

图 1.7 安装 DevEco Studio

步骤二：安装 Node.js

如果需要使用 JavaScript(以下简称 JS)语言开发 HarmonyOS 应用，则还需要下载和安装 Node.js。如果使用其他语言开发，则不用安装 Node.js，可跳过步骤二。

如果已安装 Node.js，可打开命令行工具，输入 node -v 命令，检查版本号信息，建议使用 v12.0.0 及以上版本。

接下来安装 Node.js。登录 Node.js 官方网站（网址 https://node.js.org/en/download/），下载 Node.js 软件包。选择 LTS 版本，64 位 Windows 系统对应的软件包，如图 1.8 所示。

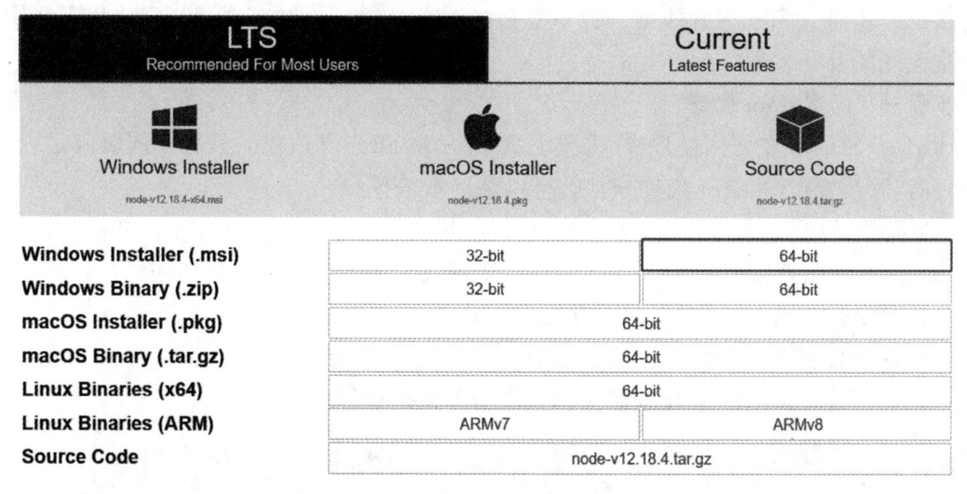

图 1.8 Node.js 安装版本

单击下载后的软件包 node-vxxx-x64.msi 进行安装，根据默认设置单击 Next 按钮，直至单击 Finish 按钮完成安装，如图 1.9 所示。

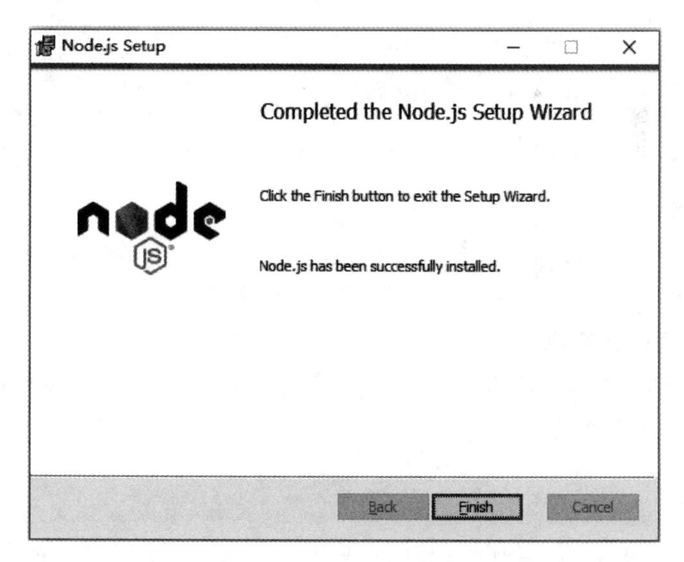

图 1.9 Node.js 安装完成

Node.js 会自动在系统的 path 环境变量中配置 node.exe 的目录路径。如果安装 Node.js 时没有选择默认的安装目录，则需要在系统变量 path 中手工添加环境变量信息（笔者的计算机→属性→高级系统设置→高级→环境变量），增加 Node.js 的安装路径。

1.2.3 搭建开发环境

DevEco Studio 的开发环境，需要保证连接 Internet 网络。若可以直接访问 Internet，则直接执行步骤二和步骤五即可。若不能直接访问 Internet，则需要逐步完成以下步骤，包括配置代理服务器。

步骤一：配置 npm 代理

DevEco Studio 在使用过程中，需要下载 JS SDK 第三方依赖。只有在同时满足以下两个条件时，才需要配置 npm 代理，否则，可以跳过本步骤。

(1) 需要使用 JS 语言开发 HarmonyOS 应用。

(2) 网络不能直接访问 Internet，而是需要通过代理服务器才可以访问。

打开命令行工具，然后根据如下方式进行 npm 代理配置。

(1) 如果使用的代理服务器需要认证，按照如下方式进行设置(将其中的 user、password、proxyserver 和 port 按照实际代理服务器进行修改)。

```
npm config set proxy http://user:password@proxyserver:port
npm config set https-proxy http://user:password@proxyserver:port
```

(2) 如果使用的代理服务器不需要认证(不需要账号和密码)，则应按照如下方式进行设置。

```
npm config set proxy http:proxyserver:port
npm config set https-proxy http:proxyserver:port
```

代理设置完成后，可在命令行工具中执行如下命令进行验证。若执行结果如图 1.10 所示，则说明代理设置成功。

```
npm info express
```

步骤二：设置 npm 仓库

下载 JS SDK 时，为提升使用 npm 安装 JS 依赖的速度，建议在命令行工具中执行如下命令，重新设置 npm 仓库地址。

```
npm config set registry https://mirrors.huaweicloud.com/repository/npm/
```

步骤三：配置 Gradle 代理

若网络不能直接访问 Internet，而是需要通过代理服务器才可以访问，在这种情况下，需要设置 Gradle 代理，以便访问和下载 Gradle 所需的依赖。否则，可以跳过本步骤。

首先，打开“此计算机”，在文件夹网址栏中输入%userprofile%，进入个人数据界面，如图 1.11 所示。

```
C:\Users\      >npm info express

express@4.17.1 | MIT | deps: 30 | versions: 264
Fast, unopinionated, minimalist web framework
http://expressjs.com/

keywords: express, framework, sinatra, web, rest, restful, router, app, api

dist
.tarball: https://registry.npmjs.org/express/-/express-4.17.1.tgz
.shasum: 4491fc38605cf51f8629d39c2b5d026f98a4c134
.integrity: sha512-mHJ9O79RqluphRrcw2X/GTh3k9tVv8YcoyY4Kkh4WDMUYKRZUq0h1o0w2rrrxBqM7VoeUVqgb27xlEMXTnYt4g==
.unpackedSize: 208.1 kB

dependencies:
accepts: ~1.3.7              cookie: 0.4.0          finalhandler: ~1.1.2         path-to-regexp: 0.1.7
array-flatten: 1.1.1         debug: 2.6.9           fresh: 0.5.2                 proxy-addr: ~2.0.5
body-parser: 1.19.0          depd: ~1.1.2           merge-descriptors: 1.0.1     qs: 6.7.0
content-disposition: 0.5.3   encodeurl: ~1.0.2      methods: ~1.1.2              range-parser: ~1.2.1
content-type: ~1.0.4         escape-html: ~1.0.3    on-finished: ~2.3.0          safe-buffer: 5.1.2
cookie-signature: 1.0.6      etag: ~1.8.1           parseurl: ~1.3.3             send: 0.17.1
(...and 6 more.)

maintainers:
- dougwilson <doug@somethingdoug.com>
- jasnell <jasnell@gmail.com>
- mikeal <mikeal.rogers@gmail.com>

dist-tags:
latest: 4.17.1       next: 5.0.0-alpha.8

published a year ago by dougwilson <doug@somethingdoug.com>
```

图 1.10 npm 代理配置成功

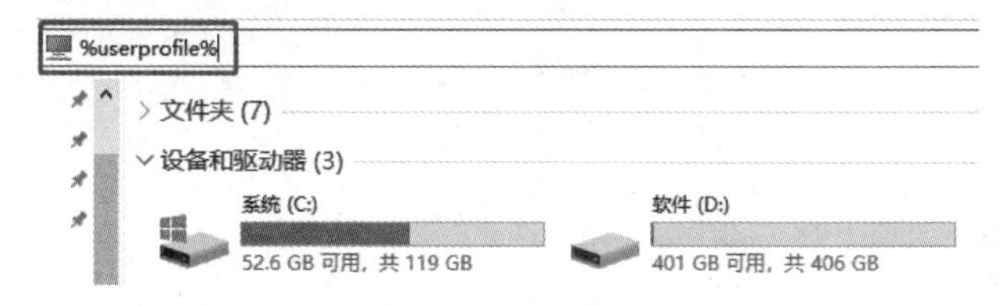

图 1.11 进入个人数据界面

创建一个文件夹，并命名为.gradle。如果已存在.gradle文件夹，则跳过此操作。

进入.gradle文件夹，新建一个文本文档，命名为gradle，并修改后缀为.properties。打开gradle.properties文件，添加如下脚本，然后保存，如图1.12所示。其中代理服务器、端口、用户名、密码和不使用代理的域名，需要根据实际代理情况进行修改。其中不使用代理的nonProxyHosts的配置间隔符是“|”。脚本代码如下：

```
systemProp.http.proxyHost = proxy.server.com
systemProp.http.proxyPort = 8080
systemProp.http.nonProxyHosts = *.company.com|10.*|100.*
systemProp.http.proxyUser = userId
systemProp.http.proxyPassword = password
systemProp.https.proxyHost = proxy.server.com
systemProp.https.proxyPort = 8080
systemProp.https.nonProxyHosts = *.company.com|10.*|100.*
systemProp.https.proxyUser = userId
systemProp.https.proxyPassword = password
```

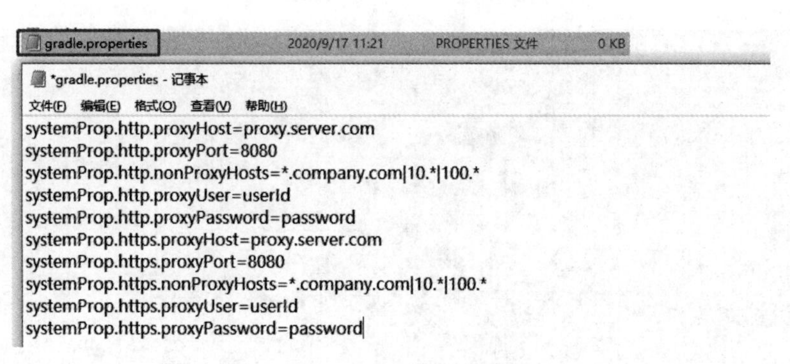

图 1.12 配置 Gradle 代理

步骤四：配置 DevEco Studio 代理

若网络不能直接访问 Internet，而需要通过代理服务器才可以访问，在这种情况下，需要设置 DevEco Studio 代理，以便访问和下载外部资源，如图 1.13 所示。否则，可以跳过本步骤。

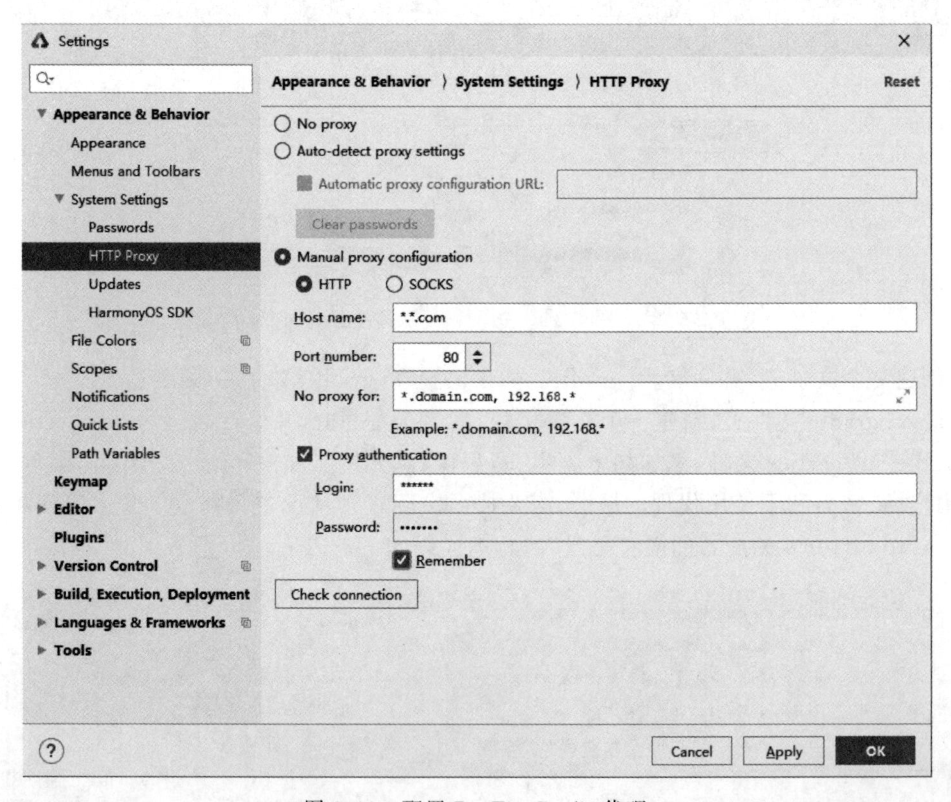

图 1.13 配置 DevEco Studio 代理

首先，运行已安装的 DevEco Studio，首次打开，选择 Do not import settings，单击 OK 按钮。

进入 DevEco Studio 欢迎界面后，单击 Configure→Settings→Appearance&Behavior→System Settings→HTTP Proxy，进入 HTTP Proxy 的设置界面，设置代理信息。其中包括 HTTP 配置项和 Proxy authentication 配置项。

（1）HTTP 配置项：设置代理服务器信息。

- Host name：代理服务器主机名或 IP 地址。
- Port number：代理服务器对应的端口号。
- No proxy for：不需要通过代理服务器访问的 URL 或者 IP 地址（地址之间用英文逗号分隔）。

（2）Proxy authentication 配置项：如果代理服务器需要通过认证鉴权才能访问，则需要设置。否则，可以跳过该配置项。

- Login：访问代理服务器的用户名。
- Password：访问代理服务器的密码。
- Remember：勾选，记住密码。

配置完成后，单击 Check connection 按钮，输入网络地址（如 https://developer.harmonyos.com），检查网络连通性。提示 Connection successful 表示代理设置成功。单击 OK 按钮完成 DevEco Studio 的代理配置。

DevEco Studio 代理配置完成后，会提示安装 HarmonyOS SDK，若同意安装至默认路径，则可以单击 Next 按钮进行下载安装。若想更改 SDK 的存储目录，则单击 Cancel 按钮，并根据下方步骤五进行 HarmonyOS SDK 下载操作，如图 1.14 所示。

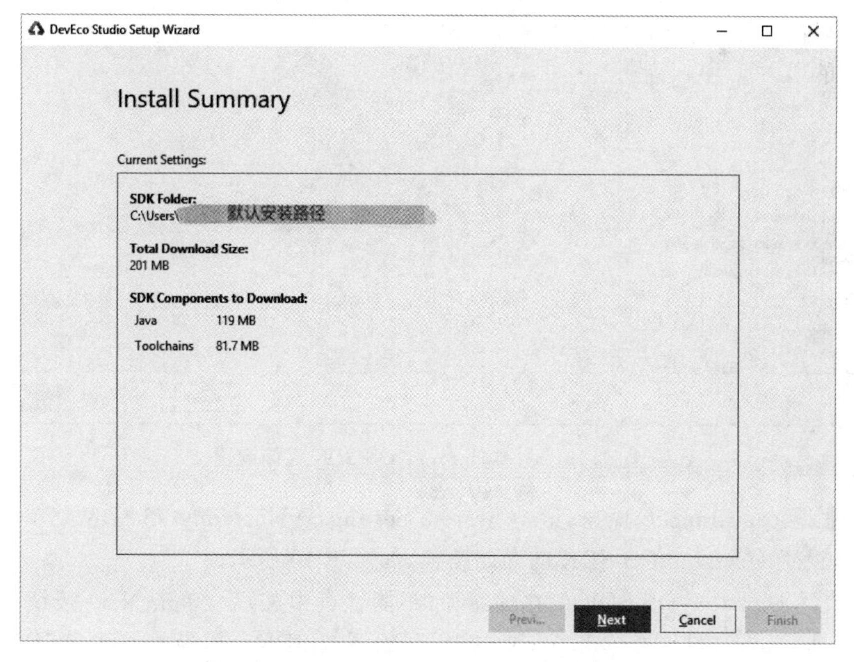

图 1.14 HarmonyOS SDK 默认安装

步骤五：下载 HarmonyOS SDK

DevEco Studio 提供 SDK Manager 统一管理 SDK 及工具链，当下载各种编程语言的 SDK 包时，SDK Manager 会自动下载该 SDK 包所依赖的工具链，因此只需下载所需编程语言对应的 SDK 包。

SDK Manager 提供多种编程语言的 SDK 包，包括 Native（C/C++语言 SDK 包）、JS（JS 语言 SDK 包）和 Java（Java 语言 SDK 包）。其中，Java SDK 在首次下载 Harmony SDK 时会默认下载，Native SDK 和 JS SDK 默认不自动下载，需要进行手动勾选下载，因此，如果需要使用 JS 或 C/C++语言开发应用，则需手动下载对应的 SDK 包。

下载 Harmony SDK 的步骤如下。

(1) 在菜单栏选择 Configure→Settings 或者按快捷键 Ctrl＋Alt＋S，打开 Settings 配置界面，如图 1.15 所示。

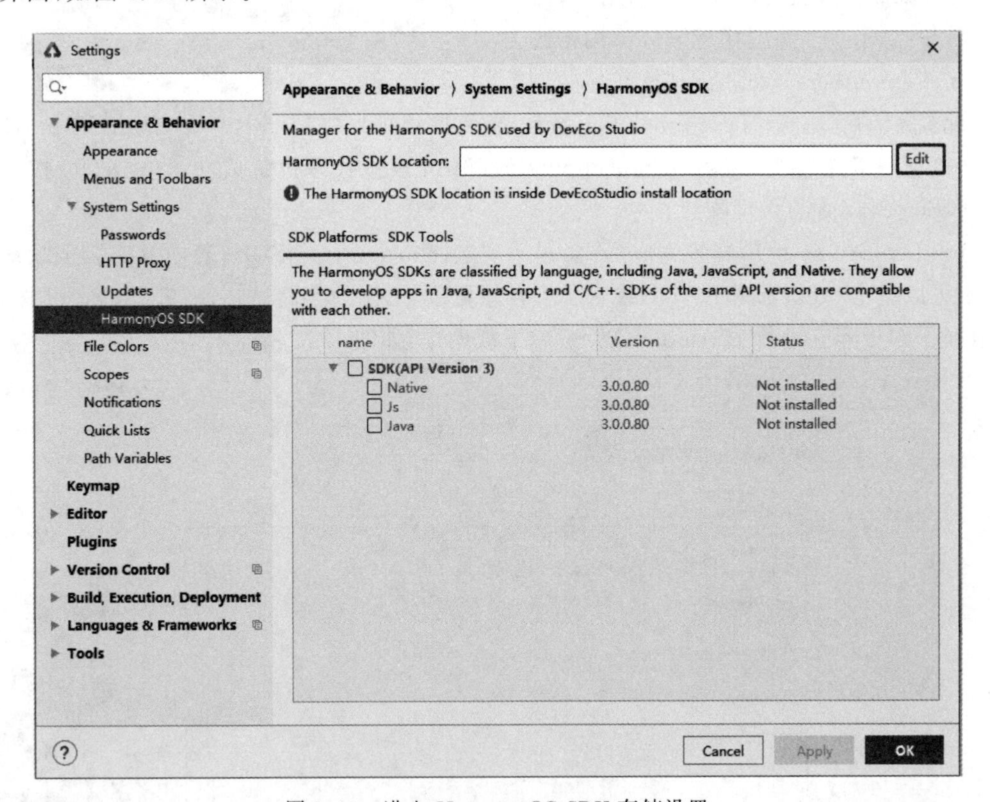

图 1.15 进入 HarmonyOS SDK 存储设置

(2) 进入 Appearance&Behavior→System Settings→HarmonyOS SDK 菜单界面，单击 Edit 按钮，设置 HarmonyOS SDK 的存储路径，如图 1.16 所示。

(3) 选择 HarmonyOS SDK 的存储路径（不能含有中文）后，单击 Next 按钮，在弹出的 License Agreement 窗口中，单击 Accept 按钮开始下载 SDK。如果本地已有 SDK 包，则选择本地已有 SDK 包的存储路径，DevEco Studio 会更新 SDK 及工具链，如图 1.17 所示。

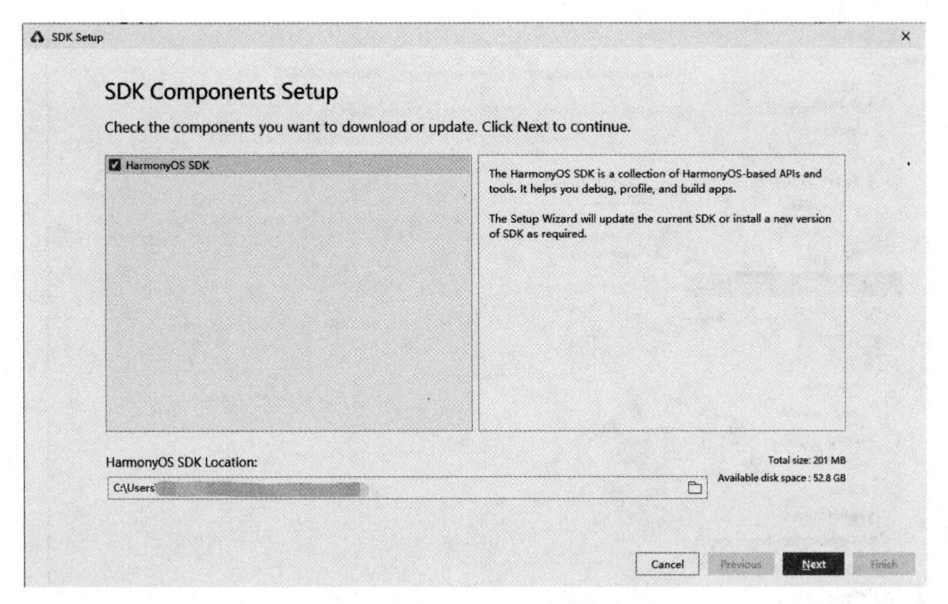

图 1.16　设置 HarmonyOS SDK 存储路径

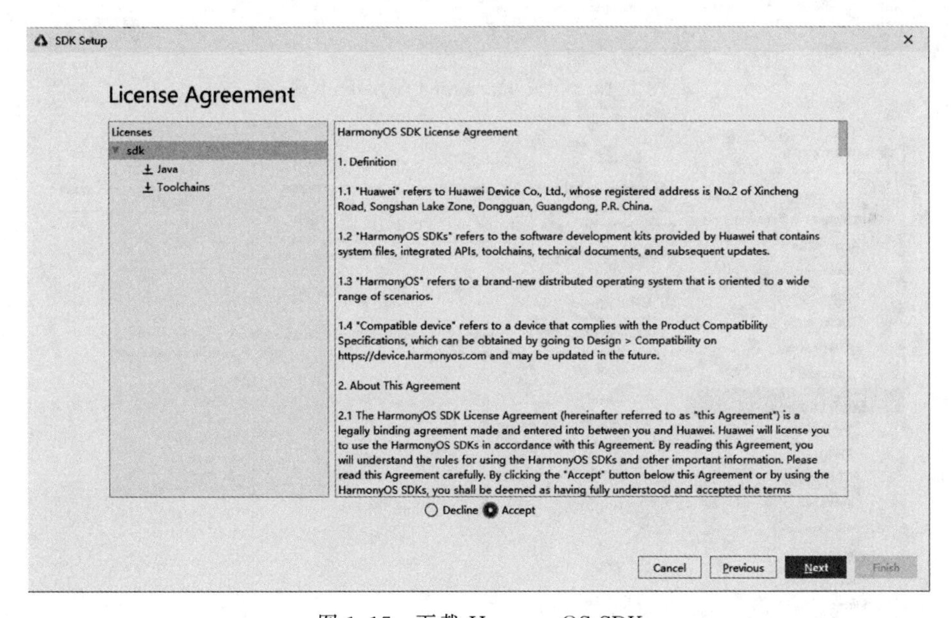

图 1.17　下载 HarmonyOS SDK

（4）待 HarmonyOS SDK 及工具下载完成后，单击 Finish 按钮，可以看到默认的 SDK Platforms→Java SDK 及 SDK Tools→Toolchains 已完成下载，如图 1.18 所示。

（5）如果需要使用 C/C++或者 JS 语言，则应在 SDK Platform 中勾选对应的 SDK 包，单击 Apply 按钮，SDK Manager 会自动将 SDK 包和工具链下载到所设置的 SDK 存储路径中，如图 1.19 所示。

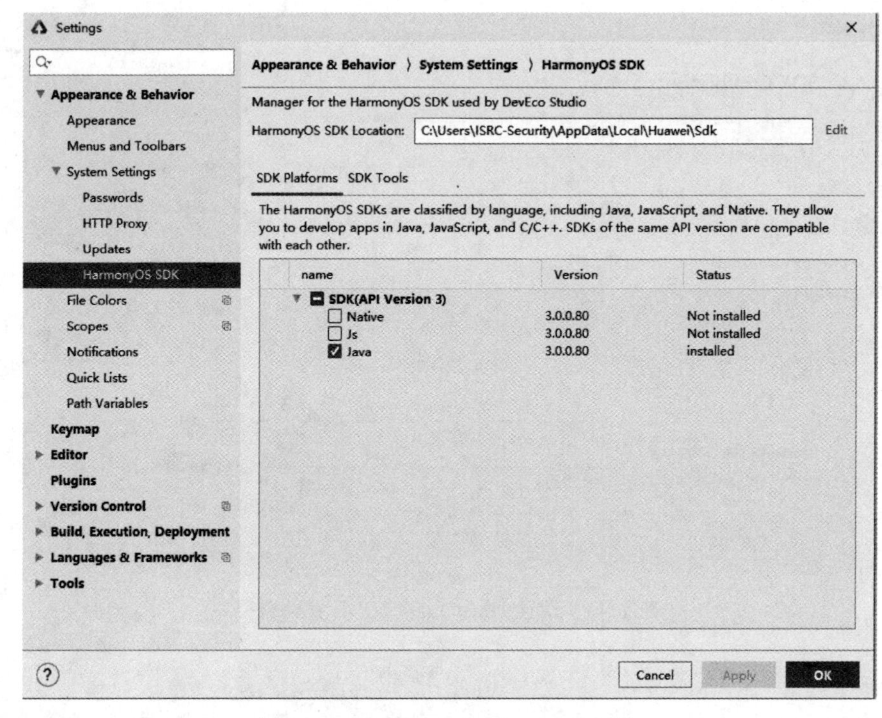

图 1.18　完成 HarmonyOS SDK 下载

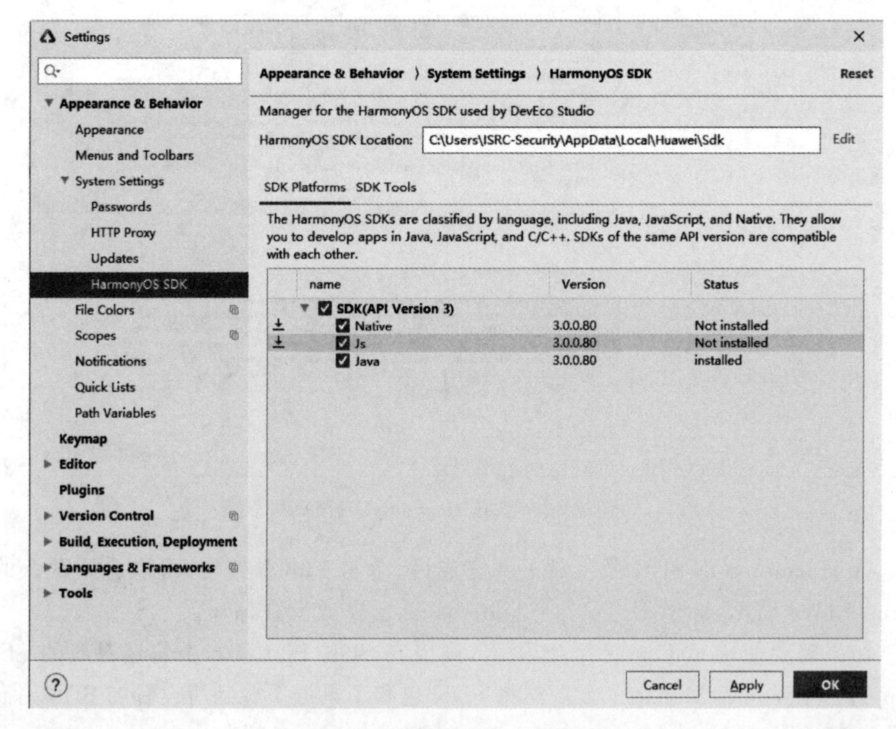

图 1.19　下载 C/C++/JS SDK

至此，开发环境已配置完成，随后我们会带领大家熟悉 DevEco Studio IDE 开发环境，并创建一个 HarmonyOS 项目，通过运行 Hello World 工程来验证环境配置。

1.3　关于 DevEco Studio

DevEco Studio 是由华为官方推出的，基于 IntelliJ IDEA Community 开源版本打造，面向华为终端全场景多设备的一站式集成开发环境（IDE），为开发者提供工程模板创建、开发、编译、调试、发布等 E2E 的 HarmonyOS 应用开发服务。通过使用 DevEco Studio，开发者可以更高效地开发具备 HarmonyOS 分布式能力的应用，进而提升创新效率。

作为一款开发工具，除了具有基本的代码开发、编译构建及调测等功能外，DevEco Studio 还具有以下特点。

(1) 多设备统一开发环境：支持多种 HarmonyOS 设备的应用开发，包括手机（Phone）、平板（Tablet）、车机（Car）、智慧屏（TV）和智能穿戴（Wearable）等设备。

(2) 支持多语言的代码开发和调试：包括 Java、XML（Extensible Markup Language）、C/C++、JavaScript、CSS（Cascading Style Sheets）和 HML（HarmonyOS Markup Language）。

(3) 支持 FA（Feature Ability）和 PA（Particle Ability）快速开发：通过工程向导快速创建 FA/PA 工程模板，一键式打包成 HAP（HarmonyOS Ability Package）。

(4) 支持 JS 应用和 Java 应用的跨设备预览器功能，在应用开发阶段，可以使用跨设备预览器查看应用在不同设备上的运行效果。

(5) 支持跨设备分布式应用调试，基于 HarmonyOS IDL 实现的跨设备的分布式场景，DevEco 提供了跨设备的 HarmonyOS 分布式应用的调试功能，方便开发者调试分布式应用。

1.3.1　DevEco Studio 界面及配置

DevEco Studio 启动后主界面结构如图 1.20 所示，主要包括位于最上方的菜单栏、导航栏（显示 IDE 打开的文件，在项目文件夹中的具体位置）、工具栏，中间的项目文件及编辑器，以及最下方的工具窗口。

这里，大家可以根据自己的喜好及习惯，配置 IDE 的主题、快捷键及代码编辑器。选择菜单栏中的 File→Settings，或者按快捷键 Ctrl+Alt+S，可以进行 Appearance&Behavior、KeyMap、Editor、Plugins 等配置。

1. Appearance&Behavior 配置

Appearance 可以配置 IDE 的背景色调，如喜欢亮色调的可以选择 HUAWEI Light Theme，喜欢暗色调的可以选择 HUAWEI Dark Theme，还可以配置工具界面的显示字体，以及字体的大小，如图 1.21 所示。

Menus and Toolbars 可以设置主菜单和快捷工具。

System Settings 可以根据网络情况配置 IDE 的 HTTP Proxy 和 HarmonyOS SDK 等，并检查 Update 版本、密码等信息。

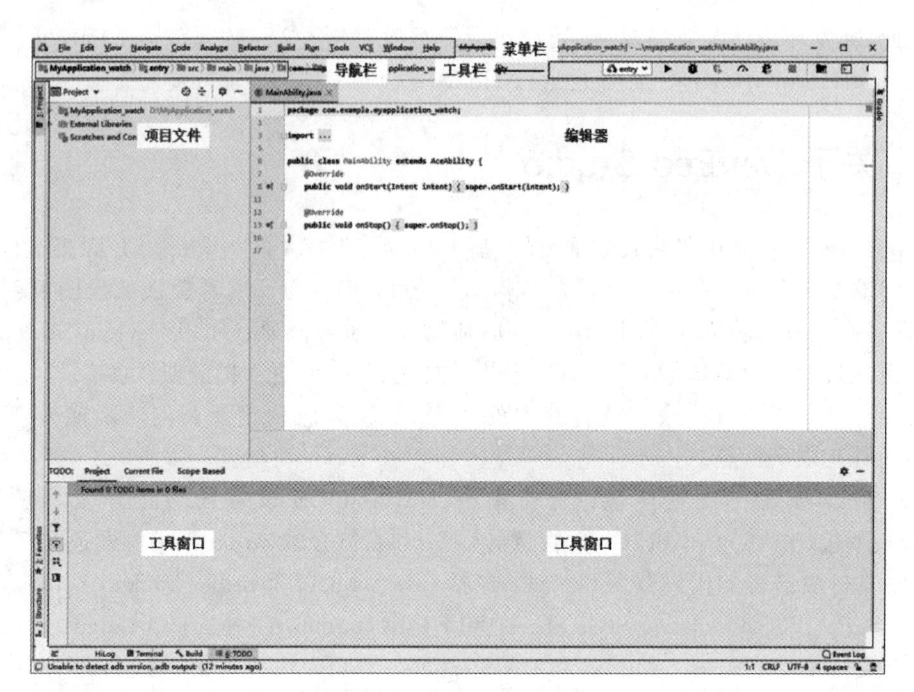

图 1.20　DevEco Studio 界面

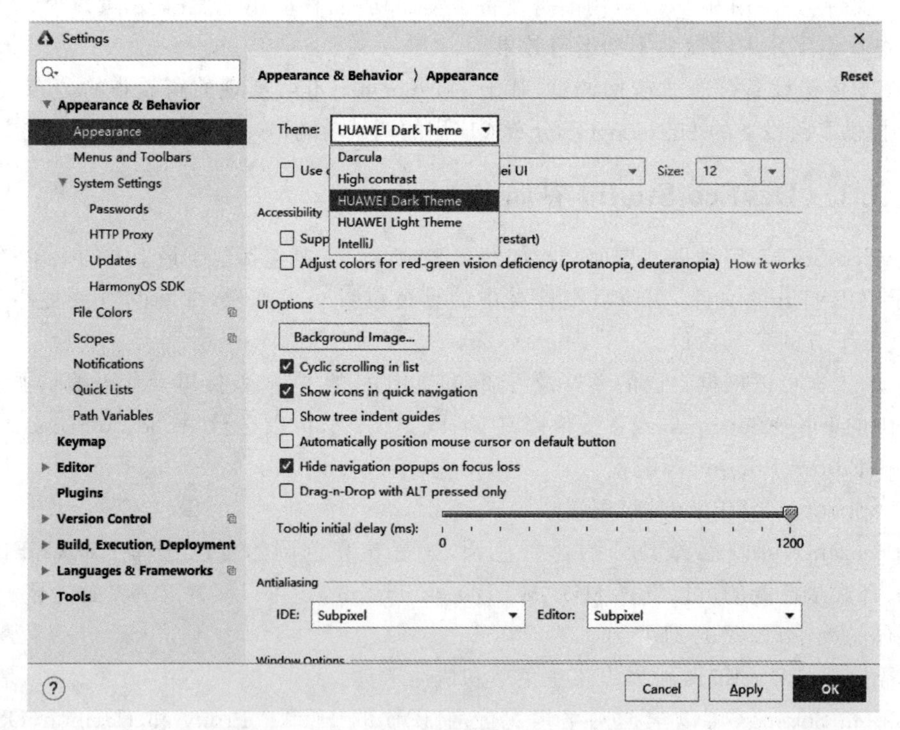

图 1.21　Appearance 配置 IDE 背景色调

2. Keymap 配置

Keymap 配置，可以查看或编辑 IDE 的快捷键，如图 1.22 所示。

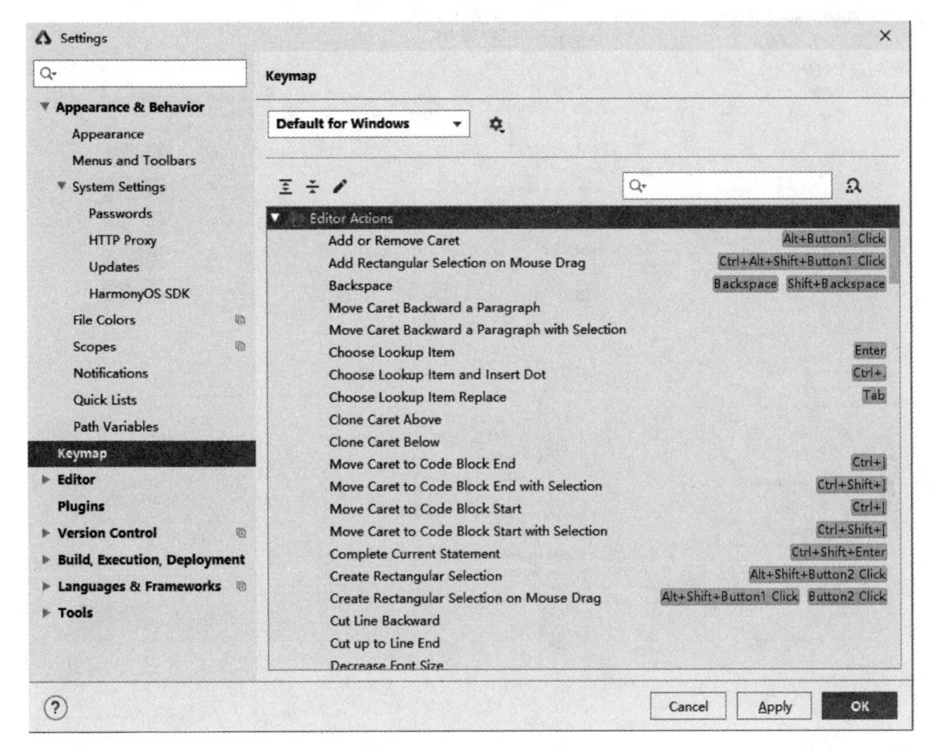

图 1.22　Keymap 配置 IDE 快捷键

3. Editor 配置

Editor 配置，可以设置编辑器的代码样式等。

选择 Editor→Color Scheme→Language Defaults→Semantic Highlighting，通过勾选 Semantic Highlighting 启用代码高亮功能，如图 1.23 所示。

进入 Editor→Code Style，单击 General Formatter Control，勾选 Enable formatter markers in comments 后，可以设置代码格式化的方式，如图 1.24 所示。若不勾选，则默认所有的代码都格式化。

1.3.2　DevEco Studio SDK 管理

通过 DevEco Studio，可以实现 HarmonyOS SDK 的下载安装及管理，方便开发者使用 SDK 中的 API 和各种工具，以便快速完成开发。这里讲解一下如何通过 DevEco Studio 进行 SDK 的管理。

首先，在 DevEco Studio 主界面上方的菜单栏，选择 Tool→SDK Manager，或者选择 File→Settings→Appearance & Behavior→System Settings→HarmonyOS SDK，进入 SDK 管理界面，如图 1.25 所示。

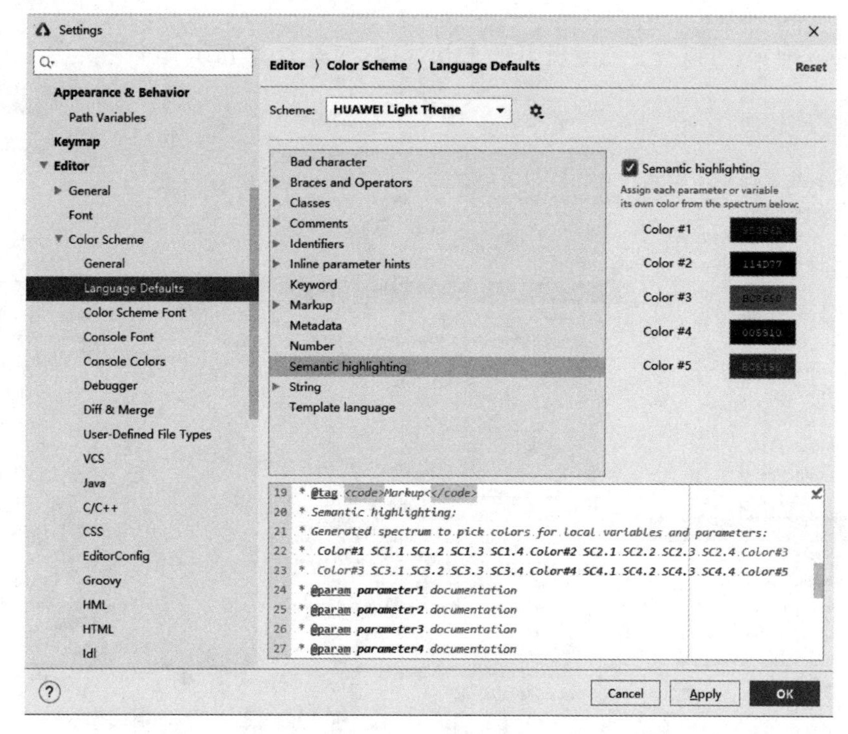

图 1.23　Editor 配置代码高亮

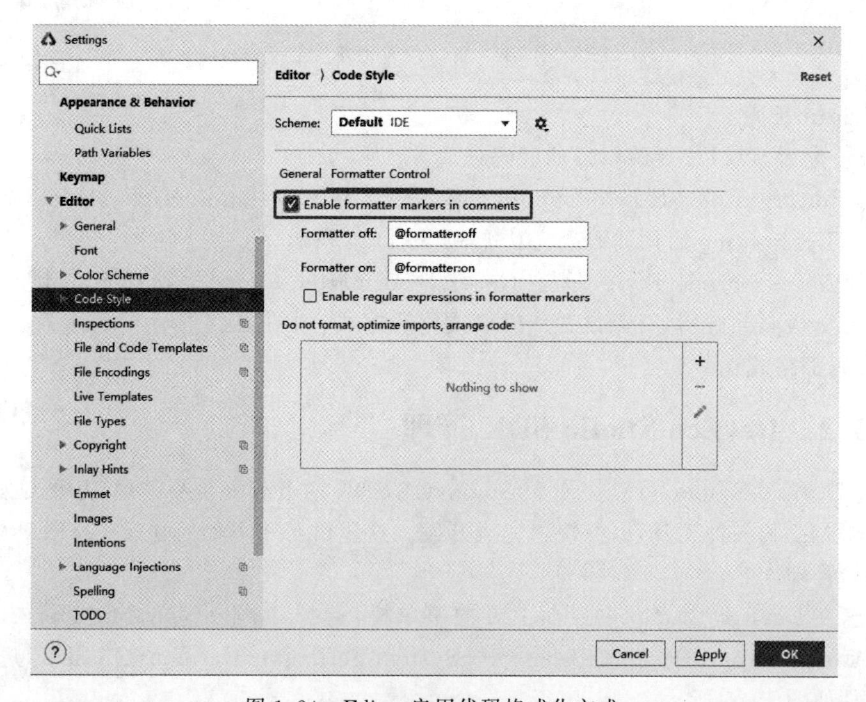

图 1.24　Editor 启用代码格式化方式

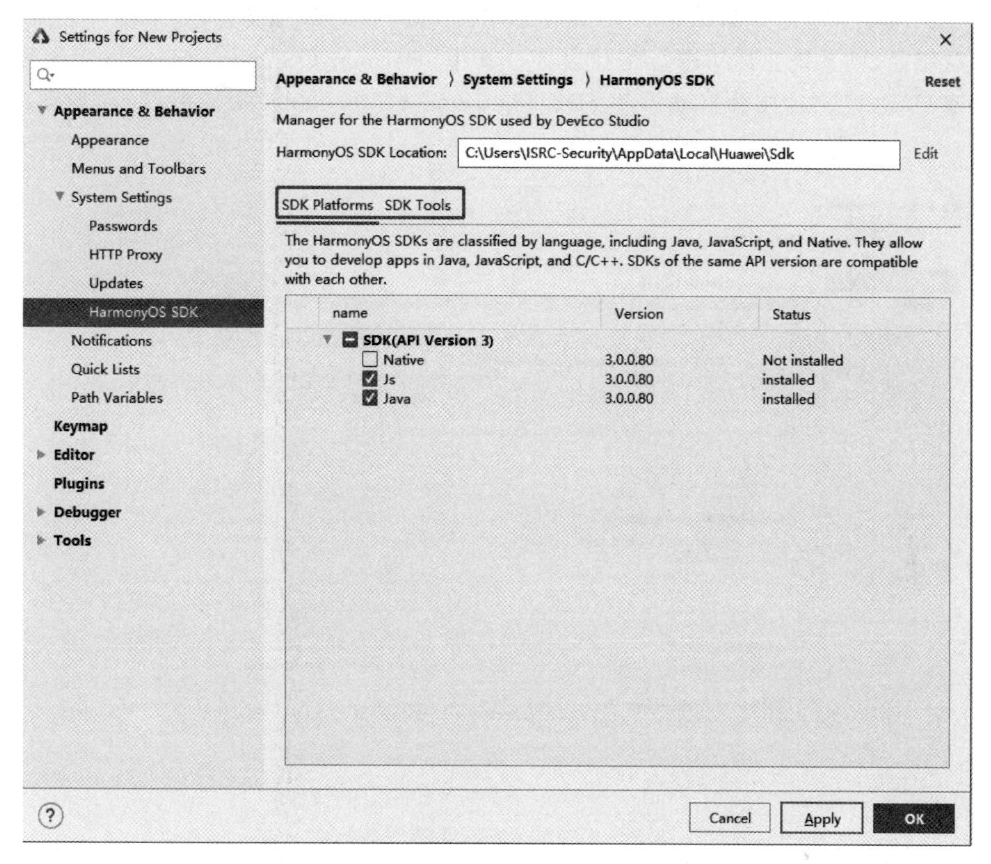

图 1.25　HarmonyOS SDK 管理

其中，配置主要包括 SDK Platforms 和 SDK Tools，作用如下。

（1）SDK Platforms→Native：C/C++开发需要的 API 和工具链，包括 Native API，编译工具链等。

（2）SDK Platforms→Js：JS 开发需要的 API 和工具链。

（3）SDK Platforms→Java：开发需要的 API 和工具链，包含了 HarmonyOS 基础 API，车机、大屏、穿戴设备的 API，以及与 Java 相关的编译构建工具等。

（4）SDK Tools→Toolchains：打包所需最小集工具链及 API。

（5）SDK Tools→Previewer：公共的内容，包括帮助，以及各种工具（打包、签名等）。

注意，与 HarmonyOS 相关的 SDK（包括 SDK-Java、Sdk-Js、Sdk-Native），需要与 HarmonyOS 版本对应一致。与 HarmonyOS 无关的 SDK Tools（Common、Install-Assisant），可以采用独立版本号。

其次，我们通过设置和检查 HarmonyOS SDK、JDK、Node.js 的本地路径，以确保项目所需的各个 SDK、JDK、Node.js 均已正确安装并进行关联。在 DevEco Studio 主界面上方菜单栏选择 File→Project Structure，或按快捷键 Ctrl＋Alt＋Shift＋D，进入 SDK Location

界面。在该界面可进行 HarmonyOS SDK、JDK、Node.js 的路径设置，如果都安装到了系统默认路径，则系统会自动进行路径设置。其中，HarmonyOS Native location 无须配置，在创建 C/C++项目时，会自动添加 SDK Platforms→SDK-Native 对应的路径，如图 1.26 所示。

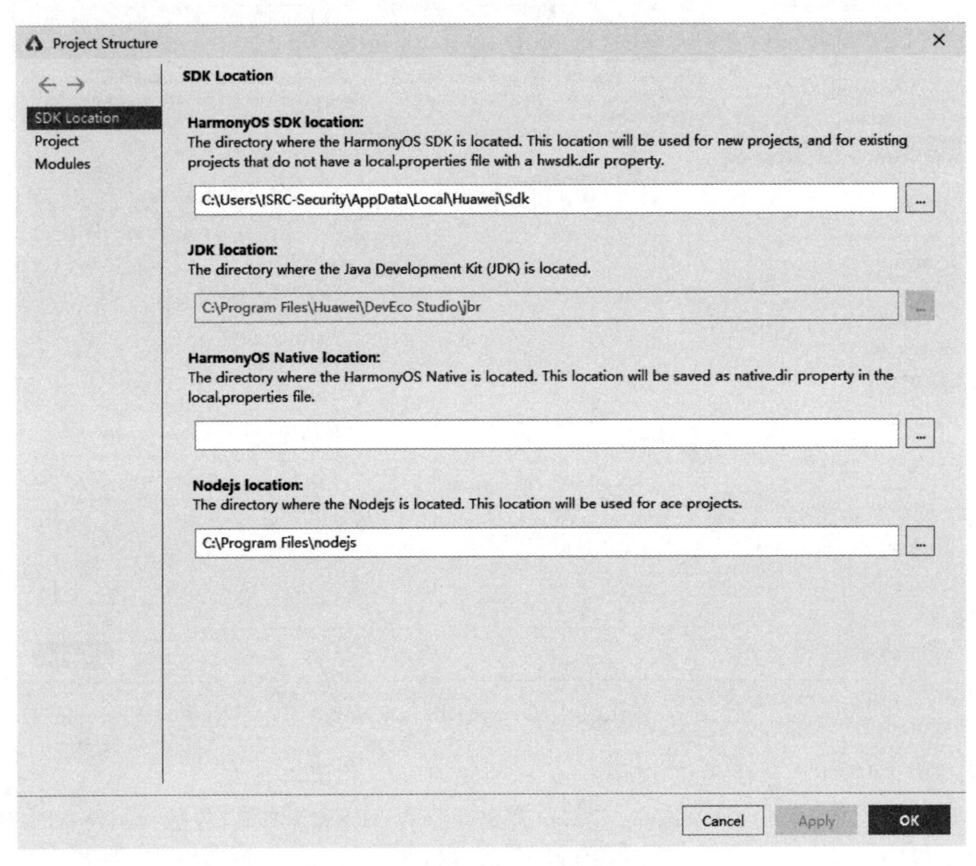

图 1.26　SDK Location 管理

1.4　创建第一个 HarmonyOS 项目

至此，HarmonyOS 的开发环境全部搭建完成了。接下来我们创建第一个 Harmony 项目。

DevEco Studio 支持包括手机、车载、智慧屏、智能穿戴、轻量级智能穿戴等多设备下的 HarmonyOS 应用开发，提供了包括 Java、JS、C/C++等多种编程语言，并支持多种语言的混合开发场景。在新建工程时，可以在 DevEco Studio 中选择并创建适用于各种设备的工程，并自动生成对应的工程模板和代码。具体的设备类型和支持的工程模板及开发语言的对应关系如表 1.1 所示。

表 1.1 各设备类型和支持的工程模板及开发语言

设 备 类 型	支持工程模板及语言
Phone	Empty Feature Ability(JS)
	Empty Feature Ability(Java)
Tablet	Empty Feature Ability(JS)
	Empty Feature Ability(Java)
Car	Empty Feature Ability(Java)
	Native C++
TV	Empty Feature Ability(JS)
	Empty Feature Ability(Java)
	List Feature Ability(JS)
	Tab Feature Ability(JS)
Wearable	Empty Feature Ability(JS)
	Empty Feature Ability(Java)
	List Feature Ability(JS)
Lite Wearable	Empty Feature Ability(JS)
	List Feature Ability(JS)
Smart Vision	Empty Feature Ability(JS)
	List Feature Ability(JS)

同时,除上述所示,手机也包含 7 个应用模板,其中覆盖了新闻、购物等场景,有相应开发需求的开发者可以直接使用应用模板。

下面以 TV 为例,创建一个 Java 项目并在模拟器上运行。

1.4.1 创建第一个项目

首先,打开工程创建向导界面。如果当前没有打开任何工程,则在 DevEco Studio 欢迎页选择 Create HarmonyOS Project。若已经打开了工程,则在上方菜单栏选择 File→New→New Project。工程创建向导界面如图 1.27 所示。

其中,Device 包含各种设备类型,Template 包含各种模板及支持的语言。首先选择需要进行开发的设备类型,然后选择对应的 Ability 模板类型。这里以智慧屏 TV 为例,选择一个空的 Ability 模板 Empty Feature Ability(Java),其支持 Java 语言。单击 Next 按钮对工程进行配置,如图 1.28 所示。

其中,Project Name 表示工程的名称,可以自定义,此应用安装到设备之后也会在设备上显示该名称。Package Name 表示软件包的名称,默认情况下应用的 ID 也会使用该名称,应用发布时,软件包名需要保持唯一性。Save Location 表示工程文件的本地存储路径,无特殊要求则保持默认即可,注意存储路径中不能包含中文字符。Compatible SDK 指兼容的 SDK 版本。配置完成后单击 Finish 按钮,工程即创建完成,如图 1.29 所示。

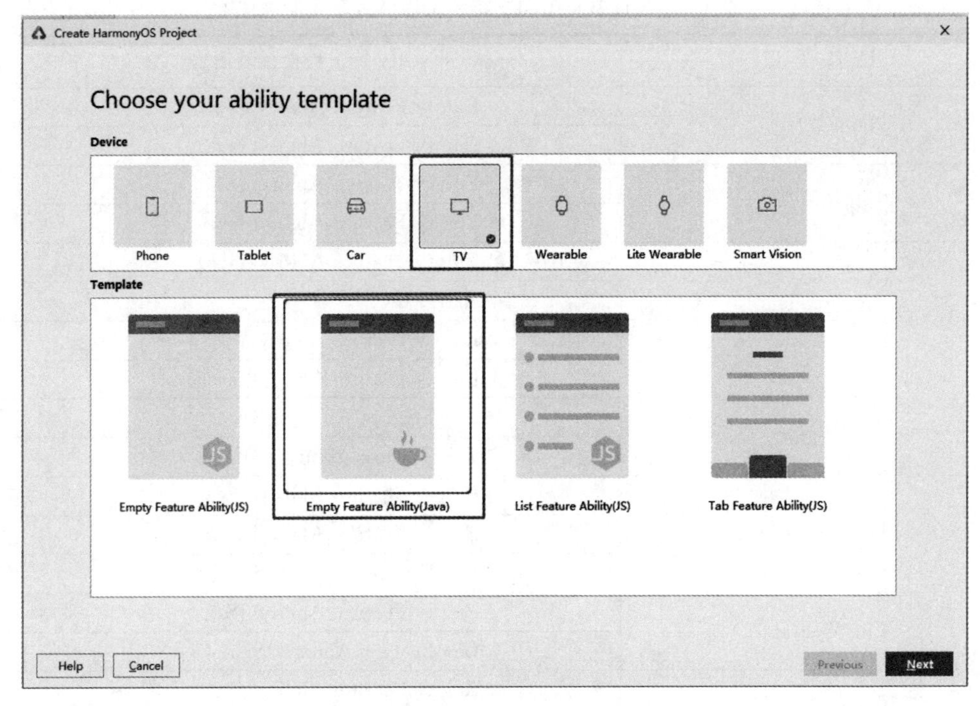

图 1.27　工程创建向导

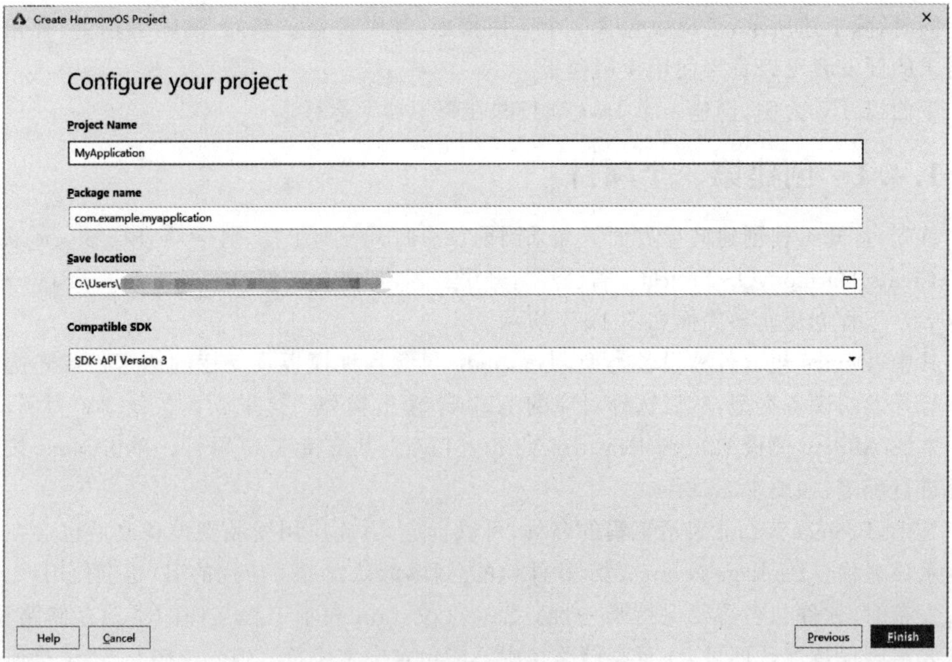

图 1.28　工程配置

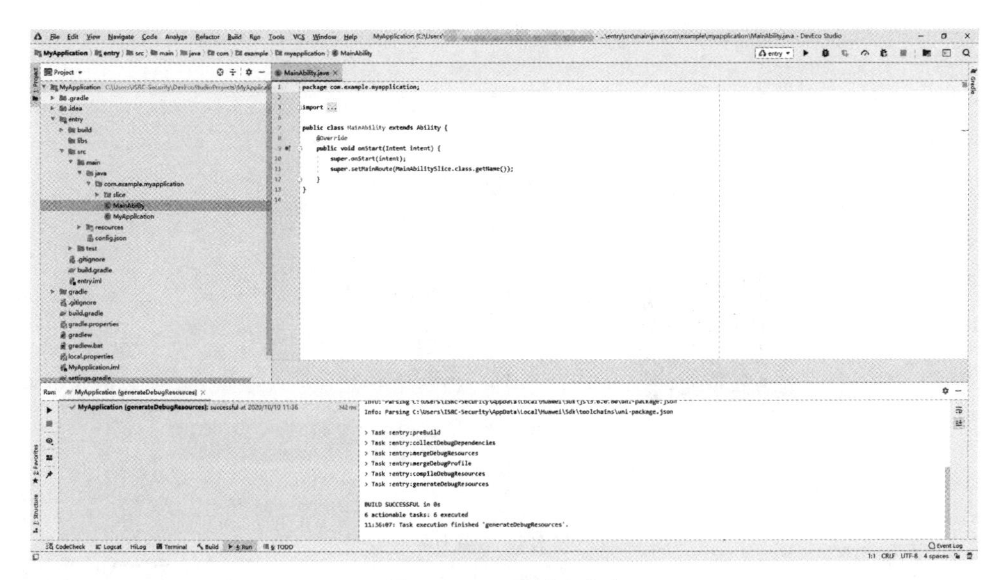

图 1.29 工程创建成功

1.4.2 模拟器运行及预览

创建完上述项目后，DevEco Studio 会自动生成 Hello World 项目中所需的代码，因此该项目不用自行编写代码，就可以正常运行 Hello World 程序。应用程序的运行需要基于设备，可以使用搭载 HarmonyOS 的设备，也可以使用 DevEco Studio 内置的模拟器。这里通过内置模拟器来运行该程序。

首先在 DevEco Studio 上方的菜单栏选择 Tools→HVD Manager，首次使用模拟器时，需要下载相关资源。弹出下方提示框后，单击 OK 按钮，如图 1.30 所示。

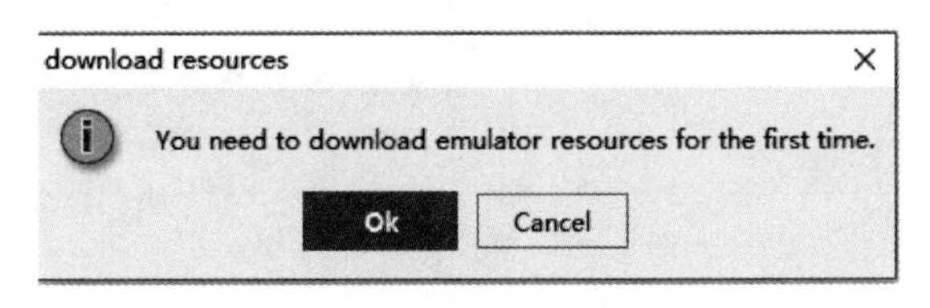

图 1.30 下载模拟器资源

随后浏览器会弹出华为账号的登录界面，如图 1.31 所示。需要先登录已完成实名认证的华为账号。这里官方推荐使用 Chrome 浏览器，如果使用 Safari 或 360 等其他浏览器，则需要取消“阻止跨站跟踪”和“阻止所有 Cookie”功能。

返回 DevEco Studio，单击 Virtual Device Manager 界面左下方的 Refresh 按钮进行授权登录，并完成下载。下载完成后会显示设备列表，如图 1.32 所示。

在设备列表中，选择 TV 设备，单击右侧的 ▷ 按钮运行模拟器，并返回 DevEco Studio 主界面，单击右上角工具栏中的 ▶ 按钮或按快捷键 Shift+F10 运行工程，在弹出的 Select

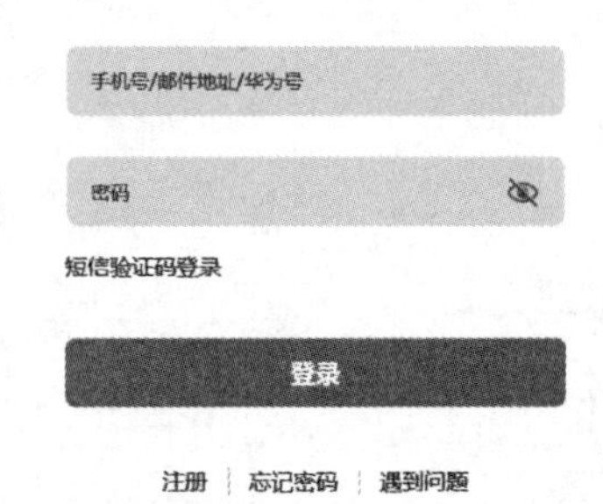

图 1.31　实名认证华为账号登录

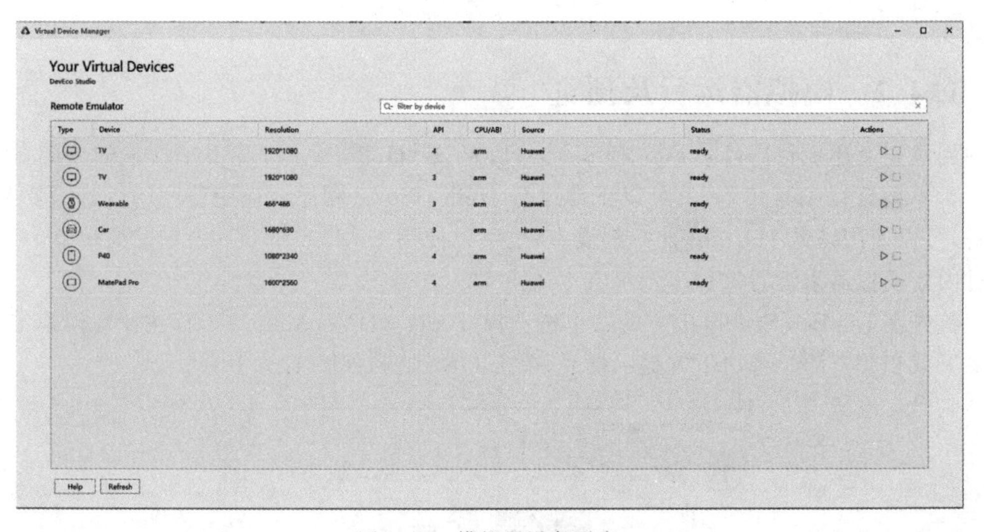

图 1.32　模拟器设备列表

Deployment Target 界面中选择相应的 Connected Devices，并单击 OK 按钮，即可看到 Hello World 程序成功运行在 TV 模拟器上，如图 1.33 所示。

同时，DevEco Studio 还支持多设备预览器和模拟器，在预览器中，可以实时查看应用的布局效果，同时还支持多设备的同时预览，查看同一个布局文件在不同设备上的呈现效果。这里新建一个 JS 工程，选择 Phone 设备下的 Empty Feature Ability(JS)，如图 1.34 所示。

打开工程目录 entry→src→main 下的 config.json 文件，在 module 配置标签下的 deviceType 字段中，增加需要支持的设备类型，如增加 TV、Wearable 设备，代码如下：

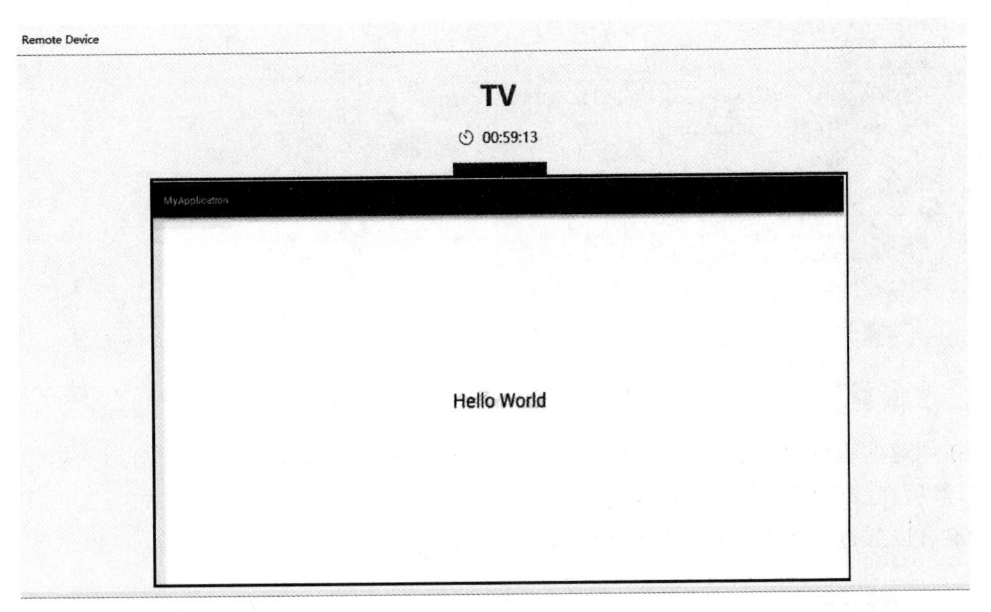

图 1.33　Hello World 程序运行在 TV 模拟器上

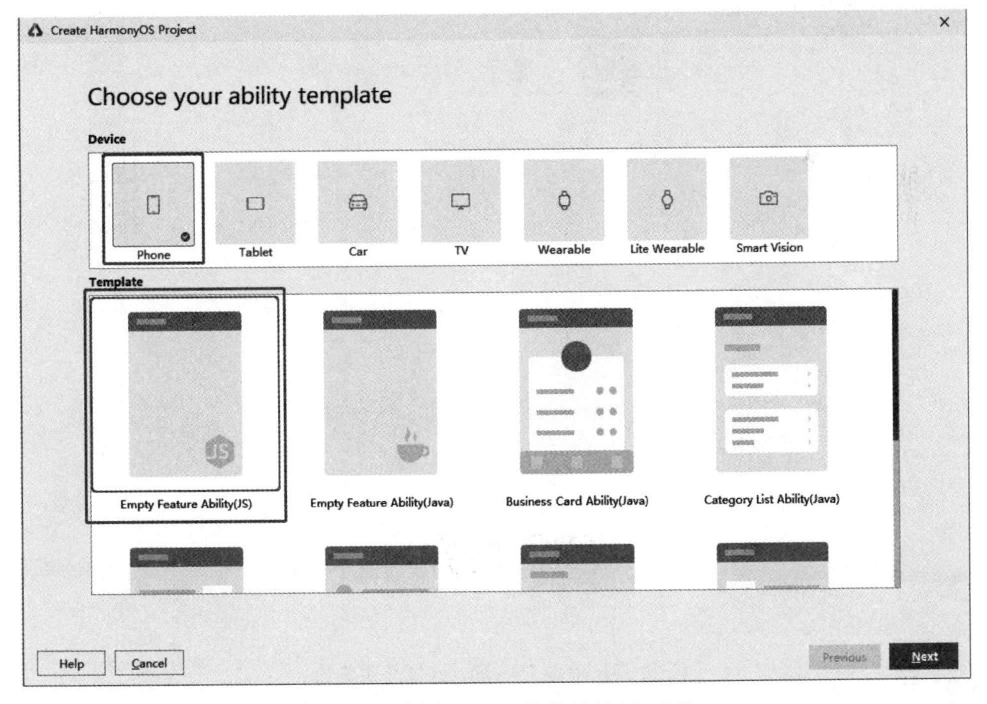

图 1.34　创建 Phone 设备下的 JS 工程

```
{
……
  "module": {
    "package": "com.example.myapplication",
    "name": ".MyApplication",
    "deviceType": [
      "phone",
"tv",
      "wearable"
    ],
  ……
  }
}
```

在创建的工程目录下，打开一个 entry→src→main→js→pages 下的 HML、CSS 或者 JS 文件，然后在编辑窗口右上角的侧边工具栏中单击 Previewer，打开预览器，如图 1.35 所示。或者也可以通过菜单栏选择 View→Tool Windows→Previewer，打开预览器。

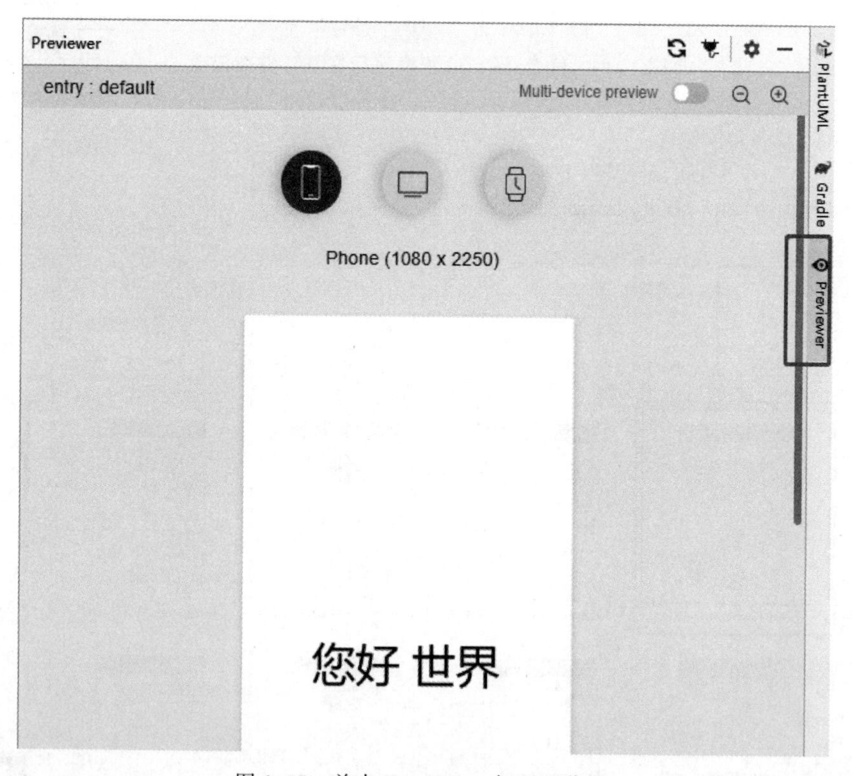

图 1.35　单击 Previewer，打开预览器

打开 index.css 文件，将.title 下的 font-size 由 100px 变为 200px，即让 Hello World 的文字变大，代码如下：

```
/* index.css */
.container {
    flex-direction: column;
    justify-content: center;
    align-items: center;
}

.title {
font-size: 200px;
}
```

修改完成后，可以立即在 Previewer 窗口的预览器中看到界面中“您好 世界”的文字实时改变，从而可以实时查看界面的布局效果。实时变化后预览器的效果如图 1.36 所示。

图 1.36　预览器实时变化效果

在预览器窗口中，可以通过预览器顶部的设备图标切换当前设备，打开 Multi-Preview 开关，可以同时查看多设备上的应用的运行效果。打开 Multi-Preview 开关，多设备预览效果如图 1.37 所示。

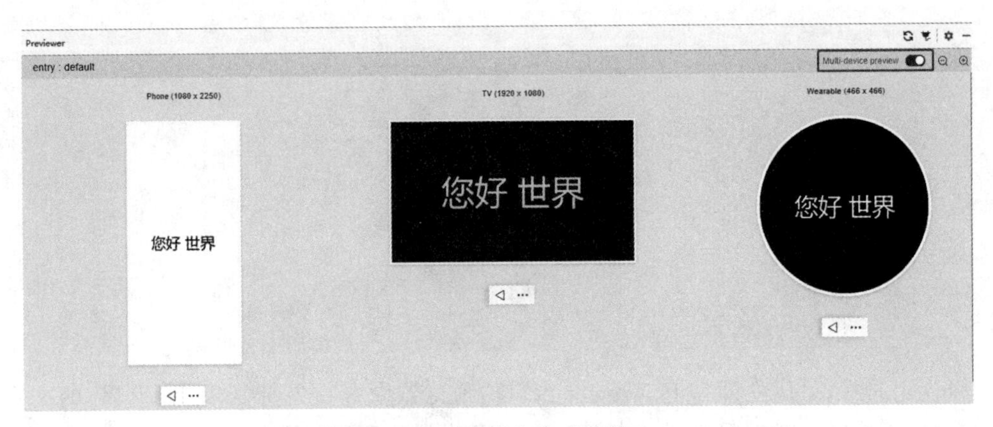

图 1.37　跨设备实时预览

1.5　HarmonyOS 项目分析

1.5.1　项目逻辑视图

首先整体了解一下一个 HarmonyOS 应用软件的整体项目逻辑视图及模块组成，如图 1.38 所示。

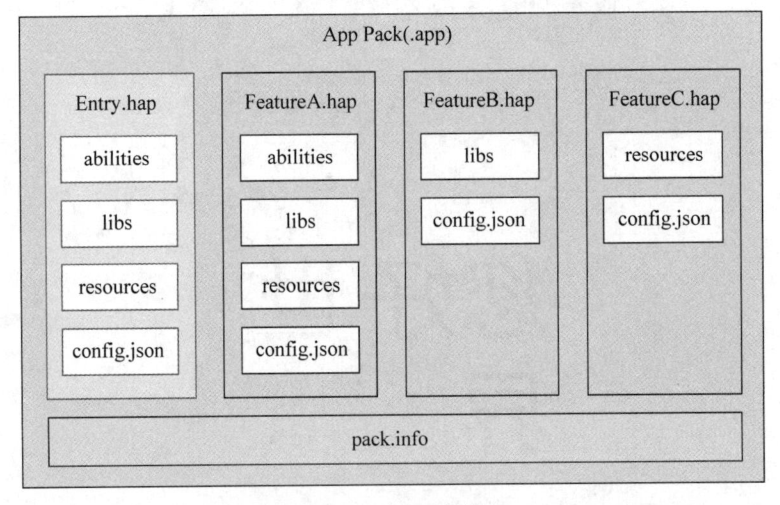

图 1.38　项目逻辑视图

HarmonyOS 的应用软件，以 App(App Pack)的形式发布，每个 App 由一个或多个 HAP(HarmonyOS Ability Package)及 pack. info 组成。其中，pack. info 用于描述每个 HAP 的属性，主要包括该 HAP 是否随应用安装(delivery-with-install)、HAP 文件名(name)、HAP 模块包类型(module-type)、支持该 HAP 运行的设备类型(device-type)等。

其中，每个 HAP 模块包由 Ability（Ability 是应用具备的能力，HarmonyOS 应用代码围绕 Ability 组件展开，第 3 章会详细讲解）、第三方库（libs）、资源文件（Resources）及应用配置文件（config.json）组成。HAP 模块包分为 entry 和 feature 两种类型。

（1）entry：应用的主模块。在一个 App 中，同一设备类型必须有且仅能有一个 entry 类型的 HAP 包，可独立安装运行。

（2）feature：应用的动态类型模块。在一个 App 中，可以包含一个或多个 feature 类型的 HAP 包，也可以不包含，但只有包含 Ability 的 HAP 包才能独立运行。

1.5.2 项目结构与文件

现在来看一下之前创建的 HelloWorld 项目，项目结构如图 1.39 所示。这里逐一介绍每个文件的内容。

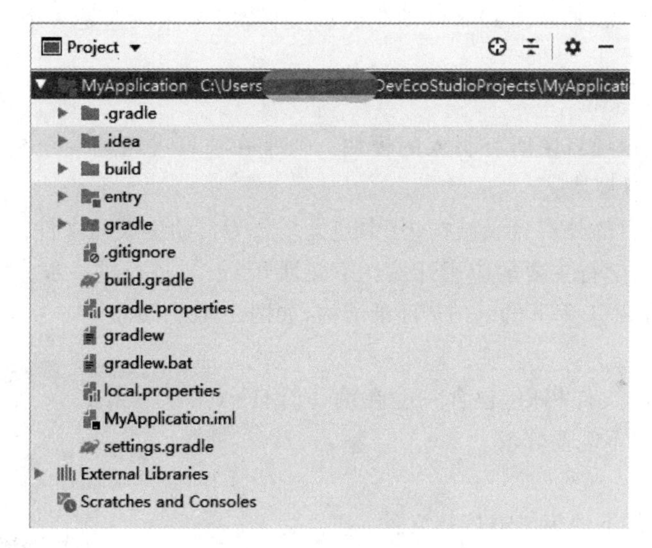

图 1.39 项目结构

1）.gradle 和 idea

这两个文件是由 DevEco Studio 自动生成的配置文件，一般情况下不需要进行修改。

2）build

build 文件夹中包含了一些在编译时自动生成的文件，一般也不需要进行修改。

3）entry

entry 是默认启动模块，是项目中的主模块，随后开发中的源码文件和资源文件均在 entry 中。

4）gradle

目录下包含了 gradle wrapper 的配置文件。

5）.gitignore

该文件用来将指定的目录或文件排除在版本控制之外。

6）build. gradle

项目全局的 gradle 构建脚本。

7）gradle. properties

项目全局的 gradle 配置文件，其中配置的属性会影响项目中所有的 gradle 编译脚本。

8）gradlew 和 gradlew. bat

这两个文件用来在命令行界面中执行 gradle 命令，其中，gradlew 用于 Linux 或 Mac 系统，gradlew. bat 用于 Windows 系统。

9）local. properties

文件用于指定本机中的 SDK 路径，通常自动生成而无须修改。若更改了本机中的 SDK 路径，则应对该文件中的路径进行修改。

10）MyApplication. iml

. iml 文件是由 DevEco Studio 自动生成的文件，用来标识这是一个 DevEco Studio 项目，无须进行修改。

11）settings. gradle

该文件用来表示项目中所有引入的模块。通常在项目中新建一个模块之后，settings. gradle 会自动引入该模块。

以上是整个项目的外层目录结构，其中绝大多数为自动生成的文件。entry 目录为整个项目的重点，在这里进行主要的应用开发。下面展开 entry 进行进一步分析。

这里先介绍 Java 工程下的 entry 目录结构，如图 1.40 所示。

1）build

与外层的 build 文件相似，包含一些在编译时自动生成的文件，一般不需要修改。

2）libs

存放 entry 模块下的第三方依赖文件。

3）src→main→java

开发者用于编写 Java 源码文件的存放目录。

4）src→main→resources

开发者用于存放开发资源文件的目录，包括图片、音视频等资源文件。

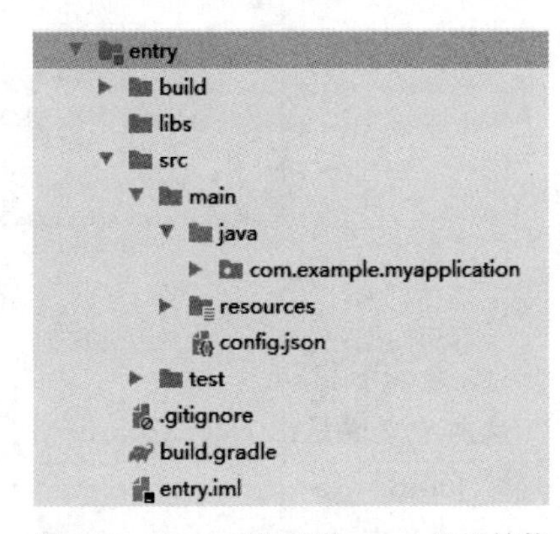

图 1.40　Java 工程下的 entry 目录结构

5）src→main→config. json

应用配置文件。用于声明应用的 Ability 信息，以及应用所需的权限等信息。

6）src→test

开发者用于编写测试文件的目录。

7）. gitignore

用于将 entry 模块内的指定目录或文件排除在版本控制之外，其作用和外层的 . gitignore 文件基本一致。

8）build.gradle

entry 模块的 gradle 构建脚本，指定与项目构建相关的配置信息。

9）entry.iml

DevEco Studio 自动生成的文件，无须进行修改。

这是 Java 项目的目录结构，这里也可以创建 JS 项目，以此来学习 JS 项目的目录结构。如选择 TV 设备下的 Empty Feature Ability(JS)，创建完成后，在项目的目录结构中，除了包括 Java 项目文件外，还多了 JS 文件夹及相应文件，如图 1.41 所示。

1）js→default→i18n

i18n 是英文单词 internationalization 的缩写，是“国际化”的简称。i18n 文件夹用于存放配置不同语言场景的资源，例如应用文本词条、图片路径等资源。

2）js→default→pages

pages 文件夹用于存放多个页面的开发文件，例如图中一个 index 表示一个页面，每个页面由 HML、CSS 和 JS 文件组成。

（1）index.hml：HML 模板文件，用来描述当前页面的布局结构，类似于网页中的 HTML 文件。

（2）index.css：CSS 样式文件，用于描述页面样式。

（3）index.js：JS 文件，用于处理页面和用户的交互。

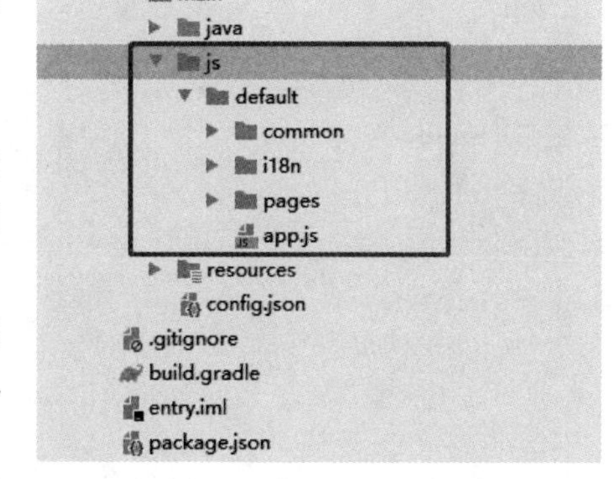

图 1.41 JS 项目目录结构

1.6 应用配置文件

1.6.1 配置文件介绍

应用的每个 HAP 的根目录下都存在一个 config.json 配置文件，如图 1.42 所示，主要涵盖以下 3 个方面：

（1）应用的全局配置信息，包含应用的包名、生产厂商、版本号等基本信息。

（2）应用在具体设备上的配置信息。

（3）HAP 包的配置信息，包含每个 Ability 必须定义的基本属性（如包名、类名、类型及 Ability 提供的能力），以及应用访问系统或其他应用受保护部分所需的权限等。

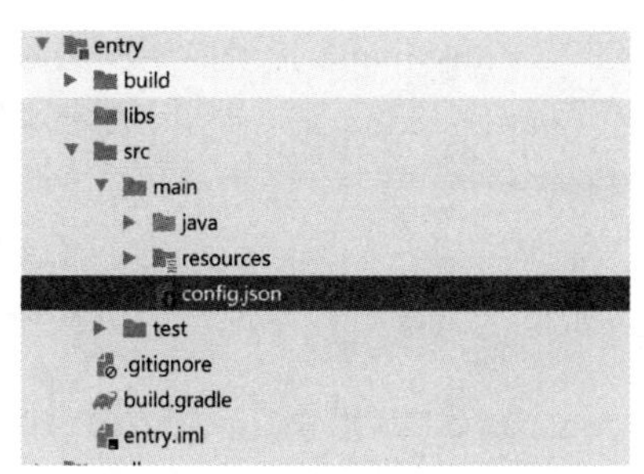

图 1.42 config.json 资源文件

config.json 由属性和值两部分构成，属性出现顺序不分先后，且每个属性最多只允许出现一次。每个属性的值为 JSON 的基本数据类型(数值、字符串、布尔值、数组、对象或者 null 类型)。如果属性值需要引用资源文件，可参见 1.7.3 节资源文件的使用。

1.6.2 配置信息 App

应用的配置文件 config.json 中由 app、deviceConfig 和 module 3 个部分组成，缺一不可。app 表示应用的全局配置信息。同一个应用的不同 HAP 包的 app 配置必须保持一致。deviceConfig 表示应用在具体设备上的配置信息。module 表示 HAP 包的配置信息。该标签下的配置只对当前 HAP 包生效。app 的示例代码如下：

```
//config.json 中 app 示例代码
"app": {
    "bundleName": "com.huawei.mytestapp.example",
    "vendor": "huawei",
    "version": {
        "code": 2,
        "name": "2.0"
    }
    "apiVersion": {
        "compatible": 3,
        "target": 3
    }
}
```

其中，bundleName 表示应用的包名，用于标识应用的唯一性。包名是由字母、数字、下画线(_)和点号(.)组成的字符串，必须以字母开头。支持的字符串长度为 7～127 字节。包名通常采用反域名形式表示(例如，com. huawei. mytestapp)。建议第一级为域名后缀 com，第二级为厂商/个人名，第三级为应用名，也可以采用多级。

vendor 表示对应用开发厂商的描述。字符串长度不超过 255 字节。vendor 可以不定义，不定义就默认为空值。

version：code 表示表示应用的版本号，仅用于 HarmonyOS 管理该应用，对用户不可见。取值为大于零的整数，不可以缺省。

version：name 表示应用的版本号，用于向用户呈现。取值可以自定义，不可以缺省。

apiVersion：compatible 表示应用运行需要的 API 最小版本。取值为大于零的整数，不可缺省。

apiVersion：target 表示应用运行需要的 API 目标版本。取值为大于零的整数，可缺省，如果不填写则自动识别为应用所在设备的当前 API 版本。

1.6.3 配置信息 deviceConfig

deviceConfig 包含在具体设备上的应用配置信息，可以包含 default、car、tv、wearable、

liteWearable、smartVision 等属性。default 标签内的配置适用于所有设备，其他设备类型如果有特殊的需求，则需要在该设备类型的标签下进行配置。

deviceConfig 的示例代码如下：

```
//config.json 中 deviceConfig 示例代码
"deviceConfig": {
    "default": {
        "process": "com.huawei.hiworld.example",
        "directLaunch": false,
        "supportBackup": false,
        "network": {
            "usesCleartext": true,
            "securityConfig": {
                "domainSettings": {
                    "cleartextPermitted": true,
                    "domains": [
                        {
                            "subDomains": true,
                            "name": "example.ohos.com"
                        }
                    ]
                }
            }
        }
    }
}
```

deviceConfig 中基本有 6 类信息：

（1）default 表示所有设备通用的应用配置信息。

（2）car 表示车机特有的应用配置信息。

（3）tv 表示智慧屏特有的应用配置信息。

（4）wearable 表示智能穿戴特有的应用配置信息。

（5）liteWearable 表示轻量级智能穿戴特有的应用配置信息。

（6）smartVision 表示智能摄像头特有的应用配置信息。

本例中只举了 default 中设置的应用配置信息，default 标签内的配置适用于所有设备，其他设备类型如果有特殊的需求，则需要在该设备类型的标签下进行配置。具体这 6 个对象的内部结构说明可查阅官方文档，此处不再赘述。

1.6.4 配置信息 module

module 中包含了 HAP 包的配置信息，示例代码如下：

```
//config.json 中 module 示例代码
"module": {
    "package": "com.example.myapplication.entry",
    "name": ".MyOHOSAbilityPackage",
    "description": "$string:description_application",
    "supportedModes": [
        "drive"
    ],
    "deviceType": [
        "car"
    ],
    "distro": {
        "deliveryWithInstall": true,
        "moduleName": "ohos_entry",
        "moduleType": "entry"
    },
    "abilities": [
        …
    ],
    "shortcuts": [
        …
    ],
    "js": [
        …
    ],
    "reqPermissions": [
        …
    ],
    "defPermissions": [
        …
    ]
}
```

(1) package 表示 HAP 的包结构名称，在应用内应保证唯一性。采用反向域名格式（建议与 HAP 的工程目录保持一致）。字符串长度不超过 127 字节。该标签仅适用于智慧屏、智能穿戴、车机，不可缺省。

(2) name 表示 HAP 的类名。采用反向域名方式表示，前缀需要与同级的 package 标签指定的包名一致，也可采用"."开头的命名方式。字符串长度不超过 255 字节。该标签仅适用于智慧屏、智能穿戴、车机，不可缺省。

(3) description 表示 HAP 的描述信息。字符串长度不超过 255 字节。如果字符串超出长度或者需要支持多语言，可以采用资源索引的方式添加描述内容。该标签仅适用于智慧屏、智能穿戴、车机，可以缺省，默认值为空。

(4) supportedModes 表示应用支持的运行模式。当前只定义了驾驶模式(drive)。该标签仅适用于车机，可以缺省，默认值为空。

(5) deviceType 表示允许 Ability 运行的设备类型。系统预定义的设备类型包括：tv

(智慧屏)、car(车机)、wearable(智能穿戴)、liteWearable(轻量级智能穿戴)和 default(通用)等,不可缺省。

(6) distro 表示 HAP 发布的具体描述。该标签仅适用于智慧屏、智能穿戴、车机,不可缺省。distro 示例代码如下:

```
"distro": {
    "deliveryWithInstall": true,
    "moduleName": "ohos_entry",
    "moduleType": "entry"
}
```

- deliveryWithInstall 表示当前 HAP 是否支持随应用安装,是一个布尔类型的值,不可以缺省,如果其值是 true,则表示支持随应用安装,如果其值是 false,则表示不支持随应用安装。
- moduleName 用字符串的形式表示当前 HAP 的名称,不可以缺省。
- moduleType 用字符串的形式表示当前 HAP 的类型(entry 和 feature)。entry 表示一个应用的主模块。一个 App 中,对于同一设备类型必须有且只有一个 entry 类型的 HAP,可独立安装运行。feature 表示应用的动态特性模块。一个 App 可以包含一个或多个 feature 类型的 HAP,也可以不含。只有包含 Ability 的 HAP 才能够独立运行。

(7) ability 表示当前模块内的所有 Ability。采用对象数组格式,其中每个元素表示一个 Ability 对象,可缺省,默认值为空。Ability 非常重要,每当我们创建新的 Ability,都要确认该处的配置是否正确,后面将详细介绍 Ability 的形式。

(8) JS 表示基于 JS UI 框架开发的 JS 模块集合,其中的每个元素代表一个 JS 模块的信息,不过不写则说明该模块没有使用 JS 框架,下面给出一个示例,page 中定义了 JS UI 的页面路径+页面名称,JS UI 具体用法可阅读第 5 章 JS UI 布局,示例代码如下:

```
//config.json 中 JS 配置信息示例代码
"js": [
    {
        "name": "default",
        "pages": [
            "pages/index/index",
            "pages/detail/detail"
        ],
        "window": {
            "designWidth": 750,
            "autoDesignWidth": false
        }
    }
]
```

(9) shortcuts 表示应用的快捷方式信息。采用对象数组格式,其中的每个元素表示一个快捷方式对象,可缺省,默认值为空。

(10) defPermissions 表示应用定义的权限。应用调用者必须申请这些权限,这样才能正常调用该应用,该值可缺省,默认值为空。

(11) reqPermissions 表示应用运行时向系统申请的权限,该值可缺省,默认值为空。

这里再详细讲解一下 abilities 对象的内部结构,示例代码如下:

```
//config.json 中 abilities 对象示例代码
"abilities": [
    {
        "name": ".MainAbility",
        "description": "himusic main ability",
        "icon": "$media:ic_launcher",
        "label": "HiMusic",
        "launchType": "standard",
        "orientation": "unspecified",
        "permissions": [
        ],
        "visible": true,
        "skills": [
            {
                "actions": [
                    "action.system.home"
                ],
                "entities": [
                    "entity.system.home"
                ]
            }
        ],
        "directLaunch": false,
        "configChanges": [
            "locale",
            "layout",
            "fontSize",
            "orientation"
        ],
        "type": "page",
        "formEnabled": false
    },
    {
        "name": ".PlayService",
        "description": "himusic play ability",
        "icon": "$media:ic_launcher",
        "label": "HiMusic",
        "launchType": "standard",
```

```
            "orientation": "unspecified",
            "visible": false,
            "skills": [
                {
                    "actions": [
                        "action.play.music",
                        "action.stop.music"
                    ],
                    "entities": [
                        "entity.audio"
                    ]
                }
            ],
            "type": "service",
            "backgroundModes": [
                "audioPlayback"
            ]
        },
        {
            "name": ".UserADataAbility",
            "type": "data",
            "uri": "dataability://com.huawei.hiworld.himusic.UserADataAbility",
            "visible": true
        }
    ]
```

（1）第一行的 name 表示 Ability 名称。取值可采用反向域名方式表示，由包名和类名组成，如 com.example.myapplication.MainAbility；也可采用“.”开头的类名方式表示，如.MainAbility。该标签仅适用于智慧屏、智能穿戴、车机，不可缺省。

（2）第二行的 description 则表示对 Ability 的描述。取值可以是描述性内容，也可以是对描述性内容的资源索引，以便支持多语言，可缺省，默认值为空。

（3）icon 表示 Ability 图标资源文件的索引。如果写了如下代码：$media:ability_icon 就可以使用 resource 文件夹中的 media 中的名字为 ability_icon 图片作为 App 的图标来显示。如果在该 Ability 的 skills 属性中，actions 的取值包含 action.system.home，entities 取值中包含 entity.system.home，则该 Ability 的 icon 将同时作为应用的 icon。如果存在多个符合条件的 Ability，则取位置靠前的 Ability 的 icon 作为应用的 icon。这个数值可缺省，默认值为空。

（4）label 表示 Ability 对用户显示的名称。取值可以是 Ability 名称，也可以是对该名称的资源索引，以便支持多语言。如果在该 Ability 的 skills 属性中，actions 的取值包含 action.system.home，entities 取值中包含 entity.system.home，则该 Ability 的 label 将同时作为应用的 label。如果存在多个符合条件的 Ability，则取位置靠前的 Ability 的 label 作为应用的 label。可缺省，默认值为空。

(5) uri 表示 Ability 的统一资源标识符。格式为[scheme:][//authority][path][?query][#AbilitySlice]。可缺省，但对于 data 类型的 Ability 不可缺省。

(6) launchType 表示 Ability 的启动模式，支持 standard 和 singleton 两种模式：standard 表示该 Ability 可以有多实例。standard 模式适用于大多数应用场景。singleton 表示该 Ability 只可以有一个实例。例如，具有全局唯一性的呼叫来电界面即采用 singleton 模式。该标签仅适用于智慧屏、智能穿戴、车机。可缺省，默认值为 standard。

(7) visible 表示 Ability 是否可以被其他应用调用，是布尔类型的值。

(8) permissions 表示其他应用的 Ability 调用此 Ability 时需要申请的权限。通常采用反向域名格式，取值可以是系统预定义的权限，也可以是开发者自定义的权限。如果是自定义权限，取值必须与 defPermissions 标签中定义的某个权限的 name 标签值一致。可缺省，默认值为空。

(9) skills 表示 Ability 能够接收的 Intent 的特征，可缺省，默认值为空。

(10) deviceCapability 表示 Ability 运行时要求设备具有的能力，采用字符串数组的格式表示。

(11) type 表示 Ability 的类型。page 表示基于 Page 模板开发的 FA，用于提供与用户交互的能力。service 表示基于 Service 模板开发的 PA，用于提供后台运行任务的能力。data 表示基于 Data 模板开发的 PA，用于对外部提供统一的数据访问抽象。

(12) formEnabled 和 form 是绑定使用的，formenabled 表示 FA 类型的 Ability 是否提供卡片(form)能力。只有当 formEnabled 生效时，form 才会生效，而 form 表示 AbilityForm 的属性。

(13) orientation 表示屏幕的方向，主要有 4 个选项：

- unspecified：由系统自动判断显示方向。
- landscape：横屏模式。
- portrait：竖屏模式。
- followRecent：跟随栈中最近的应用。

如果没有设置 orientation，系统会自动使用 unspecified 属性。

(14) 剩下的还有 backgroundModes、readPermission、directLaunch、configChanges、mission、targetAbility、multiUserShared、supportPipMode 等属性，具体完整的属性可查阅官方文档。

1.7 资源文件

1.7.1 Resource 目录介绍

HarmonyOS 中的应用程序会使用各种资源，例如图片、字符串等，开发者会把它们放入源码的相应文件夹下面，HarmonyOS 也支持并鼓励开发者把与 UI 相关的布局和元素，用 XML 资源实现。

HarmonyOS 中的资源文件分为两大类，一类为 base 目录与限定词目录，这一类中的资源为可直接访问资源，可以用 ResourceTable 直接访问，都保存在 resource 文件下，在编译的情况下会自动生产 ResourceTable.java 文件，其中保存着每个资源的编号索引文件供开发者使用。另一类为 rawfile 目录，rawfile 目录下的文件不能被 ResourceTable 直接访问，只能通过指定文件路径和文件名来引用。

1.7.2　Resource 文件编写

resources 文件夹中的结构如图 1.43 所示。

其中，base 文件夹和限定词文件夹按照两级目录形式来组织，目录命名必须符合规范，以便根据设备状态去匹配相应目录下的资源文件，一级子目录为 base 目录和限定词目录。

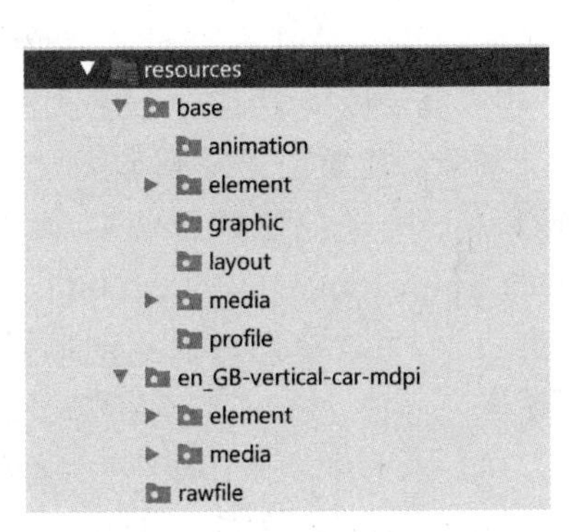

图 1.43　resources 文件结构

base 目录是默认存在的目录。当应用的 resources 目录中没有与设备状态匹配的限定词目录时，会自动引用该目录中的资源文件。

限定词目录需要开发者自行创建。目录名称由一个或多个表征应用场景或设备特征的限定词组合而成，而二级子目录为资源目录，用于存放字符串、颜色、布尔值等基础元素，以及媒体、动画、布局等资源文件，下面详细介绍一下。

首先介绍一下限定词目录的创建规则。限定词目录可以由一个或多个表征应用场景或设备特征的限定词组合而成，包括语言、文字、国家或地区、横竖屏、设备类型和屏幕密度等 6 个维度，限定词之间通过下画线(_)或者中画线(-)连接。开发者在创建限定词目录时，需要掌握限定词目录的命名要求及与限定词目录与设备状态的匹配规则。之所以需要用到限定词，就是为了让资源限定在特定的情况下使用，如果限定词目录中包含语言、文字、横竖屏、设备类型限定词，则对应限定词的取值必须与当前的设备状态完全一致，这样该目录才能够参与设备的资源匹配。例如，限定词目录 zh_CN-car-ldpi 不能参与 en_US 设备的资源匹配。

下面详细说明下限定词的创建规则。限定词目录的命名规则为语言_文字_国家或地区-横竖屏-设备类型-屏幕密度。开发者可以根据应用的使用场景和设备特征，选择其中的一类或几类限定词组成目录名称。语言、文字、国家或地区之间采用下画线(_)连接，除此之外的其他限定词之间均采用中画线(-)连接。例如 zh_Hant_CN、zh_CN-car-ldpi。

然后详细说明一下限定词的类型。上述讲到限定词目录的命名规则为语言_文字_国家或地区-横竖屏-设备类型-屏幕密度。其中第 1 个限定词语言表示设备使用的语言类型，由 2 个小写字母组成。例如 zh 表示中文，en 表示英语。具体详细取值范围，可以参见 ISO 639-1(ISO 制定的语言编码标准)。

第 2 个限定词文字表示设备使用的文字类型，由 1 个大写字母（首字母）和 3 个小写字母组成。例如 Hans 表示简体中文，Hant 表示繁体中文。具体详细取值范围，参见 ISO 15924（ISO 制定的文字编码标准）。

第 3 个限定词表示用户所在的国家或地区，由 2～3 个大写字母或者 3 个数字组成。例如 CN 表示中国，GB 表示英国。具体详细取值范围，参见 ISO 3166-1（ISO 制定的国家和地区编码标准）。

第 4 个限定词表示横竖屏，其中，vertical 代表竖屏，horizontal 代表横屏，再后面是设备类型和屏幕密度，具体规则可以查阅官方文档。

在为设备匹配对应的资源文件时，限定词目录匹配的优先级从高到低依次为区域（语言_文字_国家或地区）>横竖屏>设备类型>屏幕密度。

资源组目录创建规则如下，base 目录和限定词目录下可以创建资源组目录，包括 element、media、animation、layout、graphic、profile 6 种文件夹目录，可以分别存放不同的资源，如图 1.44 所示。

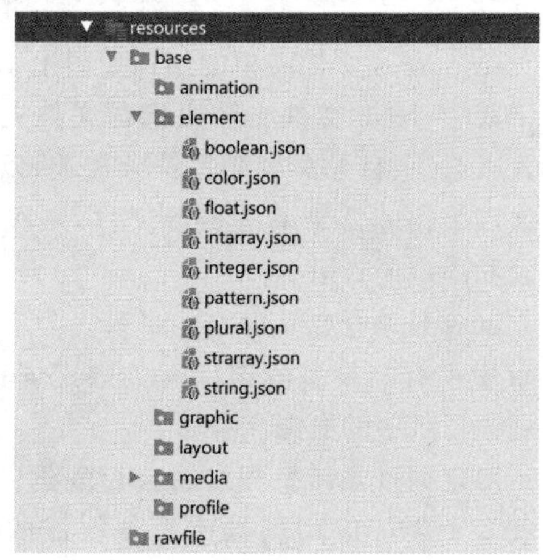

图 1.44　资源组目录结构

（1）element：element 表示元素资源，其下的每个资源文件都用相应的 JSON 文件来表征。

boolean 表示布尔型，boolean.json 的示例代码如下：

```
//boolean.json
{
    "boolean":[
        {
            "name":"boolean_1",
            "value":true
        },
        {
            "name":"boolean_ref",
            "value":"$boolean:boolean_1"
        }
    ]
}
```

color 表示颜色，color.json 的示例代码如下：

```
//color.json
{
    "color":[
```

```
        {
            "name":"red",
            "value":"#ff0000"
        },
        {
            "name":"red_ref",
            "value":"$color:red"
        }
    ]
}
```

float 表示浮点型，float.json 的示例代码如下：

```
//float.json
{
    "float":[
        {
            "name":"float_1",
            "value":"30.6"
        },
        {
            "name":"float_ref",
            "value":"$float:float_1"
        },
        {
            "name":"float_px",
            "value":"100px"
        }
    ]
}
```

intarray 表示整型数组，intarray.json 的示例代码如下：

```
//intarray.json
{
    "intarray":[
        {
            "name":"intarray_1",
            "value":[
                100,
                200,
                "$integer:integer_1"
            ]
        }
    ]
}
```

integer 表示整型，integer.json 的示例代码如下：

```
//integer.json
{
    "integer":[
        {
            "name":"integer_1",
            "value":100
        },
        {
            "name":"integer_ref",
            "value":"$integer:integer_1"
        }
    ]
}
```

pattern 表示样式，pattern.json 的示例代码如下：

```
//pattern.json
{
    "pattern":[
        {
            "name":"base",
            "value":[
                {
                    "name":"width",
                    "value":"100vp"
                },
                {
                    "name":"height",
                    "value":"100vp"
                },
                {
                    "name":"size",
                    "value":"25px"
                }
            ]
        },
        {
           "name":"child",
           "parent":"base",
           "value":[
               {
                   "name":"noTitile",
                   "value":"Yes"
               }
           ]
        }
    ]
}
```

plural 表示复数形式，plural.json 的示例代码如下：

```
//plural.json
{
    "plural":[
        {
            "name":"eat_apple",
            "value":[
                {
                    "quantity":"one",
                    "value":"%d apple"
                },
                {
                    "quantity":"other",
                    "value":"%d apples"
                }
            ]
        }
    ]
}
```

strarray 表示字符串数组，strarray.json 的示例代码如下：

```
//strarray.json
{
    "strarray":[
        {
            "name":"size",
            "value":[
                {
                    "value":"small"
                },
                {
                    "value":"$string:hello"
                },
                {
                    "value":"large"
                },
                {
                    "value":"extra large"
                }
            ]
        }
    ]
}
```

string 表示字符串，string.json 的示例代码如下：

```
//string.json
{
    "string":[
        {
            "name":"hello",
            "value":"hello base"
        },
        {
            "name":"app_name",
            "value":"my application"
        },
        {
            "name":"app_name_ref",
            "value":"$string:app_name"
        },
        {
            "name":"app_sys_ref",
            "value":"$ohos:string:request_location_reminder_title"
        }
    ]
}
```

(2) media：media 表示媒体资源，包括图片、声频、视频等非文本格式的文件。媒体资源文件名可自定义，例如 icon.png，如图 1.45 所示。

(3) animation：animation 表示动画资源，其中资源采用 XML 文件格式表示。文件名可自定义，例如 zoom_in.xml。

(4) layout：layout 表示布局资源，其中资源采用 XML 文件格式表示。详细布局内容参考第 3 章 Java UI 布局。

(5) graphic：graphic 表示可绘制资源，其中资源采用 XML 文件格式表示，如图 1.46 所示。

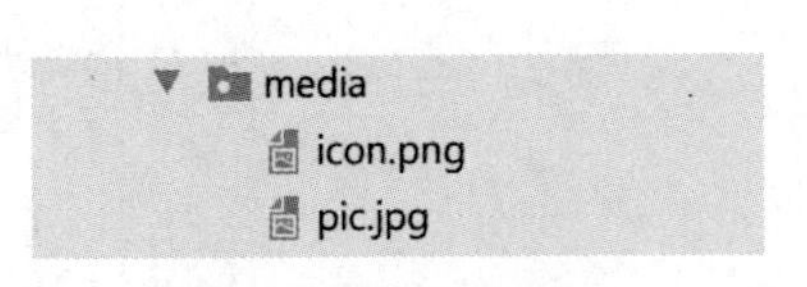

图 1.45 媒体资源文件

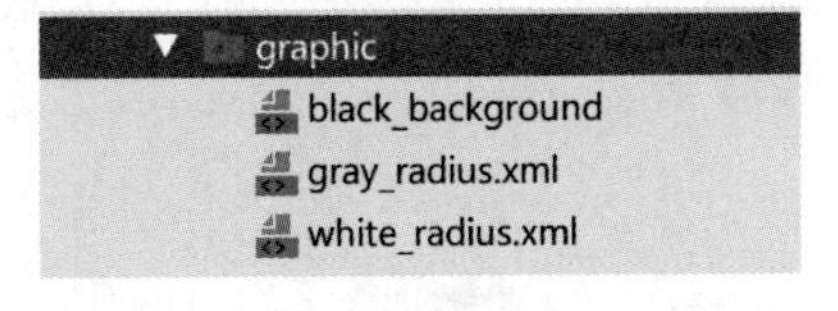

图 1.46 XML 文件

如要定义一个白色，圆角为 5dip 的椭圆，可以创建一个 white_radius 的 XML 文件，代码如下：

```
<?xml version="1.0" encoding="utf-8"?>
<shape xmlns:ohos="http://schemas.huawei.com/res/ohos"
```

```
        ohos:shape = "oval">
<solid
        ohos:color = "#88ffffff"/>
<corners ohos:radius = "5dip" />
</shape>
```

(6) profile：profile表示其他类型文件，以原始文件形式保存，文件名可以随意定义。

1.7.3 Resource文件使用

1. Java代码访问资源组文件

base目录与限定词目录中的资源文件可以通过指定资源类型(type)和资源名称(name)来引用。Java文件引用资源文件的格式为ResourceTable.type_name。特别地，如果引用的是系统资源，则采用ohos.global.systemres.ResourceTable.type_name。

如在Java文件中，引用string.json文件中类型为String、名称为app_name的资源，则代码如下：

```
ohos.global.resource.ResourceManager resManager = getAbilityContext().getResourceManager();
String result = resManager.getElement(ResourceTable.String_app_name).getString();
```

在Java文件中，引用color.json文件中类型为Color、名称为red的资源，则代码如下：

```
ohos.global.resource.ResourceManager resManager = getAbilityContext().getResourceManager();
int color = resManager.getElement(ResourceTable.Color_red).getColor();
```

2. Java代码访问原生资源文件

访问原生资源文件可以通过指定文件路径和文件名称来引用。

例如在Java文件中，引用一个路径为resources/rawfile/、名称为example.js的资源文件，则代码如下：

```
ohos.global.resource.ResourceManager resManager = getAbilityContext().getResourceManager();
ohos.global.resource.RawFileEntry rawFileEntry = resManager.getRawFileEntry("resources/
rawfile/example.js");
```

3. XML文件中引用资源文件

如果需要在XML文件中引用资源文件，需要用$type:name来引用。如果引用的是系统资源，则采用$ohos:type:name。

例如，在XML文件中，引用string.json文件中类型为String、名称为app_name的资源，则代码如下：

```
<?xml version = "1.0" encoding = "utf - 8"?>
<DirectionalLayout xmlns:ohos = "http://schemas.huawei.com/res/ohos"
```

```
    ohos:width = "match_parent"
    ohos:height = "match_parent"
    ohos:orientation = "vertical">
< Text ohos:text = " $ string:app_name"/>
</DirectionalLayout >
```

1.8 日志管理工具

HarmonyOS 为开发者提供了 HiLog 日志系统，让应用可以按照指定类型、指定级别、指定格式字符串输出日志内容，帮助开发者了解应用的运行状态，以便更好地调试程序。

输出日志的接口由 HiLog 类提供。HiLog 中定义了 5 种日志级别，用于输出不同级别的日志信息。

（1）Debug：输出 Debug 级别的日志。Debug 级别日志表示仅用于应用调试，默认不输出，输出前需要在设备的"开发人员选项"中打开"USB 调试"开关。

（2）Info：输出 Info 级别的日志。Info 级别日志表示普通的信息。

（3）Warn：输出 Warn 级别的日志。Warn 级别日志表示存在警告。

（4）Error：输出 Error 级别的日志。Error 级别日志表示存在错误。

（5）Fatal：输出 Fatal 级别的日志。Fatal 级别日志表示出现致命错误、不可恢复错误。

这里在 Hello World 项目中体验一下日志工具。

首先在输出日志前，需要调用 HiLog 的辅助类 HiLogLabel 来定义日志标签。打开 MainAbility.java 文件，在其中定义 HiLogLabel 类实例，代码如下：

```
static final HiLogLabel label = new HiLogLabel(HiLog.LOG_APP, 0x00201, "MY_TAG");
```

其中，HiLog.LOG_APP 用于指定输出日志的类型，当前 HiLog 中只提供了一种日志类型，即应用日志类型 LOG_APP。0x00201 用于指定输出日志所对应的业务领域，取值范围为 0x0～0xFFFFF，可以根据需要自定义。MY_TAG 用于指定日志标识，可以设置为任意的字符串。

这里输出一条 Warn 级别的日志信息，代码如下：

```
//输出一条 Warn 级别的日志信息
public class MainAbility extends Ability {

    static final HiLogLabel label = new HiLogLabel(HiLog.LOG_App, 0x00201, "MY_TAG");
    private String URL = "www. * * * .com";
    public int errno = 503;

    @Override
```

```
    public void onStart(Intent intent) {
        super.onStart(intent);
        super.setMainRoute(MainAbilitySlice.class.getName());
HiLog.warn(label, "Failed to visit %{private}s, reason: %{public}d.", URL, errno);
    }
}
```

其中，定义了虚拟的一个 URL 网址字符串和一个错误码 errno，在 Hilog.warn()日志输出语句中，第一个参数 label 为定义好的 HiLogLabel 标签，第二个双引号内的参数为字符串，用于日志的格式化输出，其中也可以设置多个参数，上述代码中的%s、%d 表示参数类型为 string 和 int 的变参标识，URL 和 errno 的参数类型和数量必须与标识一一对应。每个参数需添加隐私标识，分为{public}或{private}，默认为{private}，{public}表示日志打印结果可见，{private}表示日志打印结果不可见。

这时运行程序，在下方的 HiLog 窗口中可以查看日志信息，当日志信息较多时，可以通过设置设备信息、进程、日志级别和搜索关键词来筛选日志信息。如这里根据运行情况选择了设备信息和进程信息，然后选择 Warn 级别的日志，可以根据设置的 0x00201 或 MY_TAG 进行日志的筛选和查找。这里搜索 00201，就得到了对应的日志信息，如图 1.47 所示。

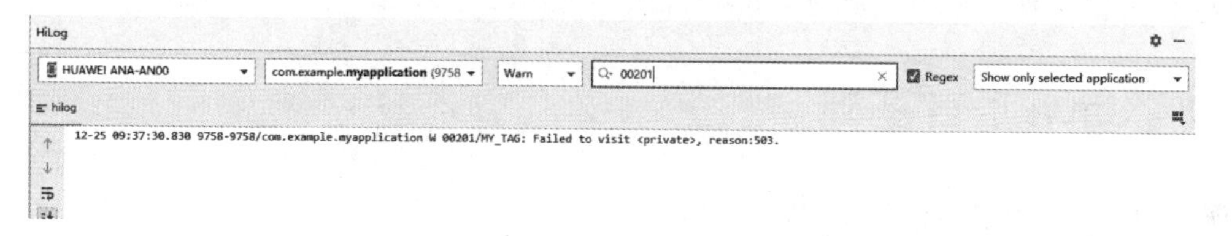

图 1.47 日志输出信息

第2章

Java

HarmonyOS之所以选择Java语言作为开发语言之一，是因为Java语言具有很多其他语言不具备的优势。首先，Java是一种面向对象的程序设计语言，较为适合大型软件的设计和实现。其次，Java还提供了丰富的类库和第三方开发包，会极大地减少开发人员的开发成本。除此之外，Java还有多线程性、安全性、分布式等多种特点，以及较好的跨平台性，在不同的平台上执行时不需要重新编译，即一次编译后，可以在多个系统平台上运行。

因此，在这一章会讲解Java语言的一些基础特性知识，便于初学者直接进行应用开发。有Java基础的读者可直接跳过本章，没有Java基础的读者可通过学习本章知识，并结合其他与Java相关的书籍，补充必要的知识基础。

17min

2.1 Java语言基础

2.1.1 面向对象编程

在软件开发流行初期，开发者普遍使用面向过程编程模式，面向过程是一种以事件为中心的编程思想，主要关注“怎么做”。开发者把需要解决的问题切割为多个步骤，然后按照一定的顺序，通过函数实现这些步骤，但是随着软件规模的不断增大，软件迭代过程中的可扩展性也变得越来越重要。当扩展一个新功能时，面向过程的编程模式可能需要重构整个项目，因此，开发者开始将另一种开发思想引入程序中，即面向对象的开发思想。

面向对象编程OOP（Object Oriented Programming），是一种以对象为中心的编程思想，主要关注“谁来做”。面向对象的思想是人类最自然的一种思考方式，程序设计过程中尽可能模拟人类的思维方式，使得软件开发方法与过程尽可能接近人类认识世界、解决现实问题的方法和过程。OOP＝对象＋类＋继承＋多态＋消息，其中对象和类是核心部分。

对象是事物存在的具体表现形式。一个人、一台计算机、一只大雁等都能称为对象，所以对象是组成系统的最基础的单位。在程序设计中，一般会思考一个对象具有哪些部分。以人为例，观察一个人一般会先观察这个人的表象属性：高矮、胖瘦、性别，然后观察其动作行为：微笑、行走、说话，如图2.1所示，因此，人们一般通过对象的属性和行为来认识一个对象。

类是对现实世界的抽象，是一个抽象的概念集合，不能将单个事物描述成一类，就好像不能把一只鸟称为鸟类。类是同一类事物的统称。同一类事物通常具有一些相似特征，例如鸟类，鸟类具有翅膀这一属性，并且可以通过这一属性实现飞行的行为。像这样具有相同属性和行为的一类实体，将它称为类。类是封装对象的属性和行为的载体，如图 2.2 描述了鸟类所封装的部分共有属性和行为。

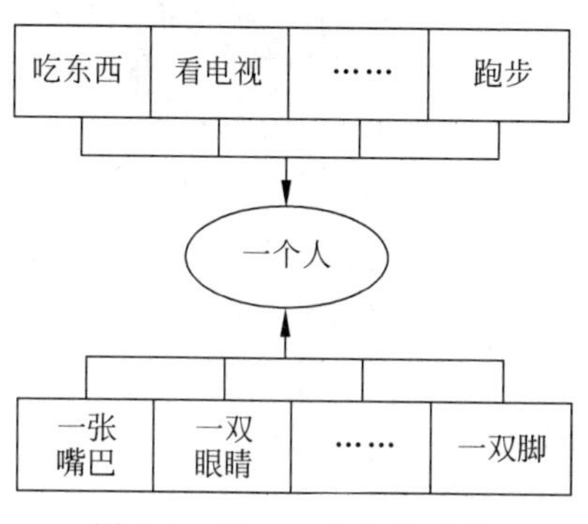

图 2.1 当一个人作为对象

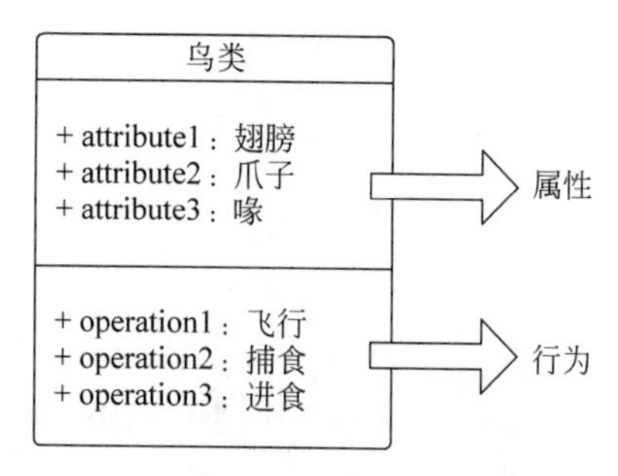

图 2.2 鸟类共有的部分属性和行为

类和对象的关系就像模具和铸件的关系一样，类的实例化结果是对象，而将对象抽象出来就是类，换句话说，类描述了一类具有相同属性和行为的对象。

2.1.2 Java 程序基本结构

这是一段简单的 Java 程序，代码如下：

```
//hello.java
package hello                                   //定义包名

public class Structure {                        //创建类
    static int num = 1;                         //定义类的成员变量
    public static void main(String[ ] args) {   //定义主方法
        String str = "这是 Java 程序";           //定义局部变量
        System.out.println(num);                //输出成员变量的值
        System.out.println(str);                //输出局部变量的值
    }
}
```

一段简单的 Java 程序基本上可以由以下几个部分组成。

1. 包名

Java 文件的第一行应先声明本文件所属的包。包可以简单地理解为“文件夹目录”。当新建一个 Java 工程时，工程中有一个 src 文件夹，这个文件夹可以简单地理解为“源码文件夹”，package 包名就是在这个文件夹下的路径，Java 中通过 package 关键字定义包名。例如在某个 Java 文件里，package 包名为 com.huawei.myapplication，如图 2.3 所示。

则它在 src 文件夹中的路径如图 2.4 所示。

```
package com.huawei.myapplication;

import ...

/**
 * Created by ve on 2020/10/30.
 */

public class mLog {
    public static void log(String TAG, String content) {
//        Log.d(TAG, content);
        HiLogLabel LABEL_LOG = new HiLogLabel(3, 0, TAG);
        HiLog.info(LABEL_LOG,content);
        }
        }
```

图 2.3　包名 com. huawei. myapplication

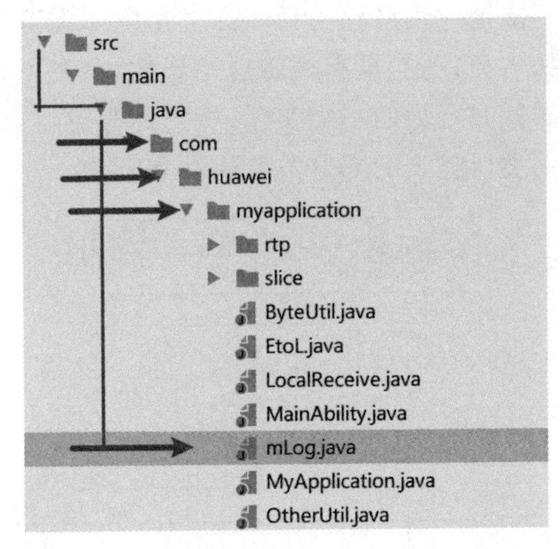

图 2.4　com. huawei. myapplication 包文件路径

2. 类和对象

类是具有共同属性和行为的对象的集合，用来描述一类对象的行为和状态，是客观世界中某类群体的一些基本特征抽象。对象则是类的实例化，指一个个具体的东西。如动物可以表示为一个类，但它只是一个概念，不存在具体的实体，而动物中每个实体，如一只猫、一条狗等，都可以称为一个对象。

在 Java 中类需要使用 class 关键字进行定义，使用 new 关键字创建对象，代码如下：

```
//定义 Animal 类
public class Animal{
    //类的主体内容
}

//创建 Animal 对象
Animalanimal = new Animal();
```

3. 类的成员变量和成员方法

对象拥有的属性称为类的属性，类的成员变量定义了类的属性，如动物类中，一般有名字、颜色、年龄等属性。定义类的成员变量示例代码如下：

```
//定义 Animal 类的成员变量
public class Animal{
    public String name;          //名字
    public String color;         //颜色
    private int age;             //年龄
}
```

对象执行的操作称为类的方法，类的成员方法定义了类的行为。如动物类中，一般有行走、进食、睡觉等行为。一个完整的方法，主要由方法名、方法参数、返回值类型和方法体组成。定义类的成员方法示例代码如下：

```
//定义 Animal 类的成员方法
public class Animal{
    public void sleep(){
        System.out.println("现在在睡觉");
    }
}
```

4. 主方法

main 方法是程序的入口，根据 Java 的语言规范，main 方法必须声明为 public，且 main 方法必须是静态的，用关键字 static 修饰。void 表示该方法不需要有返回值。

2.2 类与对象

2.2.1 类

在 2.1 节了解到类描述了一类具有相同属性和行为的对象，它是封装对象的属性和行为的载体，而在 Java 语言中，类中描述对象属性的部分以成员变量的形式存在，类中描述对象行为的部分以成员方法的形式存在，代码如下：

```
class 类名称 {
    //声明成员变量
    //声明成员方法
}
```

1. 成员变量和成员方法

成员变量对应的是类中对象的属性，以 Time 类为例，在 Time 类中设置了 3 个成员变量：hour、minute、second，分别对应时间的时、分和秒，代码如下：

```
//Time.java
class Time {
    //声明成员变量
    private int hour;
    private int minute;
    private int second;
    //声明成员方法
}
```

从上述代码中可以看出，成员变量的设置与一般变量的声明相同，可以在声明的同时设置初始值，如果不设置初始值也会赋值默认值。

成员方法对应的是类中对象的行为，同样以 Time 类为例，在 Time 类中设置了 getHour()和 setHour()两种方法，这两种方法的作用分别是获取时间和设置时间，代码如下：

```
//Time.java
class Time {
    //声明成员变量
    private int hour;
    private int minute;
    private int second;
    //声明成员方法
    public int getHour() {
        return hour;
    }
    public void setHour(int hour) {
        this.hour = hour;
    }
}
```

成员方法可以根据有无返回值分为两种，一种是有明确返回值的方法，另一种是返回 void 类型的方法。也可以根据参数分为带参数方法和无参数方法。总结出一种通用的定义成员方法的语法格式如下：

```
权限修饰符返回值类型方法名(参数类型参数名) {
    ... //方法体
    return 返回值;
}
```

一般情况下，成员方法会对每个成员变量设置 get 和 set 两个函数，用来对成员变量进行赋值和获取成员变量值。

2. 权限修饰词

Java 中主要的权限修饰词包括 private、public 和 protected，还有 default 表示没有权限修饰词，这些权限修饰词控制着对类和类中的成员变量及成员方法的访问。访问权限控制的等级从大到小依次为 public > protected > default > private。

- public：除了在本类中使用，还可以在子类和其他包中使用。
- protected：只能在本包内的类和子类中使用。
- default：当前包中所有类对该成员都有访问权限。
- private：除了本类，其他类都无法访问该成员。如果父类不允许继承它的子类访问它的成员变量，则可以使用 private 修饰词。

public、protected、default 和 private 4 个修饰词的修饰权限如表 2.1 所示。

表 2.1 public、protected、default 和 private 的修饰权限

访问权限	修饰词	同类	同包	子类	不同的包
公开	public	√	√	√	√
受保护	protected	√	√	√	×
包访问	default	√	√	×	×
私有	private	√	×	×	×

3. 类的构造方法

在一个 Java 类中，其方法可以分为成员方法和构造方法。构造方法是一个与类名相同的方法，Java 中每个类都有一个默认的构造方法，用来初始化类的一个新对象，在创建对象时会自动调用。若在代码中没有添加构造方法，编译器会自动添加一个默认的无参构造方法。含有构造方法的 Timer 类代码如下：

```
//Time.java
public class Time {
    public Time() {
        hour = 11;
        minute = 11;
        second = 11;
    }
    public Time(int hour) {}
    public Time(int hour, int minute) {}
    public Time(int hour, int minute, int second) {}
}
```

在构造方法中可以为成员变量赋值，这样当实例化一个类的对象时，相应的成员变量也会一并被初始化。

构造方法和普通的成员方法有很多不同，例如构造方法没有返回值，而且也不用添加 void 用来标明这个构造方法没有返回值。其次，构造方法在命名上非常严格，必须与类名相同，只有当名字相同时，编译器才会将这个与类名同名的方法当作构造方法。一个类中必须有一个构造方法，也可以有多个构造方法，这些构造方法名字都相同，一般通过方法中传入的参数来区分，即重载。以上面的代码为例，Time 类中定义了 4 个构造函数，其中参数数量也各有不同。那么当实例化一个对象时，编译器会根据所传递参数的数量和类型来选择对应的构造函数，所以当实例化一个 time 对象时，代码如下：

```
Time time = new Time(11,11);
```

则对应 Time 类中第 3 个构造函数，对应的构造函数代码如下：

```
Time time = new Time(11,11);      //第 3 个构造函数
```

4. this 关键词

在 Java 和 HarmonyOS 开发中，this 关键词一直都非常重要，this 关键词主要有 3 个方面的应用：

(1) this 调用本类中的属性，即类中的成员变量。

this 关键词用来表示本类对象的引用，代码如下：

```
public void setHour(int hour) {
        this.hour = hour;
}
```

从上述代码可以看到，成员变量 hour 和 setHour 中设置的形式参数名称相同。在 Java 中规定，this 关键词可以隐式地引用对象的成员变量和方法，例如在上述代码中，this. hour 表示的是 Time 类中的 hour 成员变量，而后面的 hour 即为形式参数。

(2) this 调用本类中的其他方法。

this 关键词还可以调用本类中的其他构造方法，调用时要放在构造方法的首行，代码如下：

```
//Time.java
public class Time {
    public Time() {
        this(11,12,13);
    }

    public Time(int hour,int minute,int second) {}
}
```

在上述代码中，定义了两个构造方法，一个为无参构造，另一个为有参构造。在无参构造方法中，通过 this()调用有参构造，并为 hour、minute 和 second 赋值。

(3) this 关键词除了可以引用成员变量和方法之外，还可以返回类的引用，代码如下：

```
public Time getTime() {
    return this; //返回 Time 类引用
}
```

getTime()方法的返回值是 Time 类，而使用 this 关键词作为类的引用，所以方法中使用的 return this 实际是返回一个 Time 类的对象。

2.2.2 对象

对象是类抽象出来的结果，所有的问题都由对象进行处理。在 Java 语言中通过 new 来创建新的对象，而实质的创建过程是调用前文中讲解的构造函数，因此，准确地说，在 Java 语言中使用 new 操作来调用构造函数创建对象，代码如下：

```
Time time = new Time();
Time time = new Time(11);
```

其中，Time 是类名，time 是被创建的对象名，new 是操作符，11 是传入构造函数中的参数。

每个对象都是相互独立的，创建之初，JVM 虚拟机会为它们分配独立的内存空间，互不影响，并且每个对象都有自己的生命周期，当对象的生命周期结束时，JVM 虚拟机会通过垃圾回收机制将对象回收。创建 Time 对象代码如下：

```
//Time.java
class Time {
    private int hour;
    private int minute;
    private int second;
    public Time() {
        System.out.println("成功创建对象");
    }

    public static void main(String [] args) {
        Time time = new Time();
    }
}
```

将上述代码保存为 Time.java 文件，并在控制台运行，结果如图 2.5 所示。

图 2.5 Time.java 控制台执行结果

创建完对象之后就可以获取对象自身的属性和行为。在 Java 中引用对象的属性和行为需要使用“.”操作符，对象名在圆点左边，属性或者方法在圆点右边，代码如下：

```
对象名.属性              //访问对象的属性
对象名.方法名()          //访问对象的方法
```

还是以 Time 类为例，分别创建两个对象 t1 和 t2，在构造函数中初始化时间，在主函数中用“.”引用 t2 的属性并赋上新的值，分别打印两个实例的时间，代码和运行结果如下：

```
//Time.java
class Time {
```

```
    private int hour;
    private int minute;
    private int second;
    public Time() {
        hour = 11;
        minute = 11;
        second = 11;
    }

    public static void main(String [] args) {
        Time t1 = new Time();
        Time t2 = new Time();
        t2.hour = 0;
        t2.minute = 0;
        t2.second = 0;
        System.out.println("第一个实例对象的时间: " + t1.hour + ":" + t1.minute + ":" +
t1.second);
        System.out.println("第二个实例对象的时间: " + t2.hour + ":" + t2.minute + ":" +
t2.second);
    }
}
```

运行结果：

```
第一个实例对象的时间: 11:11:11
第二个实例对象的时间: 0:0:0
```

从运行结果可以看出，不同对象的属性值互不影响，虽然参数名称相同，但第一个对象t1在打印时间时仍然是成员变量的初始值，而第二个对象t2打印出的是设置的属性值，改变t2的hour属性值并不会影响t1的hour属性值。

2.3 继承、接口、抽象类与多态

2.3.1 继承

在前面学习了面向对象编程中类的概念，一个工程可以由单个或者多个类组成，但是当类的数目非常庞大，或者需要使用多个相似的类时，分别把每个类都从头到尾实现一遍的做法是非常低效的，因此Java为类添加了继承的特性，其基本的用法就是从一个现有的类(称为父类)扩展(extend)出一个新的类(称为子类)，而继承的意思就是指子类从父类继承父类的全部成员，这些成员包括父类所含的变量、方法及内嵌类。如子类Son继承自父类Parent，示意图如图2.6所示。

如图2.6所示，父类Parent拥有成员变量a、b及方法func1()和func2()。子类Son拥有成员变量a、b、c及方法func1()、func2()和func3()，其中，a、b、func1()和func2()继承于

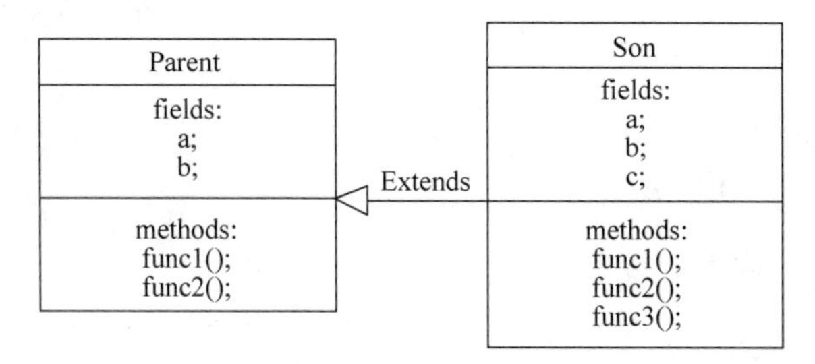

图 2.6 子类继承父类示意图

父类 Parent，而 c 和 func3()是子类 Son 所独有的。简而言之，也就是父类所拥有的成员、方法，子类都有，并且子类还可以同时拥有更多其他的成员和方法。

在定义子类时，只需让子类继承于父类，并且在子类中加上需要的新成员，这样不仅节省了开发时间，提高了开发效率，而且还使代码看起来更加简洁。继承通过 extends 关键字实现，首先定义一个父类 Parent，代码如下：

```
//Parent.java
public class Parent {
    //fields
    private int a;
    public int b;

    //methods
    private void func1(){...};
    public void func2(){...};
}
```

此时创建一个子类 Son 继承于 Parent，则在子类名称后接 extends 关键字，extends 后是要继承的父类名称，代码如下：

```
//Son.java
public class Son extends Parent {
    //fields
    public int c;

    //methods
    public son();
    public void func3(){...};
}
```

此时的子类 Son 看起来非常简洁，但是不代表其内部只有一个成员变量和一个成员方法，事实上 Son 内部的代码应该如下：

```
//Son.java
public class Son extends Parent {
    //fields
    private int a;
    public int b;
    public int c;

    //methods
    public son(){...};
    private void func1(){...};
    public void func2(){...};
    public void func3(){...};
}
```

在定义子类的新方法时还可以使用父类已经有的方法，这时要使用 super 关键字来代表父类，例如在定义 Son 的构造函数时，代码如下：

```
public class Son(){
    super();
    ...
};
```

super()的意思是首先构造出父类所拥有的所有特性，再在省略号处构造子类的新特性。

类之间的继承其实是非常自由的，而除了上面所讲的子类可以在继承时新增其他的成员外，还可以不使用父类已经定义好的方法，而是根据需求将父类的方法重写。即在父类的方法不再适用于子类的情况下，子类可以改写继承自父类的方法。例如在自然界中，人和鲸都是哺乳动物，将哺乳动物定义为一个父类，而将人和鲸定义为继承于哺乳动物的子类。最初的哺乳动物是在陆地上用四肢爬行进行移动的，因此将哺乳动物的“移动”方法定义为“使用四肢爬行”，但是子类人和鲸在各自的生存环境内均不是使用四肢爬行来移动的，因此在子类“人”和“鲸”中需要重写“移动”这种方法。在“人”子类中，将“移动”定义为“双腿步行”，而在“鲸”子类中，将“移动”定义为“鲸尾划水”，示意图如图 2.7 所示。

将上面的例子转移到子类 Son 中，假设在定义 Son 的时候需要重写父类 Parent 中的 func2()方法，则代码如下：

```
//Son.java
public class Son extends Parent {
    public Son(){
        super();
        ...
    };
    @Override
    public void func2(){...};
}
```

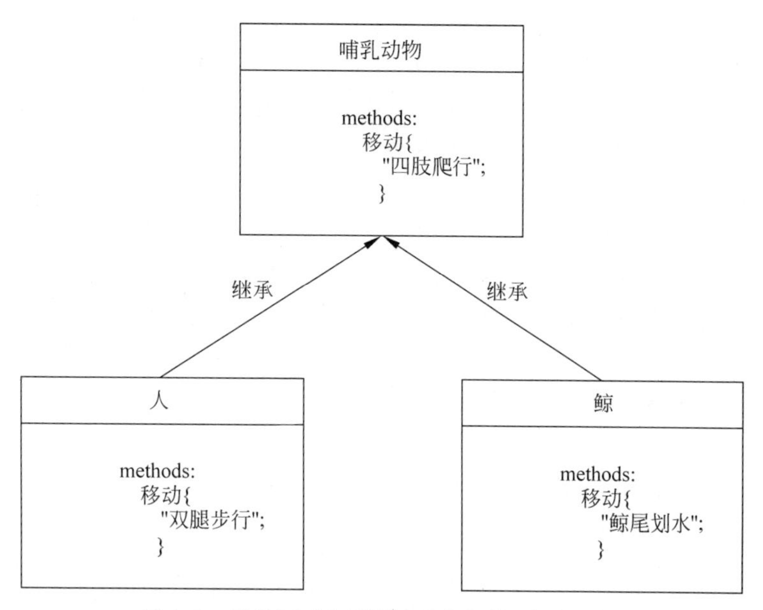

图 2.7　子类"人"和"鲸"继承自父类"哺乳动物"

其中的@Override是Java中的一种注解，代表在子类Son中重写了父类的func2()方法。在重写父类方法时，必须保证父类中有这种方法，也就是必须保证前面的Parent类中有func2()方法。

2.3.2 抽象类

当一个类中所包含的信息(成员、方法)不足以描述一个完整的类时，这样的类就叫作抽象类，因此抽象类不能直接被实例化，而当需要子类继承它时，由子类进行实例化。

例如定义3个类："人类""教师"和"司机"，其中，"人类"为抽象类，而"教师"和"司机"继承于"人类"，代码如下：

```
//定义 Taecher 和 Driver 类继承 Human 类
public abstract class Human{
    public void work(){};
}

public class Teacher extends Human{
    @Override
    public void work(){System.out.println("授课");}
}
public class Driver extends Human{
    @Override
    public void work(){System.out.println("开车");}
}
```

在 Human 类中的 work()方法不能直接实现,因为不同的人工作方式是不同的,因此 work()方法需要在具体的子类当中实现,例如教师的工作就是授课,而司机的工作就是开车。

抽象类一般用在项目开始阶段,通过抽取很多类的相似特征,将这些相似的特征统一声明在一个抽象类里,这样会使类间关系变得更加明确,从而增加代码的可读性。

2.3.3 接口

目前的 Java 特性不支持多重继承,即一个类最多只能继承于一个父类。有时必须从几个类中派生出一个子类,继承它们所有的属性和方法,此时通过接口即可达到多重继承的效果。

接口其实是一种特殊的抽象类,是由全局常量和公共的抽象方法所组成,而没有变量和具体的实现方法体。

假设定义一个类叫作"飞机",代码如下:

```
//飞机.java
public class 飞机 {
    private string wing;
    private string head;
    private string tail;

    void 起飞(){System.out.println("起飞")};
    void 降落(){System.out.println("降落")};
    void 飞行(){System.out.println("飞行")};
}
```

定义一个子类"战斗机"继承自"飞机",则"战斗机"继承了"飞机"的所有成员和方法,但是"战斗机"还需要拥有战斗功能,而不同装备的战斗方式均不同,无法统一具体的战斗方式,因此将"战斗"定义为一个接口,"攻击"和"防御"为抽象方法,代码如下:

```
public interface 战斗{
    void 攻击(){};
    void 防御(){};
}
```

接口定义采用关键字 interface,接口"战斗"中没有"攻击"和"防御"方法的执行体,在接口的不同实现类中,会对"攻击"和"防御"方法依据自身特性进行具体实现。例如定义"战斗机"和"坦克"类,分别实现它们关于战斗的方法,代码如下:

```
//"战斗机"类和"坦克"类实现"战斗"接口
public class 战斗机 extends 飞机 implements 战斗{
    @Override
    void 攻击(){System.out.println("发射导弹")};
```

```
    @Override
    void 防御(){System.out.println("发射干扰弹")};
}

public class 坦克 implements 战斗{
    @Override
    void 攻击(){System.out.println("发射穿甲弹")};
    @Override
    void 防御(){System.out.println("加速逃离")};
}
```

上述代码中,“战斗机”不但具有“飞机”的特性,还具有自身的“战斗”特性,而坦克也拥有自身的“战斗”特性。值得注意的是,在继承一个接口时,子类(抽象类除外)需要实现接口中的所有抽象方法,且一个类可以继承一个父类,并同时实现多个接口,但是 implements 部分必须放在 extends 部分之后。

这里需要注意几点接口与抽象类之间的区别:

(1) 接口里的方法均只是声明,而没有实现;抽象类里面的方法可以只是声明,也可以在抽象类里实现。

(2) 当一个类实现接口时需要实现接口所声明的所有方法,而继承于抽象类的类可以不用实现其父类的所有方法。

(3) 一个类在实现接口时,可以同时实现多个接口,而在继承抽象类时只能继承一个抽象类。

2.3.4 多态

多态即一个对象可以拥有多种不同的表现形式,也就是当父类中定义的属性和方法被子类继承后,可以具有不同的数据类型或表现出不同的行为,使得同一属性或方法在父类及其各个子类中具有不同的含义。

引用之前的例子,一个人的工作方式可以有多种,如果他是教师则工作方式为授课,如果他是司机则工作方式为开车,多态特性的示例代码如下:

```
public static void main(String[] args) {
    Human human;
    human = new Teacher();
    human.work();
    human = new Driver();
    human.work();
}
```

运行结果如下:

```
授课
开车
```

在上述代码中，声明了 Human 类的变量 human，当 human 指向的对象是 Teacher 时，调用 human. work()方法，执行的是“授课”，当 human 指向的对象是 Driver 时，调用 human. work()方法，则执行的是“开车”。无论 human 变量的对象是 Teacher 还是 Driver，它们都是 Human 类的子类，因此可以向上转型为该类，从而实现多态。

值得注意的是，在多态中引用的方法，必须在父类里有所声明或实现，即多态中只能使用在父类中已经声明的方法。

在本例中，父类 Human 为抽象类，实际上这里的 Human 可以是抽象类，可以是接口，也可以是一个普通类，均可以实现多态的效果。

第3章 Java UI

应用开发中,界面设计和 UI 布局是必不可少的。任何一个应用都需要在屏幕上显示用户界面,包含用户可查看并与之交互的所有内容。在 HarmonyOS 中,可通过 Java 和 JS 两种方式布局 UI。本章先讲述 Java UI 布局,在第 5 章将讲述 JS UI 布局。

在学习使用 Java 对 UI 进行设计之前,首先需要理解一下 HarmonyOS 的 UI 逻辑。依据功能和特点,可将 HarmonyOS 的 UI 组件分为两类:单体组件 Component 和可用于装载其他组件的容器组件 ComponentContainer。

UI 组件根据一定的层级结构进行组合形成整体的 UI。单体组件在未被添加到容器组件中时,既无法显示也无法交互,因此一个用户界面至少包含一个容器组件。此外,容器组件中不但可以添加单体组件,也可以添加其他的容器组件,这样就可以形成丰富多样的 UI 样式。HarmonyOS 的 UI 视图树范例如图 3.1 所示。

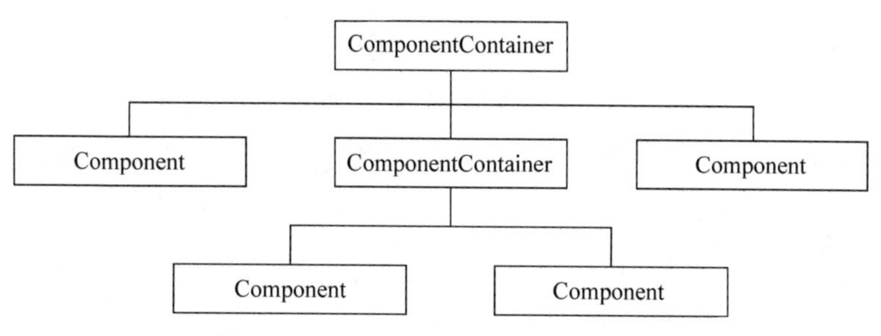

图 3.1　HarmonyOS 的 UI 视图树范例

单体组件和容器组件以树状的层级结构进行组织,这样的布局被称为视图树。视图树的特点是仅有一个根视图,其他视图有且仅有一个父节点,视图之间的关系受到父节点的规则约束。

3.1　Java UI 单体组件

本章主要讲解使用 Java 语言设置单体 UI 组件。

HarmonyOS 常用的单体 UI 组件全部继承自 Component 类,Component 类包含了对

UI 组件的全部常用方法，例如创建、更新、缩放、旋转及设置各类事件监听器。其他的 UI 组件，如 Text、Button 等，都是在 Component 类的基础上添加了对应的功能实现的。以下是常用单体组件的开发流程。

3.1.1 Text 组件

Text 是用来显示字符串的组件，是应用开发中最基础的组件，在界面上显示为一块文本区域。下面学习 Text 的用法。

1. Java 代码创建 Text 组件

首先按照第 1 章的介绍，创建一个 Phone 设备的 Java 模板，用来学习使用代码创建 UI 布局。在新建项目中的界面左侧找到 Project 项目栏，双击打开 entry→src→main→项目 Package 名称(本书中为 com. huawei. mytestapp)→Slice 中的 MainAbilitySlice. java 文件，如图 3.2 所示。

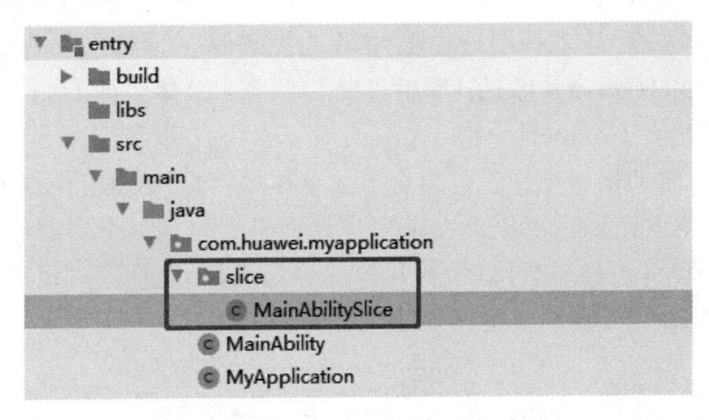

图 3.2 MainAbilitySlice 文件路径

在打开后的 Java 文件中找到 onStart()方法，修改其中的内容，代码如下：

```
//MainAbilitySlice.java
public void onStart(Intent intent) {
    super.onStart(intent);

    //容器
    DirectionalLayout myLayout = new DirectionalLayout(this);
    myLayout.setWidth(DirectionalLayout.LayoutConfig.MATCH_PARENT);
    myLayout.setHeight(DirectionalLayout.LayoutConfig.MATCH_PARENT);
    myLayout.setAlignment(LayoutAlignment.HORIZONTAL_CENTER);
    myLayout.setPadding(32,32,32,32);

    //Text 组件
    Text text = new Text(this);
    text.setWidth(DirectionalLayout.LayoutConfig.MATCH_CONTENT);
    text.setHeight(DirectionalLayout.LayoutConfig.MATCH_CONTENT);
```

```
        text.setText("My name is Text.");
        text.setTextSize(24, Text.TextSizeType.VP);
        myLayout.addComponent(text);

        super.setUIContent(myLayout);
    }
```

下面对代码的具体内容进行讲解。

第一个代码块对容器组件进行了设置，页面容器组件的相关说明会在下一节中进行。第二个代码块设置并添加了一个 Text 组件，其中前 5 行代码创建并设置了一个 Text 组件，每一句的具体意义如下：

第 1 行：实例化一个 Text 类的实例，实例名为 text。

第 2 和第 3 行：为 text 设置了宽和高属性，这样可以让组件的大小自动适配其内容(文本)的大小。

第 4 行：为 text 添加要显示的文字信息 My name is Text。

第 5 行：为 text 的文本字体设置大小 24vp。

至此，text 的属性就设置完毕了，为了使 text 能够在手机屏幕上显示，还需要执行 super.setUIContent(myLayout)，将创建好的 Text 组件实例 text 放入容器组件中，相关知识在后面讲到容器组件时会具体说明。运行上述代码，效果如图 3.3 所示。

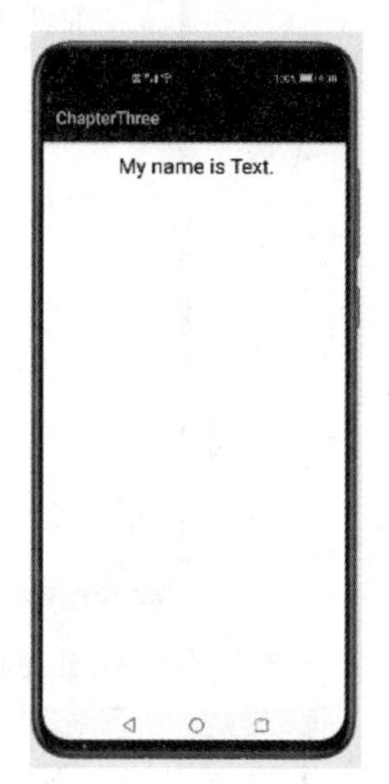

图 3.3　Text 组件的运行效果

从图 3.3 可以看出，页面顶部出现了一条黑色的文本信息，显示 My name is Text，说明上述设置均已生效。

还可以使用 Java 代码修改或添加一些属性，例如，使用 setTextColor()方法可以修改 text 的字体颜色，使用 setFont()方法可以修改 text 的字体。在 Text 组件代码块中添加代码如下：

```
//MainAbilitySlice.java 中的 Text 组件
Text text = new Text(this);
text.setWidth(DirectionalLayout.LayoutConfig.MATCH_CONTENT);
text.setHeight(DirectionalLayout.LayoutConfig.MATCH_CONTENT);
text.setText("My name is Text.");
text.setTextSize(24, Text.TextSizeType.VP);
text.setTextColor(Color.BLUE);
text.setFont(Font.DEFAULT_BOLD);
myLayout.addComponent(text);
```

上述代码将 text 的文本设置为蓝色，字体加粗。重新运行程序，运行结果如图 3.4 所示。

2. xml 创建 Text 组件

在实际应用中，为了代码的简洁美观，通常会使用 xml 进行布局，这里学习通过 xml 进行 UI 布局，实现相同的效果。在界面左侧的 Project 项目栏选择 entry→src→main→resources→base→layout，打开 ability_main. xml，如图 3.5 所示。

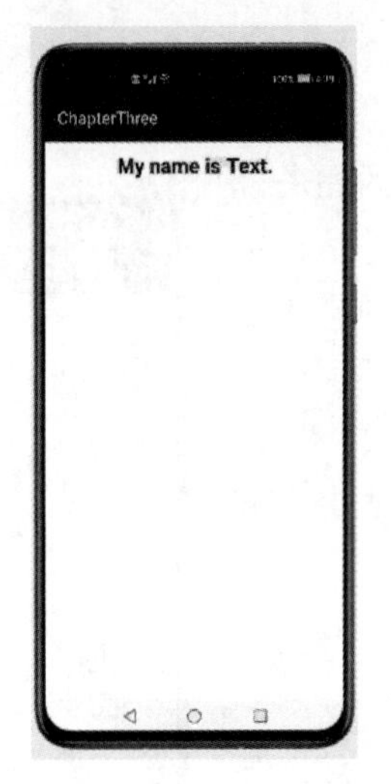

图 3.4　Text 组件的属性设置效果

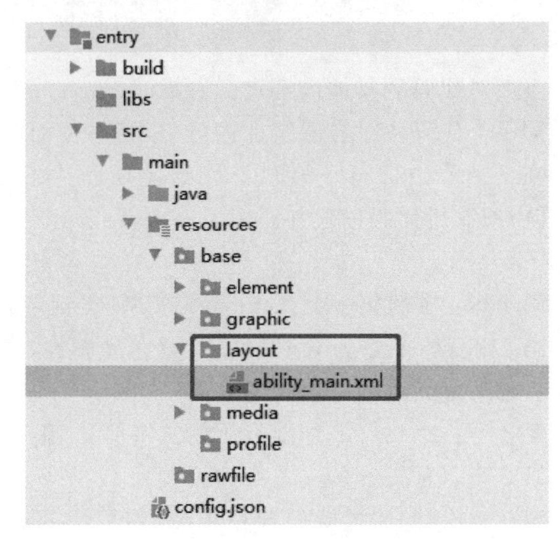

图 3.5　xml 布局目录结构

打开 ability_main. xml 文件，对布局和组件进行描述，代码如下：

```
<! -- ability_main.xml -- >
<?xml version = "1.0" encoding = "utf - 8"?>
< DirectionalLayout
        xmlns:ohos = "http://schemas.huawei.com/res/ohos"
        ohos:width = "match_parent"
        ohos:height = "match_parent"
        ohos:orientation = "vertical"
        ohos:padding = "32">

< Text
            ohos:id = " $ + id:text"
            ohos:width = "match_content"
            ohos:height = "match_content"
            ohos:layout_alignment = "horizontal_center"
            ohos:text = "My name is Text. "
            ohos:text_size = "24vp"/>

</DirectionalLayout >
```

其中，最外层的 DirectionalLayout 指页面的整体竖向布局。在 Text 中，使用 ohos：id 为当前控件定义一个唯一的标识 id，布局中的组件通常都需要设置独立的 id，以便在程序中查找该组件。若布局中有不同的组件设置了相同的id，则通过id查找会返回查找到的第一个组件，因此尽可能保证id的唯一性。使用ohos:width和ohos:height为控件指定了宽度和高度，其属性值match_parent表示组件大小将扩展为父容器允许的最大值，占据父组件方向上的剩余大小，即由父组件决定当前控件的大小。属性值match_content表示组件大小与它的内容占据的大小范围相适应，即由控件内容决定当前控件的大小。除此之外也可以通过具体的数值设置宽和高，例如24(以像素为单位)或24vp(以屏幕相对像素为单位)。使用ohos:layout_alignment指定文字的对齐方式，其中horizontal_center指水平方向居中。通过ohos:text设置Text的具体内容。

随后，在MainAbilitySlice.java中，注释掉通过Java创建Text组件的代码，并通过setUIContent()加载该xml布局。修改MainAbilitySlice.java中的代码如下：

```
//MainAbilitySlice.java
public class MainAbilitySlice extends AbilitySlice {

    @Override
    public void onStart(Intent intent) {
        super.onStart(intent);
        …                                              //注释通过 Java 创建 Text 组件代码
        super.setUIContent(ResourceTable.Layout_ability_main);   //加载 xml 布局
    }
}
```

现在运行程序，效果如图3.6所示。

在xml中也可以对Text设置字体颜色及字重。在ability_main.xml中，增加属性代码如下：

```
<!-- ability_main.xml -->
<?xml version="1.0" encoding="utf-8"?>
<DirectionalLayout
        xmlns:ohos="http://schemas.huawei.com/res/ohos"
        ohos:width="match_parent"
        ohos:height="match_parent"
        ohos:orientation="vertical"
        ohos:padding="32">

<Text
            ohos:id="$+id:text"
            ohos:width="match_content"
            ohos:height="match_content"
            ohos:layout_alignment="horizontal_center"
            ohos:text="My name is Text."
            ohos:text_size="24vp"
```

```
            ohos:text_color = "blue"
            ohos:text_weight = "700"/>

</DirectionalLayout>
```

通过 ohos:text_color 属性可以设置文本的颜色，通过 ohos:text_weight 属性可以设置文本的字重。重新运行后效果如图 3.7 所示，可见 text 文本变为蓝色，且字体加粗。

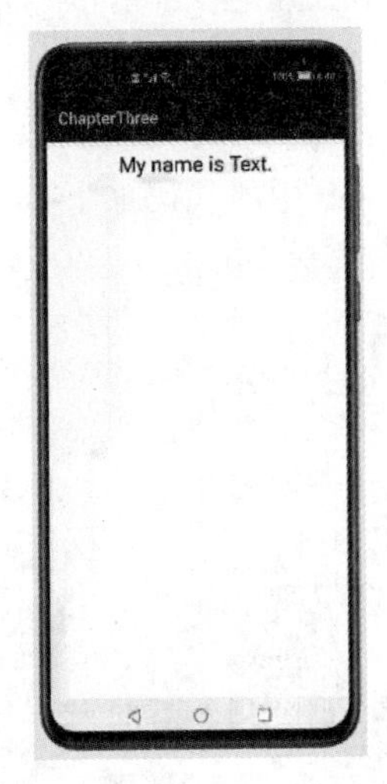

图 3.6 xml 创建 Text 组件的运行效果

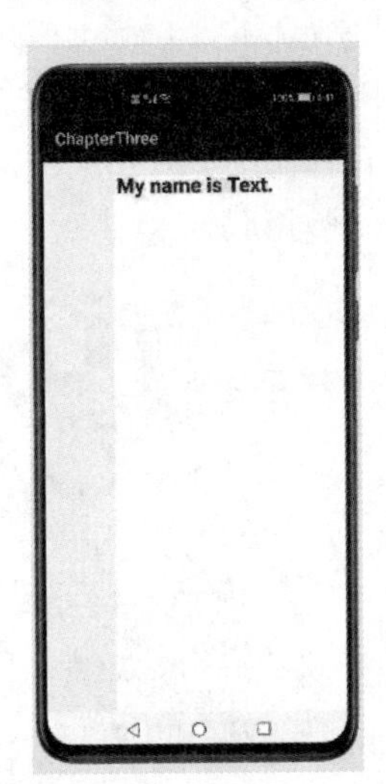

图 3.7 xml 中设置 Text 颜色及字重运行效果

除此之外，Text 还有很多其他的属性，在开发过程中根据需求查阅相关文档即可。

3.1.2 Button 组件

Button(按钮)是一种常见的与用户进行交互的组件，单击可以触发对应的操作，可以由图标和文本共同组成。

1. Java 代码创建 Button 组件

同样地，还是先尝试使用 Java 代码来添加一个 Button 组件。打开 MainAbilitySlice.java，在 onStart()函数中添加如下代码：

```
//MainAbilitySlice.java 中添加 Button 组件
Button button = new Button(this);
button.setWidth(DirectionalLayout.LayoutConfig.MATCH_PARENT);
button.setHeight(DirectionalLayout.LayoutConfig.MATCH_CONTENT);
button.setText("button");
button.setTextSize(28, Text.TextSizeType.VP);
ShapeElement backgroundElement = new ShapeElement();
backgroundElement.setRgbColor(new RgbColor(0xDC,0xDC,0xDC));
```

```
button.setBackground(backgroundElement);
myLayout.addComponent(button);
```

值得注意的是，此时如果直接运行程序会发现 UI 界面没有变化，这是因为在 3.1.1 节最后使用了 setUIContent()加载了 xml 布局 ability_main.xml，而修改的代码是创建的布局 myLayout，这是两套互不干涉的布局，所以如果想看到刚刚在代码布局中添加的 button，需要将 setUIContent()函数中的变量改为 myLayout，然后运行程序，这样就可以看到的 text 下方多出了 button。Button 组件也有丰富的自定义属性，大家同样可以根据自己的需求来查阅相关文档。

2. 使用 xml 创建 Button 组件

同样也可以使用 xml 来添加 Button 组件，在刚才编写的 ability_main.xml 文件中增加 Button，代码如下：

```
//ability_main.xml
<?xml version = "1.0" encoding = "utf-8"?>
<DirectionalLayout
    xmlns:ohos = "http://schemas.huawei.com/res/ohos"
    ohos:width = "match_parent"
    ohos:height = "match_parent"
    ohos:orientation = "vertical"
    ohos:padding = "32">

    ...

<Button
    ohos:id = "$ + id:button"
    ohos:width = "match_parent"
    ohos:height = "match_content"
    ohos:text = "button"
    ohos:text_size = "28vp"
    ohos:background_element = "$ graphic:buttonelement"/>

</DirectionalLayout>
```

Button 可配置属性与 Text 差不多一致。其中，ohos:background_element 为对 Button 引用背景色，属性值 $graphic:buttonelement 指为 Button 创建的背景色文件。在左侧 Project 窗口中，打开 entry→src→main→resources→base→graphic 文件夹，右击选择 New→File，命名为 buttonelement.xml，并在 buttonelement.xml 中定义 Button 的背景，代码如下：

```
<!-- buttonelement.xml -->
<?xml version = "1.0" encoding = "utf-8"?>
```

```
<shape xmlns:ohos = "http://schemas.huawei.com/res/ohos"
       ohos:shape = "rectangle">
<solid
           ohos:color = "#DCDCDC"/>
</shape>
```

完成上述添加 Button 的操作后，在 MainAbilitySlice.java 中，注释通过 Java 添加布局的代码，并通过 super.setUIContent(ResourceTable.Layout_ability_main)引用 xml 布局文件，运行后界面如图 3.8 所示。

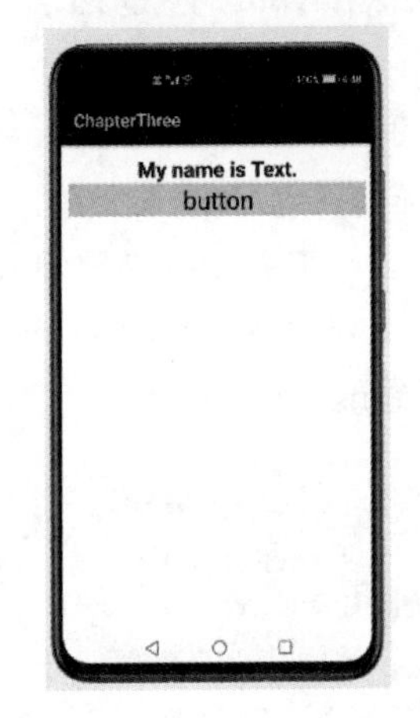

图 3.8　xml 中创建 Button 组件运行效果

Button 组件的一个重要功能是当用户单击 Button 组件时，会执行相应的操作。这就需要让系统为 Button 组件设置一个单击事件监听器，监听器会在一个新的线程中不断地检测这个 Button 组件的单击事件。当用户单击按钮时，Button 对象会收到一个单击事件，开发者可以自定义响应单击事件的方法。

例如，通过创建一个 Component.ClickedListener 对象，然后通过调用 setClickedListener 将其分配给 Button 对象，在 onClick()方法中可以添加单击 Button 后的事件逻辑。通过上述方法实现对 Button 的事件监听，示例代码如下：

```
//Button 单击事件代码示例
public class MainAbilitySlice extends AbilitySlice {

    @Override
    public void onStart(Intent intent) {
        super.onStart(intent);
        super.setUIContent(ResourceTable.Layout_ability_main);

        //从定义的 xml 中获取 Button 对象
        Button button = (Button) findComponentById(ResourceTable.Id_button);
        if(button != null) {
            //为按钮设置单击事件监听器
            button.setClickedListener(new Component.ClickedListener() {
                public void onClick(Component v) {
                 //此处添加单击按钮后的事件处理逻辑
                }
            });
        }
    }
}
```

下面通过一个实例展示一下按钮的功能。打开 MainAbilitySlice.java，在 onStart()函数中添加如下代码：

```
//MainAbilitySlice.java
super.onStart(intent);
super.setUIContent(ResourceTable.Layout_ability_main);
Text text = (Text) findComponentById(ResourceTable.Id_text);
    Button button = (Button) findComponentById(ResourceTable.Id_button);
    if(button != null && text!= null){
        button.setClickedListener(new Component.ClickedListener() {
        @Override
        public void onClick(Component component) {
        text.setTextColor(Color.RED);
        text.invalidate();
        }
        });
    }
```

在这段代码中首先从 xml 中取出了 text 和 button 两个实例，然后为 button 设置了单击事件监听器，并将修改 text 的文字颜色作为单击响应事件。再次运行程序后，单击 button，会看到 text 中的文本变为了红色，运行效果如图 3.9 所示。

图 3.9 使用 Button 监听逻辑对 Text 颜色进行更改，左图为修改前，右图为修改后

3.1.3 Image 组件

Image 是用于在屏幕上展示图像资源的组件，是软件开发中非常常用的 UI 组件。接下来讨论 Image 的具体使用方法。

Image 可以通过 setPixelMap(PixelMap)展示 PixelMap 类组件，或者通过 setImageElemen(Element)展示 Element 组件，也可以通过 setImageAndDecodeBounds(int)显示资源文件里的图片。

首先将图片资源 picture.jpg 放入 entry→src→main→resources→base→media 文件夹中，

修改 MainAbilitySlice.java 文件，初始化一个 DirectionalLayout 来承载 Image，代码如下：

```
//MainAbilitySlice.java
public class MainAbilitySlice extends AbilitySlice {
    private DirectionalLayout myLayout = new DirectionalLayout(this);
    private DirectionalLayout.LayoutConfig layoutConfig = new DirectionalLayout.LayoutConfig
(ComponentContainer.LayoutConfig.MATCH_PARENT, ComponentContainer.LayoutConfig.MATCH_PARENT);
    @Override
    public void onStart(Intent intent) {
        super.onStart(intent);
    }
}
```

随后在 onStart()方法中，设置 LayoutConfig 的属性，并使用 setImageAndDecodeBounds(int)方法加载 resources 资源文件夹中 media 文件中的图片资源，代码如下：

```
//MainAbilitySlice.java 中加载图片资源
@Override
public void onStart(Intent intent) {
super.onStart(intent);
myLayout.setLayoutConfig(layoutConfig);
    layoutConfig.alignment = LayoutAlignment.HORIZONTAL_CENTER;
    layoutConfig.width = ComponentContainer.LayoutConfig.MATCH_CONTENT;
    layoutConfig.height = ComponentContainer.LayoutConfig.MATCH_CONTENT;
Image image = new Image(this);
image.setImageAndDecodeBounds(ResourceTable.Media_picture);
image.setLayoutConfig(layoutConfig);
myLayout.addComponent(image);
super.setUIContent(myLayout);
}
```

运行上述代码，展示效果如图 3.10 所示。

图 3.10　Image 组件的效果展示

Image 还可以通过 setPixelMap(PixelMap)展示 PixelMap 类组件，修改 onStart()中的代码如下：

```
//MainAbilitySlice.java
@Override
public void onStart(Intent intent) {
    super.onStart(intent);
    myLayout.setLayoutConfig(layoutConfig);
    layoutConfig.alignment = LayoutAlignment.HORIZONTAL_CENTER;
    layoutConfig.width = ComponentContainer.LayoutConfig.MATCH_CONTENT;
    layoutConfig.height = ComponentContainer.LayoutConfig.MATCH_CONTENT;
    ResourceManager resourceManager = this.getResourceManager();
    Resource resource = null;
    try {
    resource = resourceManager.getResource(ResourceTable.Media_picture);
    } catch (IOException e) {
        e.printStackTrace();
} catch (NotExistException e) {
        e.printStackTrace();
    }
    ImageSource imageSource = ImageSource.create(resource,new ImageSource.SourceOptions());
    PixelMap pixelMap = imageSource.createPixelmap(new ImageSource.DecodingOptions());
    Image image = new Image(this);
    image.setPixelMap(pixelMap);
    image.setLayoutConfig(layoutConfig);
    myLayout.addComponent(image);
    super.setUIContent(myLayout);
    }
```

运行上述代码，可以得到和图 3.10 相同的运行效果。

还可以通过 setImageElement(Element)令 Image 组件展示 Element 组件。构造一个蓝色矩形的 element1，然后把这个 element1 传给 Image 组件并显示。修改 onStart()中代码如下：

```
//MainAbilitySlice.java
@Override
public void onStart(Intent intent) {
    super.onStart(intent);
    myLayout.setLayoutConfig(layoutConfig);
    layoutConfig.alignment = LayoutAlignment.HORIZONTAL_CENTER;
    layoutConfig.width = 1000;
    layoutConfig.height = 1000;

    ShapeElement element1 = new ShapeElement();
    element1.setShape(ShapeElement.RECTANGLE);
```

```
        element1.setRgbColor(new RgbColor(0, 0, 255));

        Image image = new Image(this);
        image.setImageElement(element1);
        image.setLayoutConfig(layoutConfig);
        myLayout.addComponent(image);
        super.setUIContent(myLayout);
    }
```

运行上述代码，展示效果如图 3.11 所示。

还可以根据需求，通过 setBounds()方法设置 element 的位置和大小，如新增 element2，设置一定的位置，将颜色设置为红色，并通过 setBackground()方法将 element1 设置为背景，在 onStart()方法中调整代码如下：

```
    ShapeElement element2 = new ShapeElement();
    element2.setShape(ShapeElement.RECTANGLE);
    element2.setRgbColor(new RgbColor(255, 0, 0));
    element2.setBounds(300,100,800,1000);
    Image image = new Image(this);
    image.setBackground(element1);

    image.setImageElement(element2);
```

运行上述代码，效果如图 3.12 所示。

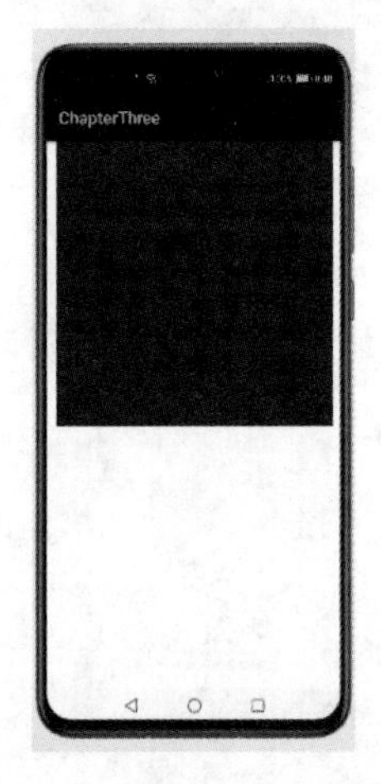

图 3.11　设置 element 效果图

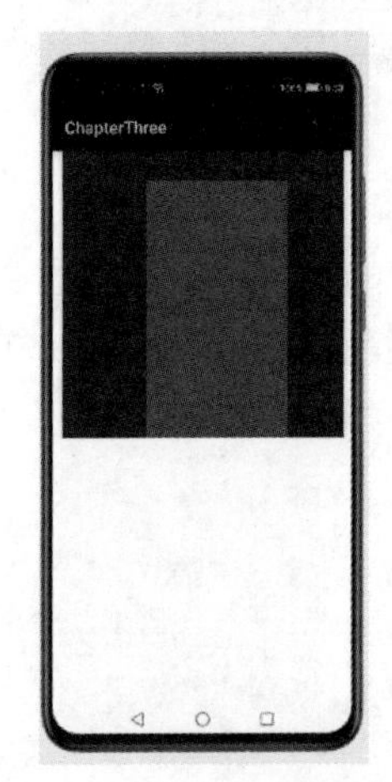

图 3.12　调整 element 大小

image 和其中放的图片内容都是有大小的，如果 image 的大小和内部图片的大小不一致，则需要使用 setscalemode()方式来调整图片大小。接下来，详细介绍一下 scalemode 的使用方法。

scalemode 有 6 种参数，分别演示一下效果。修改 onStart()方法中 Image 部分代码如下：

```
Image image1 = new Image(this);
image1.setImageAndDecodeBounds(ResourceTable.Media_picture);
Image image2 = new Image(this);
image2.setImageAndDecodeBounds(ResourceTable.Media_picture);
image2.setHeight(300);
image2.setWidth(300);
image2.setBackground(element1);

myLayout.addComponent(image1);
myLayout.addComponent(image2);
```

运行上述代码，会得到一张显示效果不全的图片，运行效果如图 3.13 所示。其中，image1 是原图，组件大小和图片大小一致。image2 把图片组件缩小到 300×300，这样就导致图片显示不全。

创建 Image 实例 image3，同样将大小设置为 300×300，并将 image3 的 scalemode 改变为 INSIDE，代码如下：

```
Image image3 = new Image(this);
image3.setImageAndDecodeBounds(ResourceTable.Media_picture);
image3.setBackground(element1);
image3.setHeight(300);
image3.setWidth(300);
image3.setScaleMode(Image.ScaleMode.INSIDE);
myLayout.addComponent(image3);
```

运行代码效果如图 3.14 所示，其中第 3 张是 INSIDE 的效果，为了直观地看到运行效果，可以将组件的背景用蓝色充满，可以看出 image3 的 scalemode 设置为 INSIDE 后，会根据比例对图像进行缩小，使得图像与组件大小相同或更小，并在中心显示图像内容。image3 大小为 300×300，但图片被按比例缩小在了 element1 中。

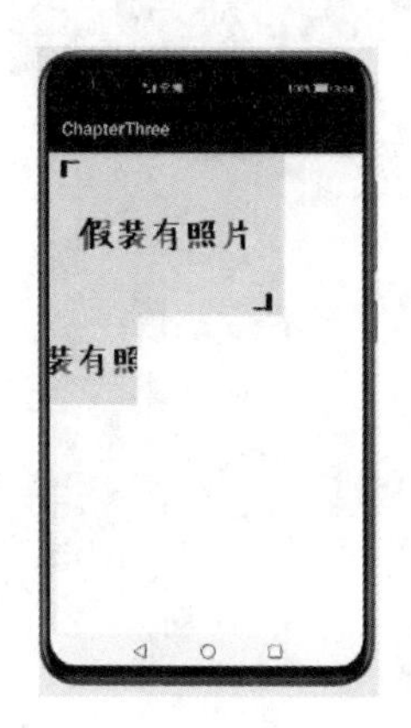

图 3.13 Image 显示不全效果图

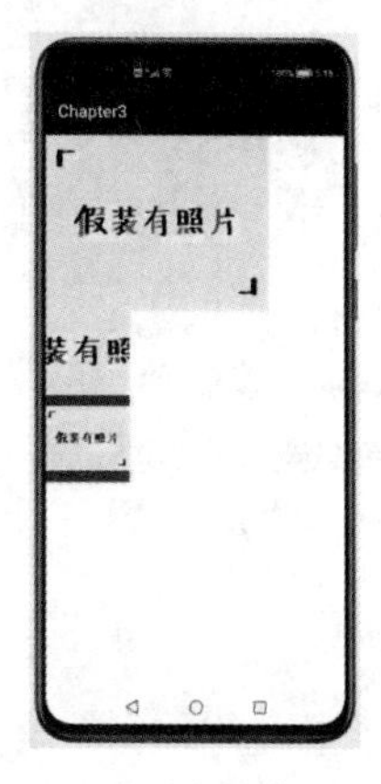

图 3.14 image3 按比例缩放效果图

若将 image3 的 scalemode 改变为 STRETCH，STRETCH 不使用任何比例对图像进行缩放。修改代码如下：

```
image3.setScaleMode(Image.ScaleMode.STRETCH);
```

运行后效果如图 3.15 所示，可以清楚地看到，图片原本的比例被破坏，被拉长后充满了 300×300 的组件。

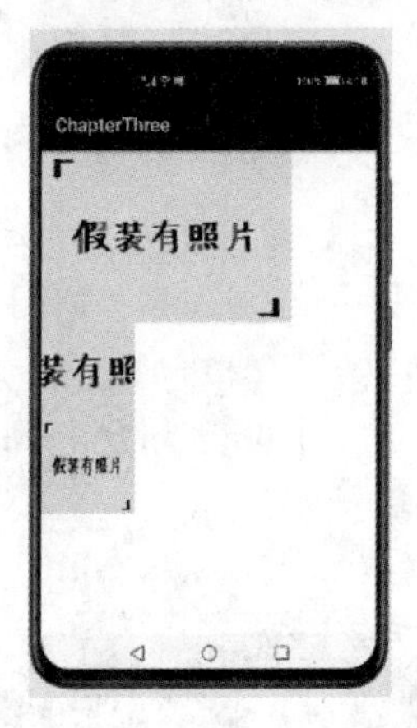

图 3.15　image3 设置为 STRETCH 效果图

ZOOM_CENTER 可以实现放大效果，将 Image 组件放大到 900×900，使得组件大于图片大小，原图片与组件 image2 间的空白位置用蓝色填充，scalemode 设置为 ZOOM_CENTER 后的图片与组件 image3 间的空白位置用青色填充，代码如下：

```
ShapeElement element1 = new ShapeElement();
element1.setShape(ShapeElement.RECTANGLE);
element1.setRgbColor(new RgbColor(0, 0, 255));

ShapeElement element2 = new ShapeElement();
element2.setShape(ShapeElement.RECTANGLE);
element2.setRgbColor(new RgbColor(0, 255, 255));

Image image1 = new Image(this);
image1.setImageAndDecodeBounds(ResourceTable.Media_picture);
image1.setLayoutConfig(layoutConfig);
image1.setHeight(900);
image1.setWidth(900);
image1.setBackground(element1);

Image image2 = new Image(this);
image2.setImageAndDecodeBounds(ResourceTable.Media_picture);
```

```
image2.setLayoutConfig(layoutConfig);
image2.setHeight(900);
image2.setWidth(900);
image2.setBackground(element2);
image2.setScaleMode(Image.ScaleMode.ZOOM_CENTER);
```

运行效果如图 3.16 所示。可以清楚地看到第一张为原图片，第二张图片根据比例将图片进行了放大，使得图片宽度与组件宽度相同，并在中心显示图片内容。

随后分别设置 image2 的 scalemode 为 ZOOM_START 和 ZOOM_END，可以看到效果如图 3.17 所示。ZOOM_START 可使图片放大并在左上角显示图片，ZOOM_END 可使图片放大并在右下角显示图片。

图 3.16　图片中心放大效果

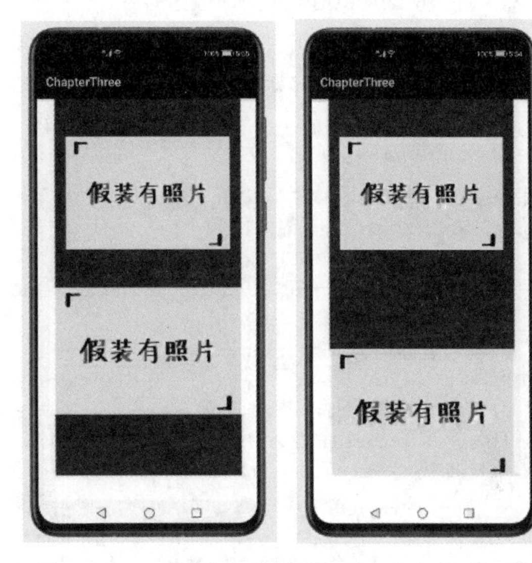

图 3.17　Image 放大并分别显示在左上角、右下角

3.2 Java UI 容器组件

13min

Java UI 框架提供了一些具有标准布局功能的容器，均继承自 ComponentContainer 类，一般以 Layout 结尾，如 DirectionalLayout、DependentLayout 等（也有例外，例如 ListContainer 类也是一个布局类）。

由图 3.1 视图树可知，完整的用户界面是一个容器组件，用户界面中的一部分也可以是一个容器组件。容器组件中可以容纳单个组件与其他容器组件。

HarmonyOS 为开发者分别提供了在 Java 代码和 xml 格式的文件中声明布局的方法。下面将分别使用代码和 XML 文件来开发 HarmonyOS 的常用布局。

3.2.1 线性布局 DirectionalLayout

DirectionalLayout 是 Java UI 中的一种重要组件布局，用于将一组组件(Component 或 ComponentContainer)按照水平或者垂直方向排布，能够方便地对齐布局内的组件。3.1.1 和 3.1.2 节就是将 Text 和 Button 等组件放在了 DirectionalLayout 中，实现了 Text 与 Button 自上而下有序排列的效果。

DirectionalLayout 的排列方向(orientation)分为水平(horizontal)和竖直(vertical)方向。使用 orientation 设置布局内组件的排列方式，默认为垂直排列。示意图如图 3.18 所示。

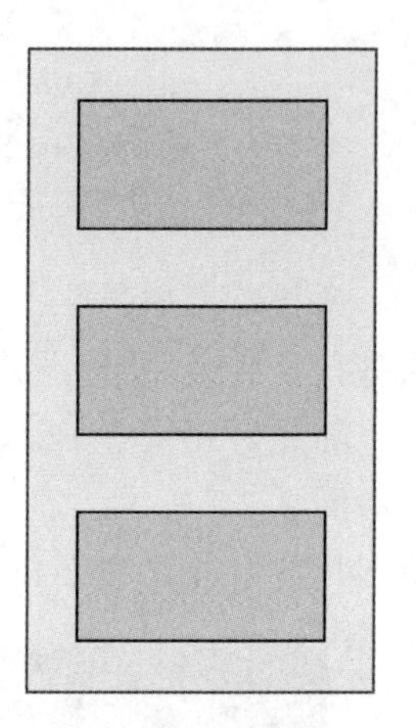

图 3.18 DirectionalLayout 竖直排列示意图

1. 代码创建 DirectionalLayout 组件

这里设置了 3 个单体组件 Button 和一个线性容器组件，并实现 Button 组件在线性容器内自上而下竖直排列。在 MainAbilitySlice.java 中修改 onStart()函数，代码如下：

```
//MainAbilitySlice.java
public void onStart(Intent intent) {
    super.onStart(intent);
    this.setDisplayOrientation(AbilityInfo.DisplayOrientation.PORTRAIT);

    //创建容器组件 DirectionalLayout 实例
    DirectionalLayout myLayout = new DirectionalLayout(this);
    //设置 DirectionalLayout 对内部组件的排列方式并设置为竖直排列
    myLayout.setOrientation(Component.VERTICAL);
    //设置 DirectionalLayout 内部的组件都排列在屏幕水平的中央位置
    myLayout.setAlignment(LayoutAlignment.HORIZONTAL_CENTER);
    //容器组件 DirectionalLayout 的 LayoutConfig
DirectionalLayout.LayoutConfig layoutConfig = new DirectionalLayout.LayoutConfig
(DirectionalLayout.LayoutConfig.MATCH_PARENT, DirectionalLayout.LayoutConfig.MATCH_PARENT);
    myLayout.setLayoutConfig(layoutConfig);

    //创建 3 个 Button 实例
    Button button1 = new Button(this);
    Button button2 = new Button(this);
    Button button3 = new Button(this);
    //单体组件 Button 的 LayoutConfig

    LayoutConfig buttonConfig = new LayoutConfig(
DirectionalLayout.LayoutConfig.MATCH_CONTENT,
DirectionalLayout.LayoutConfig.MATCH_CONTENT);
    buttonConfig.setMargins(0,100,0,100);
    //将 LayoutConfig 应用到 3 个 Button 上
    button1.setLayoutConfig(buttonConfig);
    button2.setLayoutConfig(buttonConfig);
```

```
        button3.setLayoutConfig(buttonConfig);
        //创建 Button 的背景
        ShapeElement bottomElement = new ShapeElement();
        bottomElement.setRgbColor(new RgbColor(200,200,200));
        //将背景应用到 3 个 Button 上
        button1.setBackground(bottomElement);
        button2.setBackground(bottomElement);
        button3.setBackground(bottomElement);

        //设定 3 个 Button 内显示的文本
        button1.setText("Button 1");
        button2.setText("Button 2");
        button3.setText("Button 3");

        //设定 3 个 Button 内显示的文字大小
        button1.setTextSize(80, Text.TextSizeType.VP);
        button2.setTextSize(80, Text.TextSizeType.VP);
        button3.setTextSize(80, Text.TextSizeType.VP);

        //将 3 个 Button 加入 DirectionalLayout 中
        myLayout.addComponent(button1);
        myLayout.addComponent(button2);
        myLayout.addComponent(button3);

        super.setUIContent(myLayout);

    }
```

在这段代码中首先创建了一个 DirectionalLayout 容器组件实例，然后对它的属性进行了设置，其中比较重要的属性有如下 3 种。

Orientation：代表了 DirectionalLayout 对其内部组件（子组件）的排列规则。VERTICAL 代表从上到下竖直排列，HORIZONTAL 则代表从左至右水平排列。

Alignment：代表了 DirectionalLayout 内部组件（子组件）的排列位置。HORIZONTAL_CENTER 代表水平位置的正中，常用的还有左对齐 LEFT、右对齐 RIGHT 和竖直位置的正中 VERTICAL_CENTER。

LayoutConfig：代表了对 DirectionalLayout 本身的一些设置，不过在这里只设置了宽和高为 MATCH_PARENT，这意味着这个 DirectionalLayout 的大小将充满它的上一层容器（父容器组件），在这里这个 DirectionalLayout 没有父容器组件，所以将充满整个屏幕。

尝试运行程序，可以看到 3 个 Button 组件在屏幕水平方向的正中央自上而下竖直排列，如图 3.19 所示。

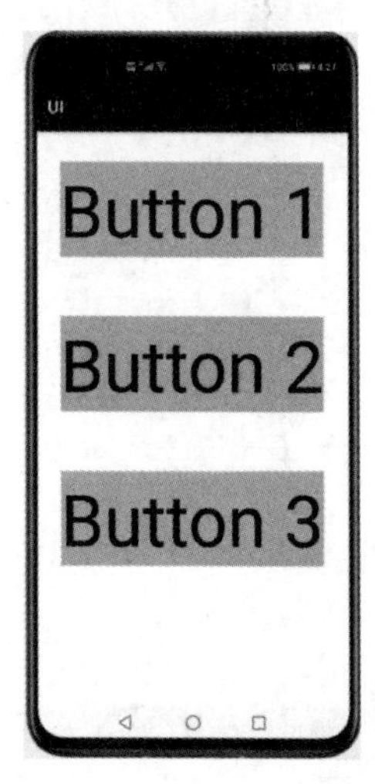

图 3.19 代码创建的 DirectionalLayout 效果展示

2. xml 创建 DirectionalLayout

现在尝试使用 xml 来创建 DirectionalLayout，以便了解更多功能。这里设置 3 个 Button，并将它们设置为垂直排列。首先在 layout 文件夹中新建布局文件 directional_layout.xml，在其中加入 3 个 Button，并将其最外层标签设置为线性布局 DirectionalLayout，代码如下：

```
<!-- directional_layout.xml -->
<?xml version = "1.0" encoding = "utf-8"?>
<DirectionalLayout
        xmlns:ohos = "http://schemas.huawei.com/res/ohos"
        ohos:width = "match_parent"
        ohos:height = "match_content"
        ohos:orientation = "vertical">
<Button
            ohos:width = "100vp"
            ohos:height = "50vp"
            ohos:bottom_margin = "13vp"
            ohos:left_margin = "13vp"
            ohos:background_element = "$graphic:buttonelement"
            ohos:text = "Button 1"
            ohos:text_size = "24vp"/>
<Button
            ohos:width = "100vp"
            ohos:height = "50vp"
            ohos:bottom_margin = "13vp"
            ohos:left_margin = "13vp"
            ohos:background_element = "$graphic:buttonelement"
            ohos:text = "Button 2"
            ohos:text_size = "24vp"/>
<Button
            ohos:width = "100vp"
            ohos:height = "50vp"
            ohos:bottom_margin = "13vp"
            ohos:left_margin = "13vp"
            ohos:background_element = "$graphic:buttonelement"
            ohos:text = "Button 3"
            ohos:text_size = "24vp"/>
</DirectionalLayout>
```

其中，ohos:orientation 的属性值设置为 vertical，即垂直排列。在 MainAbilitySlice 中修改代码，引入 directional_layout.xml 布局文件，代码如下：

```
//MainAbilitySlice.java
public class MainAbility extends Ability {

    @Override
```

```
    public void onStart(Intent intent) {
        super.onStart(intent);
        super.setUIContent(ResourceTable.Layout_directional_layout);
    }
}
```

运行效果如图 3.20 所示，可见 3 个 Button 为垂直排列。

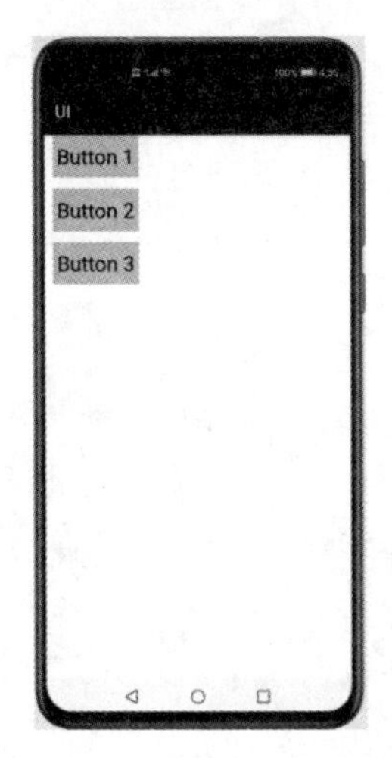

图 3.20　3 个 Button 垂直排列

垂直排列默认为左对齐，组件可以通过 ohos:layout_alignment 控制自身在布局中的对齐方式。在垂直排列中，layout_alignment 的属性值 left 表示左对齐，right 表示右对齐，horizontal_center 表示水平方向居中，center 表示水平方向和垂直方向均居中。当属性值的对齐方式与排列方式方向一致时，对齐方式不会生效，如设置了水平方向的排列方式，则左对齐、右对齐将不会生效。如修改 directional_layout.xml 布局文件，代码如下：

```
<!-- directional_layout.xml -->
<?xml version = "1.0" encoding = "utf-8"?>
<DirectionalLayout
    xmlns:ohos = "http://schemas.huawei.com/res/ohos"
    ohos:width = "match_parent"
    ohos:height = "match_content"
    ohos:orientation = "vertical">
<Button
    ohos:width = "100vp"
    ohos:height = "50vp"
    ohos:layout_alignment = "left"
    … />
<Button
    ohos:width = "100vp"
```

```
        ohos:height = "50vp"
        ohos:layout_alignment = "horizontal_center"
        …/>
    <Button
        ohos:width = "100vp"
        ohos:height = "50vp"
        ohos:layout_alignment = "right"
        …/>
    </DirectionalLayout>
```

运行后效果如图3.21所示。可见,3个Button分别实现了左对齐、水平居中和右对齐。

在DirectionalLayout布局中,可以将ohos:orientation属性值改为horizontal,这样Button则变为水平排列,修改代码如下:

```
<!-- directional_layout.xml -->
<?xml version = "1.0" encoding = "utf-8"?>
<DirectionalLayout
xmlns:ohos = "http://schemas.huawei.com/res/ohos"
ohos:width = "match_parent"
ohos:height = "match_content"
ohos:orientation = "horizontal">

…

</DirectionalLayout>
```

运行后效果如图3.22所示,可以看到3个Button变为水平排列。

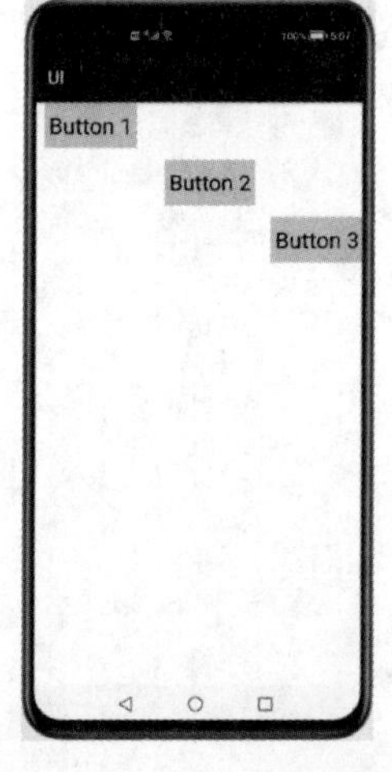

图3.21 对齐方式效果示例

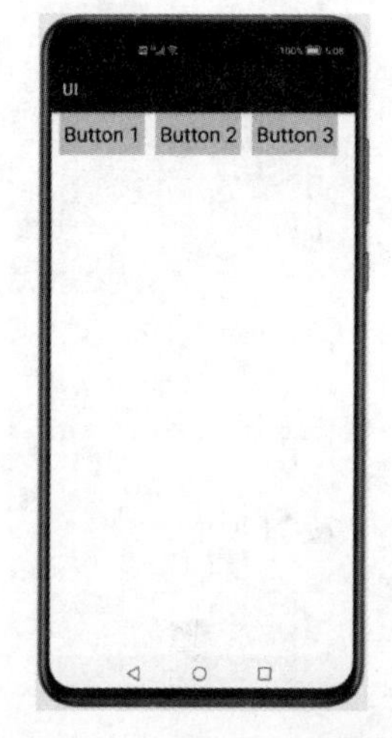

图3.22 3个Button水平排列

注意，DirectionalLayout 布局不能自动换行，子视图会按照设定的方向依次排列，若超过布局本身的大小，则超出布局大小的部分将不会被显示。修改 3 个 button 的宽度，代码如下：

```
<! -- directional_layout.xml -- >
<?xml version = "1.0" encoding = "utf - 8"?>
< DirectionalLayout
    xmlns:ohos = http://schemas.huawei.com/res/ohos
    ohos:width = "match_parent"
    ohos:height = "match_content"
    ohos:orientation = "horizontal">
< Button
    ohos:width = "150vp"
    ohos:height = "50vp"
    … />
< Button
    ohos:width = "150vp"
    ohos:height = "50vp"
    … />
< Button
    ohos:width = "150vp"
    ohos:height = "50vp"
    … />
</DirectionalLayout >
```

运行后界面效果如图 3.23 所示。

此布局包含了 3 个 Button，但是因为宽度超出了布局范围，所以界面上只完整显示了两个 Button，超出布局大小的视图部分无法正常显示，因此在开发布局中要注意避免这类问题的出现。

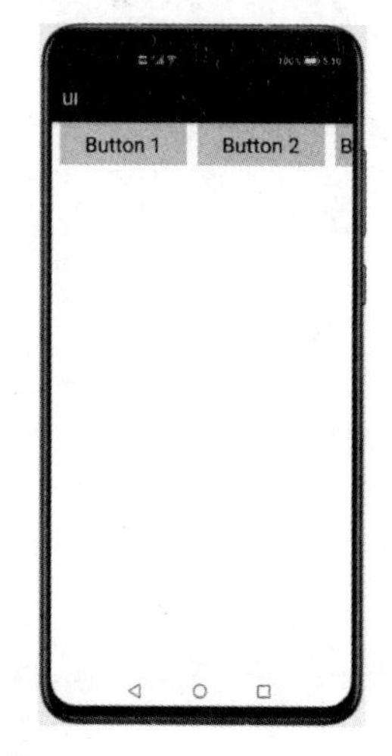

图 3.23 DirectionalLayout 不可自动换行示例

在 HarmonyOS 的 UI 组件中，布局之间也可以按层级关系互相组合，DirectionalLayout 布局和其他布局的组合，可以实现更加丰富的布局方式。具体逻辑可以参照后面章节的视图树。

3.2.2 相对布局 DependentLayout

DependentLayout 是 Java UI 里另一种常见布局，每个组件可以指定相对于其他同级元素的位置，或者指定相对于父组件的位置进行布局。与 DirectionalLayout 相比，它拥有更多的排布方式。布局的示意图如图 3.24 所示。

DependentLayout 的布局方式，有相对同级其他组件和相对父组件两种布局方式。

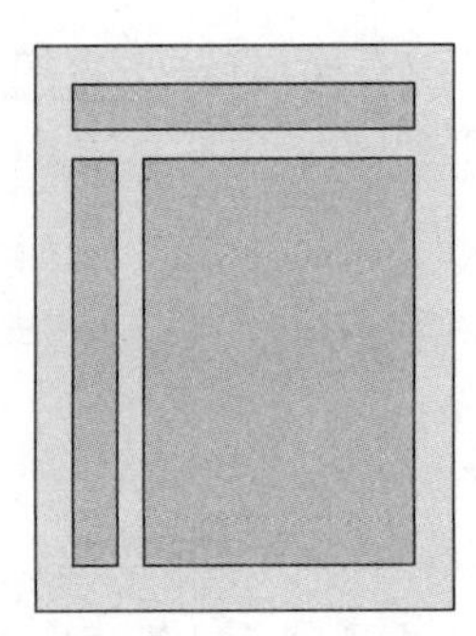

图 3.24　DependentLayout 布局示意图

1. 代码创建 DependentLayout

和 DirectionalLayout 举的例子类似，在这里设置 3 个单体组件 Button 和一个 DependentLayout 容器组件，并实现 Button 组件在 DependentLayout 容器内自上而下竖直排列。为了方便展示，还是在 MainAbilitySlice.java 中修改 onStart()函数，代码如下：

```
//MainAbilitySlice.java
public void onStart(Intent intent) {
    super.onStart(intent);
    this.setDisplayOrientation(AbilityInfo.DisplayOrientation.PORTRAIT);

    //创建容器组件 DependentLayout 实例,并为之设置 Id
    DependentLayout myLayout = new DependentLayout(this);
    int L1 = 0;
    myLayout.setId(L1);
    //容器组件 DependentLayout 的 LayoutConfig
    DependentLayout.LayoutConfig layoutConfig = new
    DependentLayout.LayoutConfig(
    DependentLayout.LayoutConfig.MATCH_PARENT,
    DependentLayout.LayoutConfig.MATCH_PARENT);
    myLayout.setLayoutConfig(layoutConfig);

    //创建 3 个 Button 实例,并为之设置 Id
    Button button1 = new Button(this);
    int B1 = 1;
    button1.setId(B1);
    Button button2 = new Button(this);
    int B2 = 2;
    button2.setId(B2);
    Button button3 = new Button(this);
    int B3 = 3;
    button3.setId(B3);

    //单体组件 Button 的 LayoutConfig
    DependentLayout.LayoutConfig buttonConfig1 = new DependentLayout.LayoutConfig(
```

```
DependentLayout.LayoutConfig.MATCH_CONTENT,
DependentLayout.LayoutConfig.MATCH_CONTENT);
    buttonConfig1.addRule(DependentLayout.LayoutConfig.CENTER_IN_PARENT);
    buttonConfig1.setMargins(0,100,0,100);
    DependentLayout.LayoutConfig buttonConfig2 = new DependentLayout.LayoutConfig(
DependentLayout.LayoutConfig.MATCH_CONTENT,
DependentLayout.LayoutConfig.MATCH_CONTENT);
    buttonConfig2.addRule(DependentLayout.LayoutConfig.BELOW, B1);
    buttonConfig2.setMargins(0,100,0,100);
    DependentLayout.LayoutConfig buttonConfig3 = new DependentLayout.LayoutConfig(
DependentLayout.LayoutConfig.MATCH_CONTENT,
DependentLayout.LayoutConfig.MATCH_CONTENT);
    buttonConfig3.addRule(DependentLayout.LayoutConfig.ALIGN_PARENT_BOTTOM);
    buttonConfig3.addRule(DependentLayout.LayoutConfig.ALIGN_PARENT_RIGHT);
    buttonConfig3.setMargins(0,100,0,100);
    //将 LayoutConfig 应用到 3 个 Button 上
    button1.setLayoutConfig(buttonConfig1);
    button2.setLayoutConfig(buttonConfig2);
    button3.setLayoutConfig(buttonConfig3);
    //创建 Button 的背景
    ShapeElement bottomElement = new ShapeElement();
    bottomElement.setRgbColor(new RgbColor(200,200,200));
    //将背景应用到 3 个 Button 上
    button1.setBackground(bottomElement);
    button2.setBackground(bottomElement);
    button3.setBackground(bottomElement);

    //设定 3 个 Button 内显示的文本
    button1.setText("Button 1");
    button2.setText("Button 2");
    button3.setText("Button 3");

    //设定 3 个 Button 内显示的文字大小
    button1.setTextSize(50, Text.TextSizeType.VP);
    button2.setTextSize(50, Text.TextSizeType.VP);
    button3.setTextSize(50, Text.TextSizeType.VP);

    //将 3 个 Button 加入 DependentLayout 中
    myLayout.addComponent(button1);
    myLayout.addComponent(button2);
    myLayout.addComponent(button3);

    super.setUIContent(myLayout);

}
```

运行效果如图 3.25 所示。

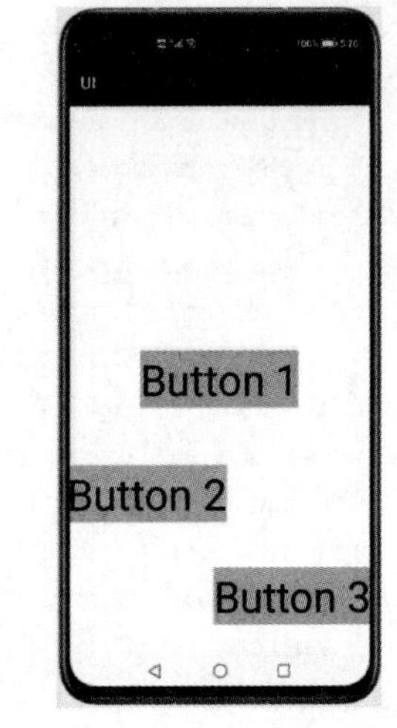

图 3.25 代码创建的 DependentLayout 效果展示

可以发现，与之前 DirectionalLayout 例子不同的是为每个组件（包括容器组件和单体组件）都设置了一个 Id，这是因为在 DependentLayout 中命令需要利用 Id 来识别组件对象，进而实现组件之间的相对布局。

为了实现相对布局，此处为每个子组件（范例中的 Button 实例）添加了一个 DependentLayout. LayoutConfig 类实例，在对应的 DependentLayout. LayoutConfig 实例中，使用 addRule()函数为这些组件添加想要的排列规则，具体的排列规则如下：

对于 button1，为其添加的规则为 LayoutConfig. CENTER_IN_PARENT，这个规则规定了 button1 将位于其父组件（范例中的 DependentLayout 实例）的正中间。

对于 button2，为其添加的规则为 LayoutConfig. BELOW 和 B1，这个规则规定了 button2 将位于 Id 为 B1（范例中的 button1 实例）的下方。值得注意的是，在效果图中可以看出 button2 并没有位于 button1 的正下方，而是下方的左侧。这是因为这个规则只规定了 button2 竖直方向的位置，而并没有规定水平方向的位置，所以水平方向的位置仍为系统的默认值，即左侧排列。这一类问题在使用时需要注意。

对于 button3，为其添加的规则为 LayoutConfig. ALIGN_PARENT_BOTTOM 和 LayoutConfig. ALIGN_PARENT_RIGHT，这两个规则分别规定了 button3 要位于其父组件的下方和右侧，所以从效果图中可以看出 button3 位于界面的右下方。这说明一个 DependentLayout. LayoutConfig 实例可以添加多个规则。如果规则之间相互"冲突"，可以尝试将 button3 添加的 2 个规则修改为 LayoutConfig. ALIGN_PARENT_RIGHT 和 LayoutConfig. ALIGN_PARENT_LEFT，运行后观察效果如何。

2. xml 创建 DependentLayout

相对于同级其他组件的位置，每个组件有 above（位于同级组件的上侧）、below（位于同级组件的下侧）、start_of（位于同级组件的起始侧）、end_of（位于同级组件的结束侧）、left_of（位于同级组件的左侧）、right_of（位于同级组件的右侧）6 种相对位置。这里通过实例直观体会一下。首先在 layout 文件夹中新建布局文件 dependent_layout. xml，代码如下：

```
<!-- dependent_layout.xml -->
<?xml version = "1.0" encoding = "utf-8"?>
<DependentLayout
    xmlns:ohos = "http://schemas.huawei.com/res/ohos"
    ohos:width = "match_parent"
    ohos:height = "match_content">
```

```
<Text
        ohos:id = "$ + id:text1"
        ohos:width = "match_content"
        ohos:height = "match_content"
        ohos:left_margin = "15vp"
        ohos:top_margin = "15vp"
        ohos:bottom_margin = "15vp"
        ohos:text = "text1"
        ohos:text_size = "20fp"
        ohos:background_element = "$ graphic:text_element"/>
<Text
        ohos:id = "$ + id:text2"
        ohos:width = "match_content"
        ohos:height = "match_content"
        ohos:left_margin = "15vp"
        ohos:top_margin = "15vp"
        ohos:bottom_margin = "15vp"
        ohos:text = "end_of text1"
        ohos:text_size = "20fp"
        ohos:background_element = "$ graphic:text_element"
        ohos:end_of = "$ id:text1"/>
<Text
        ohos:id = "$ + id:text3"
        ohos:width = "match_content"
        ohos:height = "match_content"
        ohos:left_margin = "15vp"
        ohos:text = "below text1"
        ohos:text_size = "20fp"
        ohos:background_element = "$ graphic:text_element"
        ohos:below = "$ id:text1"/>
</DependentLayout>
```

其中，ohos:end_of 属性可以让一个组件位于另一个组件的结束侧，属性值中需要指定相对控件的 id 引用，如 text2 中设置 ohos:end_of="$id:text1"，表示让该组件位于 text1 的结束侧。text3 中通过设置 ohos:below="$id:text1"，让 text3 位于 text1 的下方。其他相对位置同理。

在 graphic 文件夹中新建 text_element.xml 文件，代码如下：

```
<!-- text_element.xml -->
<?xml version = "1.0" encoding = "utf-8"?>
<shape xmlns:ohos = "http://schemas.huawei.com/res/ohos"
       ohos:shape = "rectangle">
<solid
           ohos:color = "#D5CF72"/>
</shape>
```

在 MainAbility 中更改布局文件引用，代码如下：

```
super.setUIContent(ResourceTable.Layout_dependent_layout);
```

运行代码，效果如图 3.26 所示。

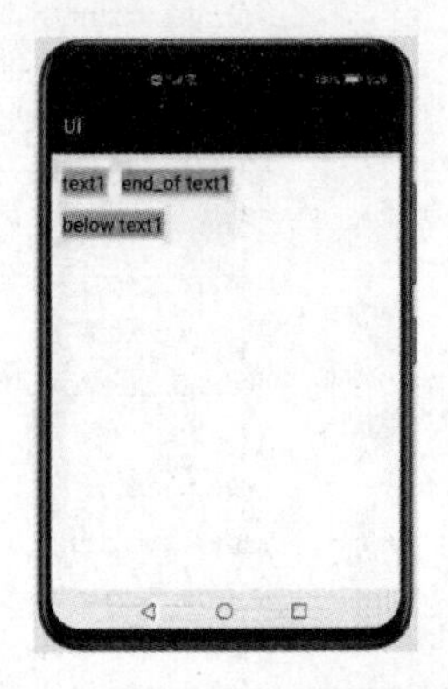

图 3.26 相对同级组件位置布局

除了可以相对于同级其他组件位置进行定位，也可以相对父组件进行布局。相对于父组件的位置，每个组件有 align_parent_left（位于父组件的左侧）、align_parent_right（位于父组件的右侧）、align_parent_start（位于父组件的起始侧）、align_parent_end（位于父组件的结束侧）、align_parent_top（位于父组件的上侧）、align_parent_bottom（位于父组件的下侧）、center_in_parent（位于父组件的中间）7 种相对位置。基于以上布局，也可以形成左上角、左下角、右上角和右下角的布局。修改 dependent_layout.xml 中的代码如下：

```
<!-- dependent_layout.xml -->
<?xml version = "1.0" encoding = "utf-8"?>
<DependentLayout
    xmlns:ohos = "http://schemas.huawei.com/res/ohos"
    ohos:width = "match_parent"
    ohos:height = "match_parent">

<Text
        ohos:id = "$+id:text1"
        ohos:width = "match_content"
        ohos:height = "match_content"
        ohos:text = "left_top"
        ohos:text_size = "20fp"
        ohos:background_element = "$graphic:text_element"
        ohos:align_parent_left = "true"
        ohos:align_parent_top = "true"/>
<Text
        ohos:id = "$+id:text2"
        ohos:width = "match_content"
        ohos:height = "match_content"
        ohos:text = "right_top"
        ohos:text_size = "20fp"
        ohos:background_element = "$graphic:text_element"
        ohos:align_parent_right = "true"
        ohos:align_parent_top = "true"/>
<Text
        ohos:id = "$+id:text3"
        ohos:width = "match_content"
        ohos:height = "match_content"
```

```
        ohos:text = "center"
        ohos:text_size = "20fp"
        ohos:background_element = " $ graphic:text_element"
        ohos:center_in_parent = "true"/>
<Text
        ohos:id = " $ + id:text4"
        ohos:width = "match_content"
        ohos:height = "match_content"
        ohos:text = "bottom_center"
        ohos:text_size = "20fp"
        ohos:background_element = " $ graphic:text_element"
        ohos:align_parent_bottom = "true"
        ohos:center_in_parent = "true"/>

</DependentLayout>
```

上述代码中，通过 ohos:align_parent_left 和 ohos:align_parent_top 进行属性设置，使 text1 位于父组件的左上角，同理令 text2 位于父组件的右上角，text3 位于父组件居中，text4 位于父容器底部居中。运行后结果如图 3.27 所示。

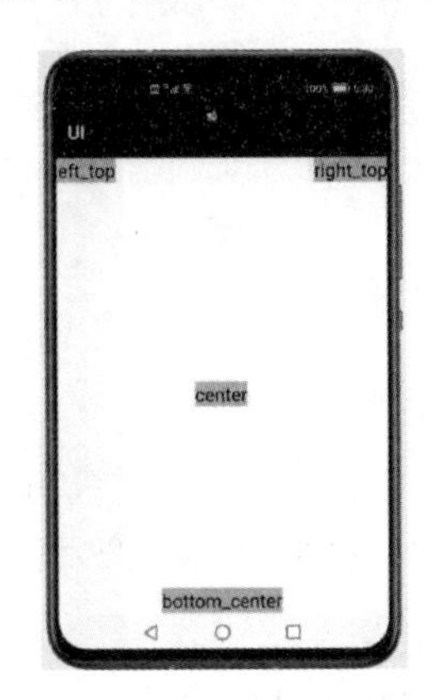

图 3.27 相对父组件位置布局

3.2.3 绝对坐标布局 PositionLayout

绝对坐标布局容器 PositionLayout 也是比较常用的基础布局组件之一，它可以依据绝对坐标对其内部的组件进行布局，示意图如图 3.28 所示。具体来讲，当 PositionLayout 作为父组件时，其在屏幕上所占有的空间可视为一个二维直角坐标系，其子组件可以依据坐标值进行布局。

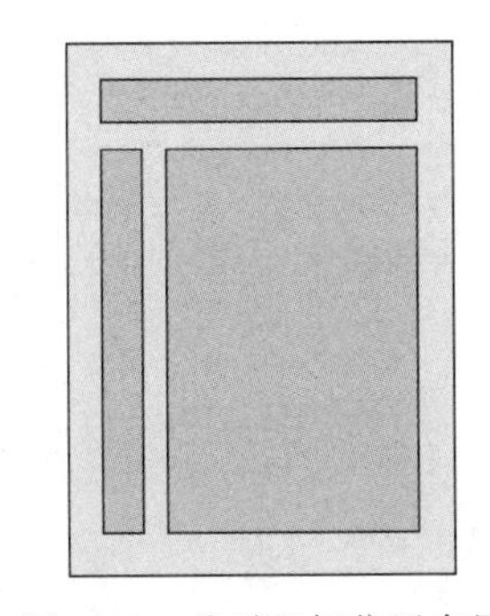

图 3.28 绝对坐标位置布局

作为最直观的容器组件，PositionLayout 的使用方法也与前两种 Layout 不同，它不需要使用 LayoutConfig 实现组件之

间的布局,而仅需要为每个子组件直接设置坐标即可,实现 PositionLayout 的代码如下:

```
//MainAbilitySlice.java
public void onStart(Intent intent) {
    super.onStart(intent);
    this.setDisplayOrientation(AbilityInfo.DisplayOrientation.PORTRAIT);

    //创建容器组件 PositionLayout 实例,并为之设置 Id
    PositionLayout myLayout = new PositionLayout(this);
    //容器组件 PositionLayout 的 LayoutConfig
    PositionLayout.LayoutConfig layoutConfig = new
    PositionLayout.LayoutConfig(
    PositionLayout.LayoutConfig.MATCH_PARENT,
    PositionLayout.LayoutConfig.MATCH_PARENT);
    myLayout.setLayoutConfig(layoutConfig);

    //创建 3 个 Button 实例,并为之设置 Id
    Button button1 = new Button(this);
    Button button2 = new Button(this);
    Button button3 = new Button(this);

    //将 LayoutConfig 应用到 3 个 Button 上
    button1.setTop(100);
    button1.setLeft(100);
    button2.setTop(1000);
    button2.setLeft(500);
    button3.setTop(1500);
    button3.setRight(500);
    //创建 Button 的背景
    ShapeElement bottomElement = new ShapeElement();
    bottomElement.setRgbColor(new RgbColor(200,200,200));
    //将背景应用到 3 个 Button 上
    button1.setBackground(bottomElement);
    button2.setBackground(bottomElement);
    button3.setBackground(bottomElement);

    //设定 3 个 Button 内显示的文本
    button1.setText("Button 1");
    button2.setText("Button 2");
    button3.setText("Button 3");

    //设定 3 个 Button 内显示的文字大小
    button1.setTextSize(50, Text.TextSizeType.VP);
    button2.setTextSize(50, Text.TextSizeType.VP);
    button3.setTextSize(50, Text.TextSizeType.VP);

    //将 3 个 Button 加入 PositionLayout 中
```

```
        myLayout.addComponent(button1);
        myLayout.addComponent(button2);
        myLayout.addComponent(button3);

        super.setUIContent(myLayout);

    }
```

在这里我们为 button1 和 button2 设置了 Left 和 Top 坐标，为 button3 设置了 Top 和 Right 坐标。运行的效果如图 3.29 所示。

可以看出，button1 和 button2 的左上角分别位于其父布局（范例中的 PositionLayout 实例）的（100，100）和（100，500）位置上，这是符合预期的，而 button3 却位于（0，1500）位置上，说明其 Right 坐标的设置是不能够改变组件位置的。经过一些测试发现对于 Bottom 坐标的设置也是不能够改变组件位置的。

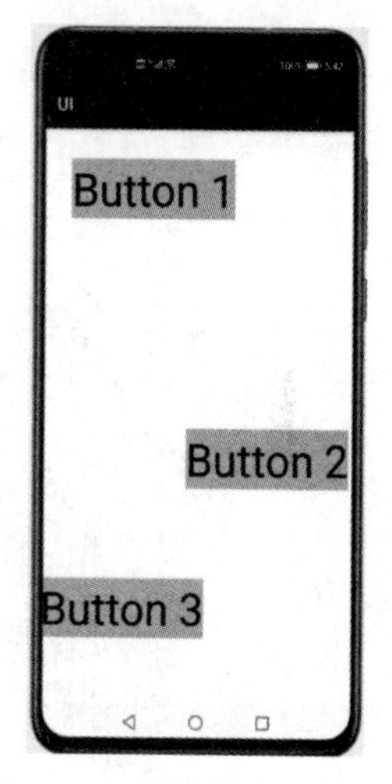

图 3.29 代码创建的 PositionLayout 效果展示

由于 PositionLayout 是依据绝对坐标进行的布局，所以代码的灵活性和扩展性都相对较差。例如，PositionLayout 在适配各种分辨率的屏幕上是比较困难的，需要对每个组件的坐标值进行修改。在切换横屏竖屏时，若要保持布局的整齐性，每个坐标值都需要重新计算，这是比较麻烦的。

3.2.4 滚动菜单 ListContainer

手机屏幕空间有限，能显示的内容不多。可以借助 ListContainer 来显示更多的内容。ListContainer 允许用户通过上下滑动来将屏幕外的数据滚动到屏幕内，同时屏幕内原有的数据滚动出屏幕，从而显示更多的数据内容。但是 ListContainer 的使用和一般的组件使用不同，ListContainer 一定要和 Provider 适配器搭配使用，HarmonyOS 里 Provider 有 BaseItemProvider 和它的子类 RecycleItemProvider。

在前面的章节中都是分别使用 Java 代码或者 xml 来创建布局容器（或单体组件），事实上还可以采用两者结合的手段进行创建，在这里将采用 xml 来创建 ListContainer，然后使用 Java 代码为其添加各种属性。首先在 resource 里定义一个带有 ListContainer 的布局，代码如下：

```
<!-- listcontainer.xml -->
<?xml version="1.0" encoding="utf-8"?>
<DirectionalLayout
        xmlns:ohos="http://schemas.huawei.com/res/ohos"
```

```
        ohos:width = "match_parent"
        ohos:height = "match_parent"
        ohos:orientation = "vertical">
<ListContainer
            ohos:id = "$ + id:list"
            ohos:width  =  "match_content"
            ohos:height = "match_content"
            ohos:margin = "50px"
            />

</DirectionalLayout>
```

然后需要设置一个适配器 provider 来为 ListContainer 提供内容，代码如下：

```
//MainAbitity Slice. java
public class SamplePagerAdapter extends BaseItemProvider {
            private List mList;
            public SamplePagerAdapter(List list) {
                mList = list;
            }
            @Override
public int getCount() {
                return mList.size();
            }

            @Override
public Object getItem(int i) {
                return mList.get(i);
            }

            @Override
public long getItemId(int i) {
                ret rn i;
            }

        @Override
public Component getComponent(int i, Component component, ComponentContainer componentContainer) {
                Text title = new Text(getContext());
                title.setText((String)mList.get(i));
                title.setTextSize(200);
                title.setTextAlignment(TextAlignment.CENTER);
                title.setLayoutConfig(new StackLayout.LayoutConfig(
            StackLayout.LayoutConfig.MATCH_CONTENT,
            StackLayout.LayoutConfig.MATCH_CONTENT));
```

```
                title.setTextColor(Color.RED);
                ShapeElement shapeElement = new ShapeElement();
                shapeElement.setShape(ShapeElement.RECTANGLE);
                shapeElement.setRgbColor(new RgbColor(0,0,255));
                title.setBackground(shapeElement);
                return title;
            }
        }
```

Provider必须继承自BaseItemProvider和它的子类RecycleItemProvider，然后需要覆写getCount()、getItem()、getItemId()和getComponent()这4种方法。

在ListContainer绘制之前，会首先调用public int getCount()方法，这种方法的返回值代表listContaimer的长度，然后根据这个值来确定getComponent()的执行次数。

在绘制每一行之前都会执行一次getComponent()，用于获取当前行需要显示的内容，其中，第一个回调参数i代表当前的行数，第二个回调参数component代表上一次绘制时本行显示的内容，所以易知首次绘制时此参数为空(因为还没有旧的component)。

另外两个函数public Object getItem(int i)的作用是返回一个子Component，即ListContainer中的一个子条目。public long getItemId(int i)的作用是返回一个item的id，由参数i决定是哪个id。

覆写完上述4个函数后，就完成了Provider的简易构建。上述Provider的业务逻辑是使getComponent()每次运行时得到列表mList中的一个字符串，然后实例化一个Text组件来展示这个字符串，所以还需要定义一个名为mList的ArrayList用于承载这些字符串，代码如下：

```
//MainAbility.java
DirectionalLayout myLayout = new DirectionalLayout(this);
DirectionalLayout.LayoutConfig layoutConfig = new
DirectionalLayout.LayoutConfig(DirectionalLayout.LayoutConfig.MATCH_PARENT, DirectionalLayout.
LayoutConfig.MATCH_PARENT);
layoutConfig.alignment = LayoutAlignment.HORIZONTAL_CENTER;
myLayout.setLayoutConfig(layoutConfig);
layoutConfig.width = ComponentContainer.LayoutConfig.MATCH_CONTENT;
layoutConfig.height = ComponentContainer.LayoutConfig.MATCH_CONTENT;
List aList = new ArrayList<>();
ListContainer listContainer = new ListContainer(this);
listContainer.setLayoutConfig(layoutConfig);
aList.add("测试 1");
aList.add("测试 2");
aList.add("测试 3");
aList.add("测试 4");
aList.add("测试 5");
aList.add("测试 6");
aList.add("测试 7");
listContainer.setItemProvider(new SamplePagerAdapter(aList));
myLayout.addComponent(listContainer);
```

上述代码创建了一个 list，用于承载 7 个字符串，把这个 list 传递给 Provider，最后对 xml 中的布局和 ListContainer 进行加载，将创建的 Provider 和布局中的 ListContainer 进行绑定，这样就能展示整个 ListContainer 了，展示效果如图 3.30 所示。

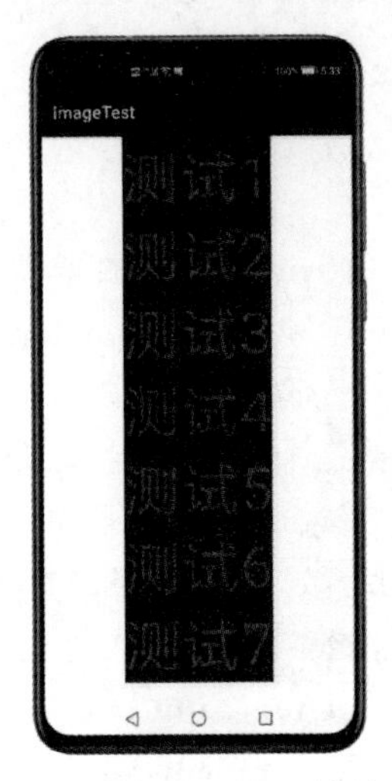

图 3.30 ListContainer 效果展示

3.2.5 滑动布局管理器 PageSlider

PageSlider 和 ListContainer 类似，是一种容器布局，需要 PageSliderProvider 作为 Provider 才能显示内容。PageSlider 可以允许用户左右或者上下滑动来翻页，其中每页都可以添加其他 Component 进行显示。

与上一节的 ListContainer 类似，首先需要定义一个 Provider，代码如下：

```java
//MainAbilitySlice.java
private class PagerAdapter extends PageSliderProvider {
    private List list;
    public PagerAdapter(List<Image> list) {
        this.list = list;
    }

    @Override
    public int getCount() {
        return list.size();
    }
    @Override
    public Object createPageInContainer( ComponentContainer componentContainer, int i) {
        componentContainer.addComponent((Component)list.get(i));
        return list.get(i);
    }
```

```
        @Override
        public void destroyPageFromContainer( ComponentContainer componentContainer, int i,
Object o) {
            componentContainer.removeComponent((Component)list.get(i));

        }
        @Override
        public boolean isPageMatchToObject(Component component, Object o) {
            return component == o;
        }
}
```

可以看到，所有的 PageProvider 都需要继承自 PageSliderProvider 类，并且需要覆写 isPageMatchToObject(Componentcomponent，Object o)、destroyPageFromContainer(ComponentContainer componentContainer，int i，Object o)、createPageInContainer(ComponentContainer componentContainer，int i)、getCount()这 4 个函数。

其中，getCount 返回要滑动的 Component 的个数，createPageInContainer 是从当前 componentContainer 的指定位置 i 中添加 Component，然后返回这个 Component。

destroyPageFromContainer 是把 component 从 componentContainer 的当前位置 i 销毁。

在示例中首先构筑布局，然后初始化一个用来存储 Component 的列表，代码如下：

```
//MainAbilitySlice.java
DirectionalLayout myLayout = new DirectionalLayout(this);
DirectionalLayout.LayoutConfig layoutConfig = new DirectionalLayout.LayoutConfig
(ComponentContainer.LayoutConfig.MATCH_PARENT, ComponentContainer.LayoutConfig.MATCH_
PARENT);
private List<Image> list = new ArrayList<>();
```

然后给 LayoutConfig 设置一些参数，这样可以保证 Component 居中展示，代码如下：

```
layoutConfig.alignment = LayoutAlignment.HORIZONTAL_CENTER;
myLayout.setLayoutConfig(layoutConfig);
layoutConfig.width = ComponentContainer.LayoutConfig.MATCH_CONTENT;
layoutConfig.height = ComponentContainer.LayoutConfig.MATCH_CONTENT;
```

接下来新建一个 PageSlider，代码如下：

```
PageSlider pageSlider = new PageSlider(this);
```

其中，Provider 中放入了包含 4 个 Image 的 list，代码如下：

```
//MainAbilitySlice.java
Image image = new Image(this);
image.setImageAndDecodeBounds(ResourceTable.Media_picture);
Image image1 = new Image(this);
```

```
image1.setImageAndDecodeBounds(ResourceTable.Media_pic);
image1.setWidth(1125);
image1.setHeight(760);
image1.setScaleMode(Image.ScaleMode.ZOOM_CENTER);

Image image2 = new Image(this);
image2.setImageAndDecodeBounds(ResourceTable.Media_pic2);
image2.setScaleMode(Image.ScaleMode.ZOOM_CENTER);
image2.setWidth(1125);
image2.setHeight(760);

Image image3 = new Image(this);
image3.setImageAndDecodeBounds(ResourceTable.Media_pic3);
image3.setScaleMode(Image.ScaleMode.ZOOM_CENTER);
image3.setWidth(1125);
image3.setHeight(760);
list.add(image);
list.add(image1);
list.add(image2);
list.add(image3);
```

然后可以把 PageSlider 和 Provider 关联起来，代码如下：

```
pageSlider.setProvider(new PagerAdapter(list));
myLayout.addComponent(pageSlider);
```

这样就可以看到翻页效果，运行上述代码，效果如图 3.31 所示，分别为第一张图片到最后一张图片的切换过程。

完成 PageSlider 后还可以初始化一个 PageSliderIndicator，然后把 Indicator 和 PageSlider 关联起来，代码如下：

```
PageSliderIndicator pageSliderIndicator = new PageSliderIndicator(this);
pageSliderIndicator.setViewPager(pageSlider);
pageSliderIndicator.setLayoutConfig(layoutConfig);
myLayout.addComponent(pageSliderIndicator);
```

这样就完成了指示器的构造，运行代码可以看到翻页效果，如图 3.32 所示。

可以看到灰色的长条表示当前所在的页面，Indicator 明确地指示了 PageSlider 的页面。

3.2.6 其他布局容器

除以上布局容器之外，HarmonyOS 还提供了多种布局容器，这里只对它们做一个简单的介绍，具体的实现可以自行尝试。

图 3.31 PageSlide 翻页展示效果

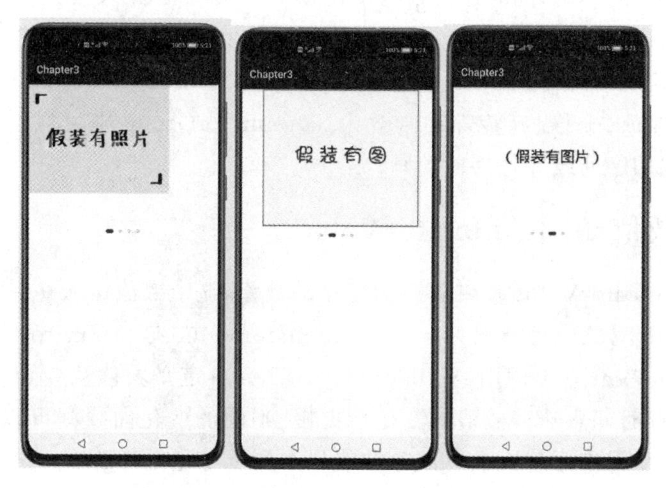

图 3.32 指示器效果运行效果图

窗口布局容器 StackLayout 提供一个窗口，提供一个框架布局，其中的元素可以重叠。StackLayout 用于在屏幕上保留一个区域来显示视图中的元素。通常，框架布局中只应该放置一个子组件。如果存在多个子组件，则显示最新添加的组件，之前添加的组件会被遮盖掉。

桌面布局容器 TableLayout 是一种像 Windows 桌面一样的排列布局容器。该布局容器用于在带有表的组件中安排组件。TableLayout 提供了对齐和安排组件的接口，以在带有表的组件中显示组件。可以配置排列方式、行数和列数及组件的位置。

还有自适应框容器 AdaptiveBoxLayout。自适应框将自动分为具有相同宽度和可能不同高度的框的行和列。框的宽度取决于布局宽度和每行中框的数量，这由它的 LayoutConfig 指定。新行仅在上一行填充后才开始。每个框都包含一个子组件。每个框的高度取决于其包含的子组件的高度。每行的高度由该行中的最高框确定。自适应框布局容器的宽度只能设置为 MATCH_PARENT 或固定值，但是开发者可以为容器中的组件自由设置长度、宽度和对齐方式。

3.3 Java UI 动画

3.3.1 动画类介绍

动画是组件的基础特性之一，精心设计的动画使 UI 变化更直观，有助于改进应用程序的外观并改善用户体验。HarmonyOS 在 Java UI 框架下提供了 Animator 类对各种组件添加动画效果，它的子类有数值动画（AnimatorValue）和属性动画（AnimatorProperty），还提供了将多个动画同时操作的动画集合（AnimatorGroup），开发者可以自由组合这些动画元素，从而构建丰富多样的动画效果。Animator 作为动画的基类，提供了与动画的启动、停止、暂停和恢复相关的 API，同时还支持为动画设定持续时间、启动延迟、重复次数和指定的曲线类型。AnimatorValue 提供随时间变化的变动数值，开发者可以自定义动画的样式，代入这个数值即可实现动画效果。AnimatorProperty 则是对 AnimatorValue 的自动化封装，原生组件提供了缩放、平移、旋转等动画效果。AnimatorGroup 则可以通过创建一个动画组，实现多个组建的序列或并行播放动画。

3.3.2 数值动画 AnimatorValue

前文说到 AnimatorValue 提供随时间变化的数值，这个数值是从 0 到 1 变化的浮点数，本身与 Component 对象或种类无关。由于 AnimatorValue 是 Animator 的子类，所以开发者可以在 AnimatorValue 中调用 Animator 中 API 并自定义数值从 0 到 1 的变化过程，例如更长的动画持续时间意味着数值的变化会更慢、不同的变化曲线则可以让数值实现各种各样的非匀速变化、设定重复次数可以让数值从 0 到 1 变化循环数次……

开发者可以利用数值随时间变化的特性，实现动画效果。例如通过值的变化改变控件的属性,从而实现控件的运动。

在本示例中主要演示一个 Text 组件的动画,并通过 Button 启动动画播放。先定义布局设置 LayoutConfig 和用于设定背景颜色的 ShapeElement 等参数,然后定义一个 Button 用于启动动画,代码如下:

```
//动画效果布局定义
layoutConfig.alignment = LayoutAlignment.CENTER;
layoutConfig.setMargins(100,600,100,100);
myLayout.setLayoutConfig(layoutConfig);
layoutConfig.width = ComponentContainer.LayoutConfig.MATCH_CONTENT;
layoutConfig.height = ComponentContainer.LayoutConfig.MATCH_CONTENT;
ShapeElement shapeElement = new ShapeElement();
shapeElement.setShape(ShapeElement.RECTANGLE);
shapeElement.setCornerRadius(80);
shapeElement.setRgbColor(new RgbColor(0,255,255));

Button button = new Button(this);
button.setLayoutConfig(layoutConfig);
button.setBackground(shapeElement);
button.setText("启动动画效果");
button.setTextSize(130);
```

接下来声明一个用于展示的 Text,此时这个 Text 可以视为动画的初始状态,代码如下:

```
Text t = new Text(this);
t.setLayoutConfig(layoutConfig);
t.setText("动画测试");
t.setTextColor(Color.RED);
t.setTextAlignment(Component.HORIZONTAL);
t.setTextSize(60);
```

然后实例化 AnimatorValue,并设置变化属性,在这里设置动画的持续时间为 2000ms、启动延时为 1000ms、循环播放 2 次,并设置变化曲线为 Bounce 型,代码如下:

```
AnimatorValue animator = new AnimatorValue();
animator.setDuration(2000);
animator.setDelay(1000);
animator.setLoopedCount(2);
animator.setCurveType(Animator.CurveType.BOUNCE);
```

接下来需要对 AnimatorValue 实例设置监听器 ValueUpdateListener,其中的回调参数 v 就是前述中从 0 到 1 变化的数值,这个数值会在设置的 Duration 时间内从 0 变化到 1,接下来将动画样式与 v 实现连接,如本例中使用了设定文本大小的函数 setTextSize((int)

(200 * v)),那么文本大小会在 2s 内从 200 * 0=0 变化到 200 * 1=200,即完成文本字体大小从 0 到 200 的变化,每一时刻的变化速度和之前设置的变化曲线有关。添加监听事件的代码如下:

```
animator.setValueUpdateListener((animatorValue, v) -> t.setTextSize((int) (200 * v)));
```

为 Button 设置监听事件,如果检测到按钮单击,则启动动画,代码如下:

```
button.setClickedListener(component -> animator.start());
```

AnimatorValue 动画效果如图 3.33 所示。

图 3.33　在 Text 组件上添加 AnimatorValue 动画效果图

可以清楚地看到，用于测试的 Text 组件已经被添加了动画效果，由于 Duration 设置值为 2000，因此这个动画效果会在 2s 内播放完毕。

3.3.3 属性动画 AnimatorProperty

因为 AnimatorProperty 是对 AnimatorValue 的自动化封装，内置实现了 Component 的平移、旋转、缩放的动画效果，使用方法较 AnimatorValue 更为简单。

首先直接对一个 Component(此处以 3.3.2 节中的 Button 为例)实例化 AnimatorProperty 对象，代码如下：

```
AnimatorProperty animator = button.createAnimator();
```

为 AnimatorProperty 实例设置变化属性，可链式调用，代码如下：

```
animator.moveFromX(50).moveToX(1000).alpha(0).setDuration(2500).setDelay(500).
setLoopedCount(5);
```

如上述代码设置了动画的起始 X 轴位置为 50、终止 X 轴位置为 1000、透明度为 0、持续时长为 2500 ms、启动延时为 500 ms 和循环次数 5 次。下面通过 start()方法启动动画，代码如下：

```
animator.start();
```

其中，AnimatorProperty 实例可以重复使用，例如可以使用 setTarget()方法改变关联的 Component 对象，代码如下：

```
animator.setTarget(t);
```

下面来看一个示例，为了便于展示，新建一个 Button 组件，并设置 LayoutConfig 和背景 ShapeElement，代码如下：

```
//动画效果布局定义
layoutConfig.setMargins(100,600,100,100);
myLayout.setLayoutConfig(layoutConfig);
layoutConfig.width = ComponentContainer.LayoutConfig.MATCH_CONTENT;
layoutConfig.height = ComponentContainer.LayoutConfig.MATCH_CONTENT;
ShapeElement shapeElement = new ShapeElement();
shapeElement.setShape(ShapeElement.RECTANGLE);
shapeElement.setCornerRadius(80);
shapeElement.setRgbColor(new RgbColor(0,255,255));
Button button = new Button(this);
button.setLayoutConfig(layoutConfig);
```

```
button.setBackground(shapeElement);
button.setText("启动动画效果");
button.setTextSize(130);
```

接下来创建一个属性动画，将这个动画和前面的 Button 实例关联起来。

设置动画的效果为 2.5s 内从 X 轴的 50 位置，移动到屏幕水平方向上 500 的位置，然后旋转 90°，代码如下：

```
AnimatorProperty animator = button.createAnimatorProperty();
animator.moveFromX(50).moveToX(500).rotate(90).alpha((float) 0.5).setDuration(2500).
setDelay(500);
animator.setCurveType(Animator.CurveType.BOUNCE);
```

运行上述代码，效果如图 3.34 所示。

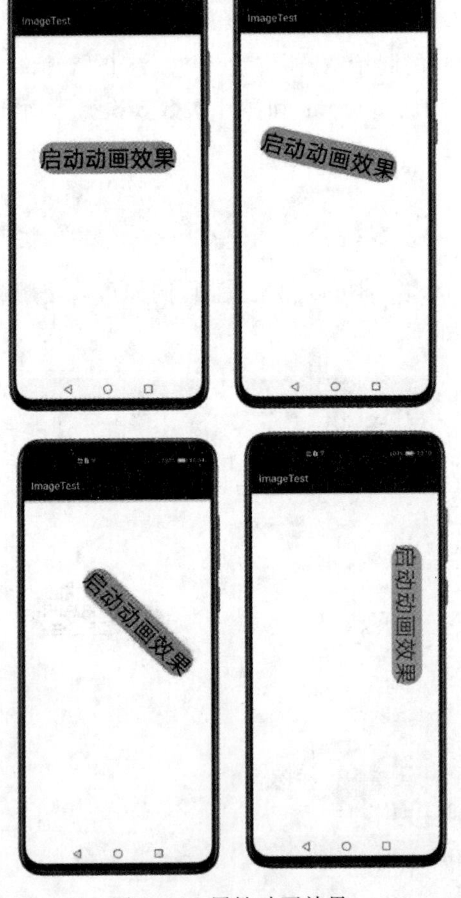

图 3.34　属性动画效果

3.3.4 动画集合 AnimatorGroup

如果需要使用一个组合动画，可以把多个动画对象添加到 AnimatorGroup 中。AnimatorGroup 提供了两种方法：runSerially()和 runParallel()，分别表示动画按序列启动和动画并发启动。下面是一个简单的示例。

首先需要声明一个动画集合 AnimatorGroup，代码如下：

```
AnimatorGroup animatorGroup = new AnimatorGroup();
```

其次，实例化多个动画，并将它们添加到动画集合中。这里实例化了 4 个动画 animator1～animator4，分别配置在 4 个按钮 button1～button4 上。动画的配置方法均参考 3.3.3 节的属性动画，animator1 只从左向右移动，animator2 从右向左移动并旋转 90°，animator3 从左向右移动并旋转 90°，animator4 与 animator2 相同，代码如下：

```
//动画的实例化与配置
AnimatorProperty animator1 = button1.createAnimatorProperty();
animator1.moveFromX(50).moveToX(500).alpha((float) 0.5).setDuration(2500).setDelay(500);
AnimatorProperty animator2 = button2.createAnimatorProperty();
animator2.moveFromX(50).moveToX(500).rotate(90).alpha((float) 0.5).setDuration(2500).
setDelay(500);
AnimatorProperty animator3 = button3.createAnimatorProperty();
animator3.moveFromX(500).moveToX(50).rotate(90).alpha((float) 0.5).setDuration(2500).
setDelay(500);
AnimatorProperty animator4 = button4.createAnimatorProperty();
animator4.moveFromX(50).moveToX(500).rotate(90).alpha((float) 0.5).setDuration(2500).
setDelay(500);
AnimatorGroup animatorGroup = new AnimatorGroup();
//4 个动画同时播放
animatorGroup.runParallel(animator1, animator2, animator3, animator4);
```

设置一个按钮并配置监听事件，用于启动动画集合，代码如下：

```
button5.setClickedListener(new Component.ClickedListener() {
            @Override
            public void onClick(Component component) {
                animatorGroup.start();
            }
        });
```

运行上述代码，可以看到动画 1 至动画 4 同时播放的效果，如图 3.35 所示。

若将上文中的 animatorGroup.runParallel (animator1，animator2，animator3，animator4)更

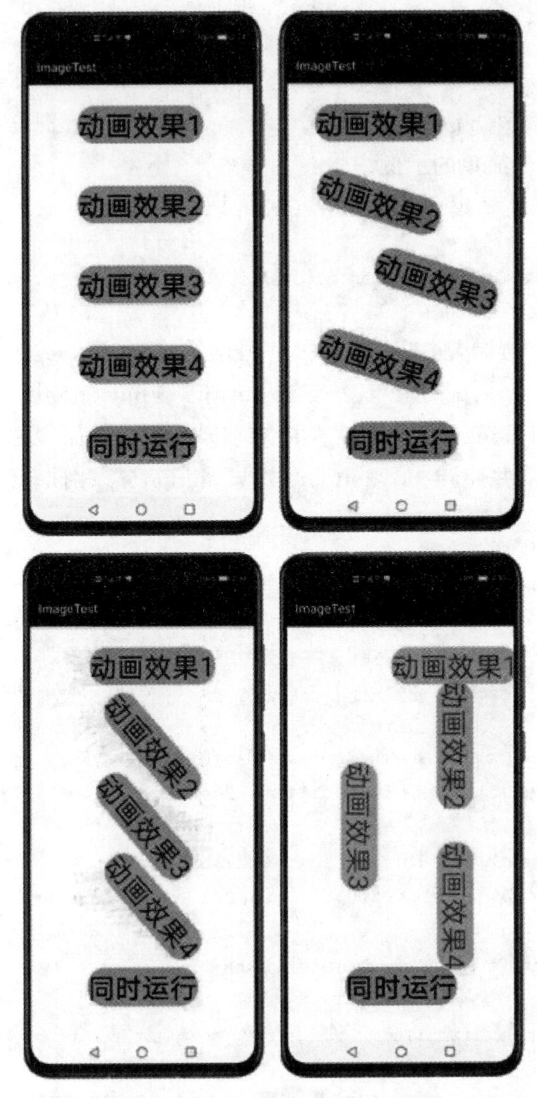

图 3.35　同时播放动画组效果

改为 animatorGroup. runSerially(animator1，animator2，animator3，animator4)，则动画将从并行播放变为序列播放，效果如图 3.36 所示。

为了更加灵活地处理多个动画的播放，例如一些动画序列播放，而另一些动画并行播放，Java UI 框架提供了更方便的动画 Builder 接口。首先声明 AnimatorGroup Builder，使用 addAnimators()方法为 Builder 添加多个动画，同一个 addAnimators()内的动画并行播放，不同的 addAnimators()中的动画按添加顺序播放。本例中动画的播放效果为首先播放完 animator1，随后并行启动 animator2 和 animator3 至两者全部播放完毕，最后启动 animator4，代码如下：

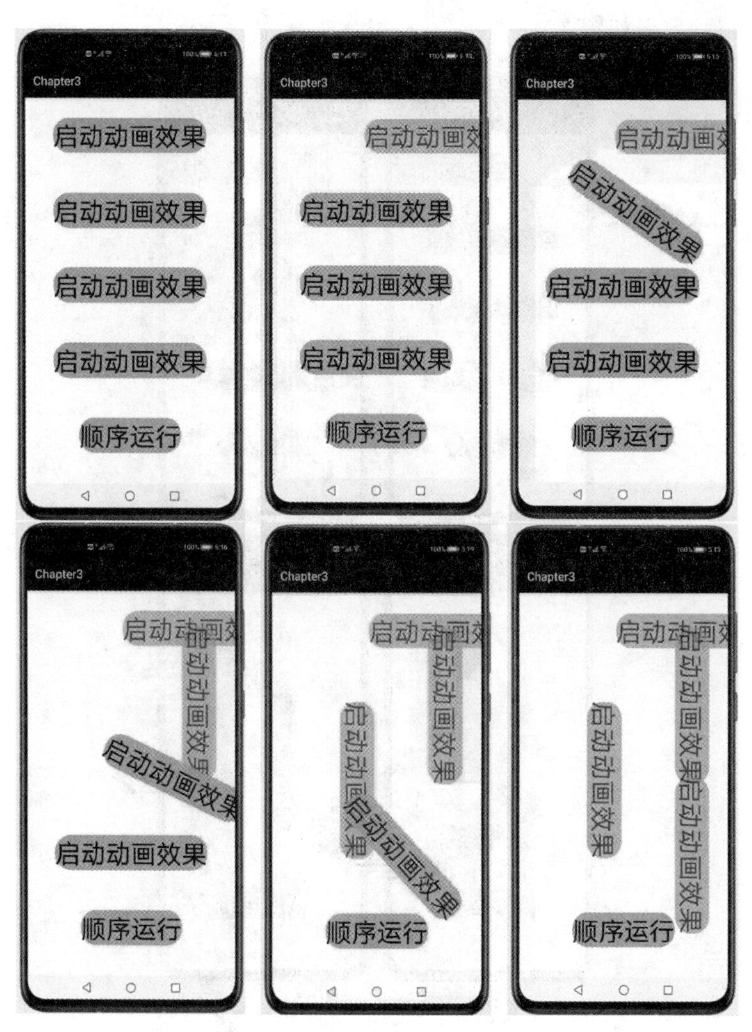

图 3.36 顺序播放动画组效果

```
//动画播放
AnimatorGroup animatorGroup = new AnimatorGroup();
AnimatorGroup.Builder animatorGroupBuilder = animatorGroup.build();
animatorGroupBuilder. addAnimators ( animator1 ). addAnimators ( animator2,  animator3 ).
addAnimators(animator4);
//4 个动画的顺序为: animator1 -> animator2/animator3 -> animator4
button5.setClickedListener(new Component.ClickedListener() {
    @Override
    public void onClick(Component component) {
        animatorGroup.start();
    }
});
```

运行代码得到效果如图 3.37 所示。

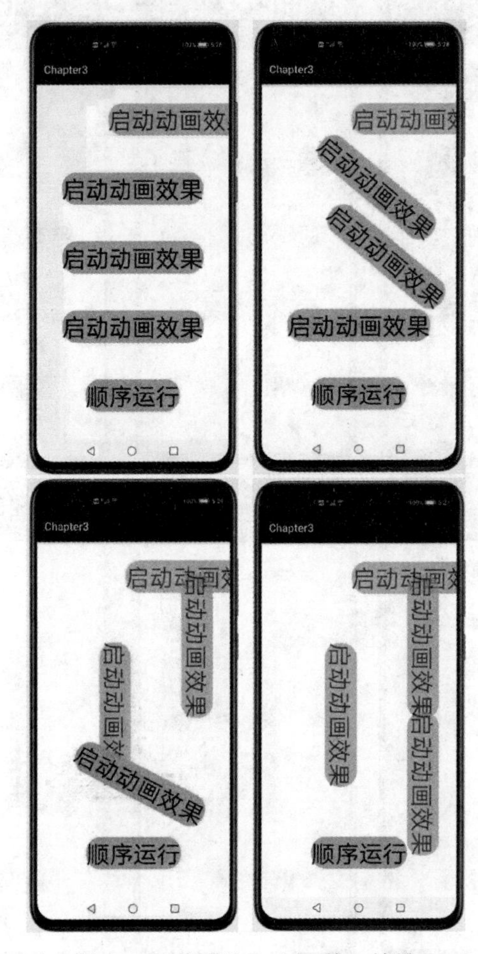

图 3.37　动画组 Builder 接口效果

第4章 JavaScript

JavaScript 简称 JS，是世界上最流行的脚本编程语言之一，它灵活轻巧，兼顾函数式编程和面向对象编程，能跨平台、跨浏览器驱动网页，以及能与用户进行交互。

JS 最初只能运行于浏览器环境，用于 Web 前端开发，后来有开发人员将 JS 从浏览器中分离出来，开发了一套独立的运行环境，所以现在的 JS 也能用于网站后台开发，如可用于游戏、桌面和移动应用程序的开发。

新兴的 Node.js 已把 JS 引入服务器端，JS 俨然变成了全能型选手。HarmonyOS 支持使用 JS 语言进行开发，为了让不太熟悉 JS 语言的读者有一个更好的开发体验，本章将对与 JS 语言相关的内容进行介绍。有一定 JS 开发基础的读者可以直接跳过本章节。

4.1 关于 JavaScript

4.1.1 JavaScript 简介

JS 是一种面向 Web 的编程语言，获得了所有网页浏览器的支持，是一种运行在客户端浏览器中的解释型的编程语言，也是网页设计和 Web 应用必须掌握的基本工具。手机、平板、计算机上支持浏览的网页，甚至于所有基于 HTML5 的手机 App，其交互逻辑均是由 JS 驱动的。

JS 语言最初于 1995 年由开启 Web 时代的最著名的第一代互联网公司——网景公司(Netscape Communications Corporation)提出，在之后的时间里，微软在浏览器中也加入了脚本编程功能，命名为 Jscript，同期，Cenvi 也推出了 ScriptEase 语言。1997 年，几个公司联合 ECMA(European Computer Manufacturers Association)组织定制了 JavaScript 语言的标准，即 ECMAScript 标准。1998 年，国际标准化组织和国际电工委员会(ISO/IEC)采用了 ECMAScript 标准(即 ISO/IEC-16262)，至此浏览器厂商就以 ECMAScript 作为各自 JavaScript 实现的规范标准，JS 语言实现了规范统一。

随着 JS 技术的不断发展，JS 开发已经不单单应用于客户端，Node.js 框架的出现成功地把 JS 应用到了服务器端。Node.js 也是开发 HarmonyOS 应用必备的软件。

4.1.2 揭开 JavaScript 面纱

简单了解了 JS 语言产生的背景后，接下来将介绍 JS 的组成、功能及特点。

一个完整的 JavaScript 主要包含以下 3 个部分：

(1) ECMAScript：JS 的语法标准。

(2) DOM(Document Object Model，文档对象模型)：网页文档操作标准，即 JS 对网页上元素进行操作的 API。

(3) BOM(Browser Object Model，浏览器对象模型)：JS 操作浏览器的部分功能的 API，即描述 JS 语言与浏览器进行交互的方法和接口。

具体地讲，ECMAScript 是 JS 的核心，描述了 JS 语言的基本语法(如：var、for、if 等)和数据类型(如数字、字符串、布尔、函数等类型)。简单来讲，ECMAScript 是 JS 的基础规范，定义了 JS 的基础语法。

实际上，Web 浏览器不仅提供基本的 ECMAScript 实现，同时也会提供各种扩展功能。

DOM 是 HTML(Hyper Text Markup Language，超文本标记语言)和 XML 的应用程序接口(API)。DOM 把整个 HTML 或 XML 页面映射成由节点层级构成的文档，页面的每个部分都是一个节点的衍生物。简单来讲，DOM 通过创建树来表示文档，从而便于开发者对文档的内容和结构进行访问和控制。此外，利用 DOM API 可以轻松地删除、添加和替换节点。

人们习惯上把所有针对浏览器的 JS 扩展算作 BOM 的一部分。BOM 提供了独立于内容而与浏览器窗口进行交互的对象，可以完成 JS 对浏览器窗口的访问和操作。例如：弹出新浏览器；移动、缩放和关闭浏览器窗口；提供详细的网络浏览器信息；提供详细的页面信息；对 Cookies 的支持等。需要注意的是，BOM 并没有相关标准的支持，每个浏览器都有自己的实现，虽然有一些非事实的标准，但还是给开发者带来一定的麻烦。

那么，一个完整的 JavaScript 能做什么呢？为什么会让无数开发者痴迷呢？事实上，JS 程序不仅可以访问浏览器和 Web 页面的元素，还可以对 Web 页面的元素进行操作、在 Web 页面创建新元素等。除此之外，JS 常见的功能还包括以下几点：

(1) 以指定尺寸、位置和样式(如：是否包含菜单栏、工具栏等)打开一个新的窗口。

(2) 检验 Web 表单输入的数据，并在提交之前确保数据格式的正确性。

(3) 能检测和响应特定的用户操作，对页面元素进行相应的操作。

(4) 检测和发现特定浏览器支持的高级功能，如：第三方插件等。

(5) 支持对浏览器新技术的原生支持。

JS 代码在用户浏览器内部运行，页面会对 JS 指令做出快速响应，有效增强了用户的体验，但 JS 作为一种脚本语言，通常也只能被嵌入到 HTML 文件中执行，因此其最大的特点就是与 HTML 的结合，只有当 HTML 文档在浏览器中被打开时，JS 代码才能被执行。若读者不了解 HTML 文件，本书将在 4.3.1 节讲解 JS 的执行方式时对其进行简单介绍。尽管 JS 可以作为单独的文件存在，但也必须通过 HTML 文档的调用才能执行。JS 扩展了标

准的 HTML,为 HTML 页面添加了交互行为。除此之外,JS 语言还具有以下特点:

(1) JS 是一种解释型脚本语言:代码不进行预编译,在程序运行过程中被逐行解释,不需要严格的变量声明。

(2) JS 语言比较简单:JavaScript 语言使用弱类型的变量类型。对使用的数据类型没有严格要求,可以进行类型转换,语法和 Java 类似,简单又灵活。

(3) JS 是基于对象的:JS 不仅可以创建对象,还可以使用现有对象。

(4) JS 是由事件驱动的:所谓事件驱动,就是指在主页中执行了某种操作所产生的动作,成为事件。当事件发生后,可能会引起相应事件的响应,例如按下鼠标、移动窗口、选择菜单等都可能会引起相应的事件响应,执行某些对应的脚本。

(5) JS 可以跨平台执行:JS 依赖于浏览器本身,与操作环境无关。即 JS 可以在任何能运行浏览器且浏览器支持 JS 的操作系统上执行(如 Windows、Linux、Mac、Android、iOS 等)。

(6) JS 可以直接在客户端浏览器中运行:可以直接对用户或者客户端的输入做出响应,无须经过 Web 服务程序。

(7) JS 语言具有一定的安全性:JS 只能通过浏览器来处理并显示信息或进行动态交互,但不能修改或删除其他文件中的内容。JS 不能访问本地硬盘且不能将数据存入服务器中,从而有效地防止数据的丢失。

4.1.3 JavaScript 与 Java 的区别

实际上,由于 JavaScript 和 Java 在命名上的相似,大部分初学者对于二者之间的联系和区别容易产生疑问。一般认为,当时 Netscape 之所以将 LiveScript 命名为 JavaScript,是因为 Java 是当时最流行的编程语言,带有 Java 的名字有助于这门新生语言的传播,且名字中的 Java 是经过 SUN Microsystems 公司授权的。然而,二者之间,不能说是毫不相干,但可以说是千差万别。

二者的区别主要体现在:

(1) 二者的开发者不同:JS 是 Netscape 公司的产品,是为了扩展 Netscape Navigator 功能而开发的一种可以嵌入 Web 页面中的解释性语言,而 Java 是 SUN Microsystems 公司推出的新一代面向对象的程序设计语言,特别适合于 Internet 应用程序开发。

(2) 基于对象和面向对象:JS 是一种基于对象和事件驱动的开发语言,是一种脚本语言,而 Java 是面向对象的,即 Java 是一种真正的面向对象的语言,即使是开发简单的程序也必须设计对象。

(3) 运行方式不同:JS 是一种解释性编程语言,不需要进行编译,由浏览器解释执行,而 Java 的源代码在执行之前,必须经过编译。

(4) 数据类型不同:JS 中变量是弱类型的,声明变量时无须指定变量的数据类型,同一个变量可以赋值不同的数据类型,而 Java 中变量是强类型的,所有变量在编译之前必须声明,且在定义变量时必须指定变量类型,如果赋值的时候类型不匹配,就会报错。

(5) 语言类型:JS 是动态类型语言,而 Java 是静态类型语言。即在 Java 中定义了一个数组，其长度就不能再改变了，但是在JS中却可以。

4.2 JavaScript 开发环境

俗话说，“工欲善其事，必先利其器”，一个好的开发工具常常能激发开发者的开发乐趣。本节将简单介绍几种JS开发环境。

4.2.1 JavaScript IDE

目前，比较知名的JS集成开发环境(IDE, Intergrated Development Environment)主要包括以下几种：

(1) Visual Studio Community：Visual Studio是微软公司的开发工具包系列产品，是一个基本完整的开发工具集，可供各个开发者、开放源代码项目、学术研究、教育和小型专业团队免费使用。提供了高级开发工具、调试功能、数据库功能和创新功能，帮助在各种平台上快速创建当前最先进的应用程序，以及开发新的程序。

下载网址：https://visualstudio.microsoft.com/zh-hans/downloads

(2) Eclipse：Eclipse是著名的跨平台的自由IDE，是一个开源的、基于Java的可扩展开发平台。就其本身而言，它只是一个框架和一组服务，用于通过插件组件构建开发环境。尽管Eclipse是使用Java语言开发的，但它的用途并不限于Java语言，支持诸如C/C++、JS、PHP等编程语言的插件。

下载网址：https://www.eclipse.org/downloads/packages/

(3) WebStorm：WebStorm是JetBrains公司旗下一款JavaScript开发工具。已经被广大中国JS开发者誉为“Web前端开发神器”“最强大的HTML5编辑器”“最智能的JavaScript IDE”等。提供JavaScript、ECMAScript 6、TypeScript、CoffeeScript、Dart和Flow代码辅助功能，支持Node.js和Web前端三大主流框架，如React、Angular、Vue.js、Meteor等。

下载网址：https://www.jetbrains.com/WebStorm/download

(4) NetBeans IDE：NetBeans包括开源的开发环境和应用平台，支持开发人员快速创建Web、桌面及移动的应用程序，NetBeans IDE已经支持PHP、Ruby、JavaScript、Groovy、Grails和C/C++等多种开发语言。

下载网址：https://netbeans.apache.org/download/index.html

(5) IntelliJ IDEA：IDEA也是JetBrains公司的产品，是Java编程语言开发的集成环境。其最突出的功能是可以对Java代码、JavaScript、JQuery、Ajax等技术进行调试。

下载网址：https://www.jetbrains.com/idea

(6) DevEco Studio：DevEco Studio是华为官方推出的，基于IntelliJ IDEA Community开源版本打造，面向华为终端全场景多设备的一站式集成开发环境(IDE)，为开发者提供工程模板创建、开发、编译、调试、发布等E2E的HarmonyOS应用开发服务。支持多语言的代码开发和调试：包括 Java、XML、C/C++、JS、CSS 和 HML。此外，不同于传统的基于 Web 进行开发，DevEco Studio 支持开发者基于 JS 进行手机、平板、智慧屏或智能穿戴设备的开发，以及对轻量级智能穿戴设备进行开发。DevEco 充分发挥了 JS 的开发潜能。

下载网址：https://developer.harmonyos.com/cn/develop/deveco-studio#download

4.2.2 浏览器

如今，大部分的网页浏览器（如 Google Chrome、Mozilla Firefox、Apple Safari 及 Microsoft Edge 等）均提供了 JS 调试工具，因此，读者完全可以使用浏览器工具进行 JS 开发。接下来，以 Chrome 浏览器为例，本节将简单演示在浏览器中的 JS 开发效果。

打开 Chrome 浏览器，Windows 系统按下 Crtl+Shift+I/J 快捷键（Mac OS 的快捷键为 Command+Option+J）即可直接打开浏览器调试模式，如图 4.1 所示。

图 4.1　Chrome 浏览器的调试模式

在 Console 区域内，键入的 JS 代码即可在当前浏览器页面中运行，在这里键入代码如下：

```
document.write("<h1>Hello World!</h1>");
```

回车后页面显示如图 4.2 所示。

图 4.2　Chrome 浏览器中运行 JS 代码

4.2.3 Node.js

Node.js 并非是一个扩展名为 js 的文档名称，而是一个开源且跨平台的运行期环境（run-time environment）。Node.js 基于 Google 的 V8 引擎（V8 引擎执行 JS 的速度非常快，性能非常好），是一个事件驱动 I/O 服务器端的 JS 环境。简单来讲，Node.js 就是运行在服务器端的 JS。只要安装了 Node.js，就可以在系统中运行 JS。在 HarmonyOS 中通过安装 Node.js 实现 JS 开发,因此在这里简要介绍一下 Node.js 的主要特点。

(1) 具备事件驱动的运行框架:事件驱动的优势在于代码在执行时无须等待某种操作完成,从而使系统资源可以用于完成其他任务,有限的资源得到了充分利用。

(2) 异步处理:在 Node.js 提供的支持模块中,许多函数甚至包括文件在内都是异步执行的。这意味着虽然在代码结构中这些函数是依次注册的,但是它们并不依赖于自身出现的顺序,而是等待相应的事件触发。Node.js 的异步特性适用于 I/O 比较密集,实时性比较强的服务。

(3) 非阻塞 I/O:在执行 I/O 操作时不阻塞,避免了由于需要等待输入或输出(数据库、文件系统等)而造成 CPU 时间损失。

(4) 单线程:事件驱动机制是 Node.js 通过内部单线程高效率的维护事件循环队列实现的,与多线程相比,不用在意一些状态同步问题,也没有资源占用和死锁问题,更没有线程进行上下文切换引起的性能开销。

(5) 跨平台:Node.js 是一个跨平台的开源软件,可以在 Linux、Windows、Android 等多个系统上运行。

综上,Node.js 基于事件驱动,使用了非阻塞 I/O 模型、单线程模式和异步编程。其轻量且高效,以及跨平台性能更让其如虎添翼。值得注意的是,尽管 Node.js 一直以来被认为是 I/O 密集型操作的特长生,但其不擅长 CPU 密集型操作。新版的 Node.js,已经在不断弥补这方面的短板。在 Node v10.5.0 中,官方给出了一个实验性质的模块 worker_threads 为 Node.js 提供了多线程能力。在 Node v12.11.0 中,worker_threads 模块正式成为稳定版。至此,Node.js 正式拥有真正的多线程能力。

4.3 走近 JavaScript

4.2 节内容主要介绍了常见的支持 JS 开发的开发工具,本节一起进入 JS 的语法世界,为后续的开发应用进行一些知识储备。

4.3.1 JavaScript 执行方式

JS 主要用于 Web 开发中。在这里首先简单介绍一下什么是 HTML。HTML 即超文本标记语言,它不是编程语言,而是一种标签语言,用于标记页面的各部分在浏览器里以何种方式展现,如作为标题还是项目列表,字体是否加粗,是否使用斜体字等。可以将 HTML 理解为使网页的"源码",浏览器可以被视为"解释和执行"HTML 源码的工具。HTML 文件一旦编写完成,就意味着无论用户何时访问这些页面,其中的标签都会以相同的方式进行解析和显示,这就决定了页面是静态的,而 JS 的主要功能是让页面"动起来",即完成与用户的交互。JS 作为一种脚本语言,通常被嵌入 HTML 文件中,JS 代码的执行取决于 HTML 文件是否被浏览器打开。当然,也有另外一种方式,可以将 JS 代码单独存放在后缀名为.js 的文件中,但也必须在 HTML 中引入该文件才能执行。

1. 使用< script >标签直接嵌入 HTML 文件中

当 JS 被嵌入 HTML 中时，JS 代码必须位于< script >与</script >标签之间。实际上，HTML 文件中可以放置任意数量的脚本。通常，JS 代码可以被放置在 HTML 的< body >或< head >标签中，或兼而有之。在 Notepad++编辑器中编写代码，使得 JS 代码嵌入在 HTML 的< body >标签，将文件保存为 test1.html 文件，并在 Chrome 浏览器中运行，代码如下：

```
<! -- test1.html -- >
< html >
< head >
< title > Hello from JS </title >
</head >
< body >
< script >
            alert("Hello, world!");
</script >
</body >
</html >
```

需要注意的是，这里的讲解重点放在对 JS 的介绍上，故对于 HTML 中各个标签的用法，感兴趣的读者可以自行了解。在浏览器中运行上述代码，运行效果如图 4.3 所示。

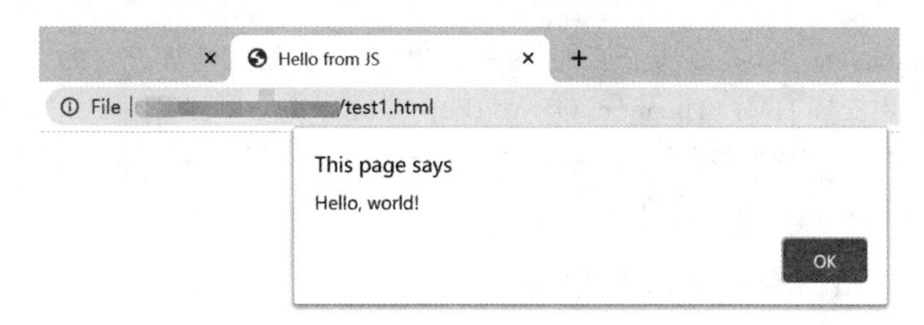

图 4.3　JS 代码嵌入 HTML 中的执行效果图

2. JS 作为单独的文件被 HTML 文件引用

JS 执行的另一种方式是将 JS 代码存为一个单独的.js 文件，在 HTML 中通过标签< script src="XX.js"></script >引入这个文件。同样，可以在 HTML 中的< head >或< body >标签中进行引用。首先，将 JS 代码单独保存为 myjs.js 文件，代码如下：

```
alert ("Hello, world!");
```

接着新建 test2.html 文件并引用该 JS 文件，并将 test2.html 文件和 myjs.js 文件保存在同一文件夹中，test2.html 的代码如下：

```
<! -- test2.html -- >
< html >
```

```
<head>
<title>Hello from JS</title>
</head>
<body>
<script src="myjs.js"></script>
</body>
</html>
```

其中,HTML 利用 src 属性对不同路径下文件的引用方式,需要读者自行了解,这里不进行详细讲述。在浏览器中的运行效果如图 4.4 所示。可以看出,以上两种方式的执行结果相同。

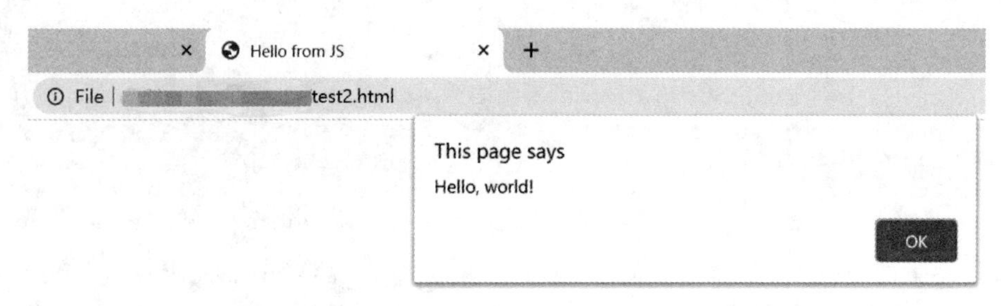

图 4.4 HTML 中引用外部 JS 文件的执行效果图

在 HarmonyOS 中,支持基于手机、平板、智慧屏和智能穿戴等设备的 JS 模板开发。HarmonyOS 中支持 HML(HarmonyOS Markup Language,HarmonyOS 标记语言),HML 为 HarmonyOS 自研的标签性语言,与 HTML 类似。下面将先讲解一些基础的语法,对 JS 有一个更深入的理解,从而更好地开发应用。

4.3.2 JavaScript 核心语法

对于 JS 核心语法,本节从以下几个方面来讲解:语法约定、数据类型、变量、运算符号、输入/输出、控制语句、注释、函数。对于这部分知识有了解的读者,可以跳过此部分内容,直接进入下一阶段的学习。

1. 语法约定

JS 语法就是指构成合法的 JS 程序的所有规则和特征的集合,定义了 JS 的语言结构,简单描述如下:

(1) 大小写:JS 严格区分大小写,如果弄错了大小写,程序将报错或者运行不正常。例如将上文所建的 test1.html 中的 alert 语句改为大写 ALERT,在浏览器中运行时,效果如图 4.5 所示。从图中可以看出,没有遵守大小写规则会导致程序报错,从而无法正常运行。

(2) 标识符:JS 中的标识符包括变量名、函数名、参数名和属性名。一个合法的标识符应遵循以下规则。

- 标识符是大小写英文、数字、$ 和_的组合,不能以数字开头。

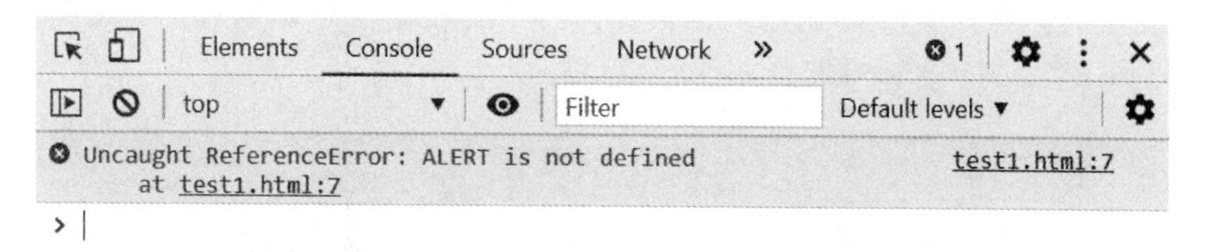

图 4.5 JS语句大小写错误效果图

- 不能与 JS 中的关键字(如 if、while 等)和保留字重名。
- 标识符区分字母大小写。

(3) 分号：JS 的每个语句以分号“;”结束，但是，JS 并不强制要求在每个语句的结尾加分号，浏览器中负责执行 JS 代码的引擎会自动将分号补在每个语句的结尾。

2. 数据类型

JS 中的数据类型主要包括以下几种，在 JS 中可以通过 typeof()检测变量的返回值。

(1) 数值(Number)：JS 不区分整数和浮点数，且可以直接进行四则运算，规则和数学中的规则一致。

(2) 字符串(String)：字符串是以单引号'或双引号"括起来的任意文本。

(3) 布尔值(Boolean)：一个布尔值只有 true、false 两种值。

(4) 数组(Array)：JS 中数组用方括号书写，数组元素之间用逗号分隔。JS 数组中可以包括任意数据类型，如新建一个数组 arr：var arr = [1, 2, 3.14, 'Hello', null, true]。数组元素可以通过索引访问，索引的起始值为 0，如：arr[0] 返回索引为 0 的元素，即 1。

(5) 对象(Object)：JS 的对象是一组由键-值组成的无序集合，由大括号{}括起来，其中，键都是字符串类型，值可以是任意数据类型，代码如下：

```
var person = {
    name:'Millie',
    age: 20,
    tags: ['cute', 'beautiful'],
    girl: true
};
```

要获取一个对象的属性，则采用“对象变量.属性名”的方式，如 person.name 可以获取 name 属性。

(6) 未定义 (Undefined)：在 JS 中，没有值的变量，其值是 undefined。任何变量均可通过设置值为 undefined 进行清空。

(7) 空 (Null)：Null 与 Undefined 不同，例如 var car=""；空的字符串变量既有值也有类型，值是""，类型是 String。在 JS 中，null 的数据类型是 Object。

3. 变量

JS 中的变量用于存储数据值。变量在 JS 中用变量名表示，在前面已经了解到了变量名的命名规则。下面将介绍变量的声明和赋值规则。

(1) 变量的声明：在JS中创建变量被称为"声明"变量。可以通过var关键词声明JS变量，代码如下：

```
var person;
```

注意，声明之后，变量是没有值的，即变量的值Undefined，但可以在声明变量的同时给它赋值。

(2) 变量的赋值：在JS中，使用等号"="对变量进行赋值，可以把任意数据类型赋值给变量，同一个变量可以反复赋值，而且可以是不同类型的变量，代码如下：

```
var a = 123;                //此时变量 a 的值是 Number 类型
a = "ABC";                  //此时变量 a 的值是 String 类型
```

(3) 变量的作用域分为全局变量和局部变量。全局变量在整个页面脚本中都是可见的，可以被自由访问。局部变量：变量仅能在声明的函数内部可见，在函数外是不允许访问的。在函数作用域或全局作用域中，通过关键字var声明的变量，无论实际上是在哪里声明的，都会被当作在当前作用域顶部声明的变量，代码如下：

```
function getName(Condition){
        if(condition){
                var name = "Bob";
                return name;
        }else{
                //此处可以访问变量 name,其值为 Undefined
                return null;
        }
         //此处可访问变量 name,其值为 Undefined
}
```

在函数getName中，有人认为只有当condition为true时才会创建变量name，然而事实上，变量name始终会被创建。在预编译阶段，上面的函数会被JS引擎修改，代码如下：

```
function getName(Condition){
        var name;
        if(condition){
                name = "Bob";
                return name;
        }else{
                return null;
        }
}
```

可以看出，变量name的声明被提升至函数顶部，而初始化操作却保留在原处执行。故在 else 语句中也可以访问变量 name，且由于在 else 中变量并没有初始化，因此其值为 Undefined。这就是在函数内部的变量提升，在全局作用域中同理。

4. 运算符号

JS 中的几类常见的运算符如表 4.1 所示。类型运算符中，typeof 用于返回变量的类型，instanceof 用于判断是否是对象实例，如果是，则返回值为 true，否则返回值为 false。

表 4.1 JS 中的运算符

类　型	运　算　符
算术运算符	+　-　*　/　%　++　--
赋值运算符	=　+=　-=　*=　/=　%=
比较运算符	>　<　>=　<=　==　!=　===　!==
逻辑运算符	&&　||　!
类型运算符	typeof　instanceof

5. 输入/输出

prompt()支持用户在页面进行输入，同时也支持多种输出方式，如使用 indow. alert()可写入警告框，使用 document. write()可写入 HTML 输出，使用 innerHTML 写入 HTML 元素，使用 console. log()写入浏览器控制台。这里编写的代码，保存为 testIO. html 文件，并在 Chrome 浏览器运行，效果如图 4.6 所示，代码如下：

```
<!-- testIO.html -->
<html>
<head>
<title>Test I/O</title>
</head>
<body>
    <script>
        var j = prompt("请输入需要输出的次数：");                    //支持用户输入
        for(var i = 0;i < j;i++) {
            document.write("<h3>Hello World!</h3>");               //在页面输出
        }
        alert("共连续：" + j + "次");                               //弹窗显示
    </script>
</body>
</html>
```

6. 控制语句

JS 中的控制语句主要包括以下几个：

(1) if else 语句：条件语句用于基于不同条件执行不同的动作。

(2) switch 语句：选择多个需要被执行的代码块之一。

(3) for 语句和 while 语句：for 循环语句在已知循环的初始和结束条件时非常有用，而

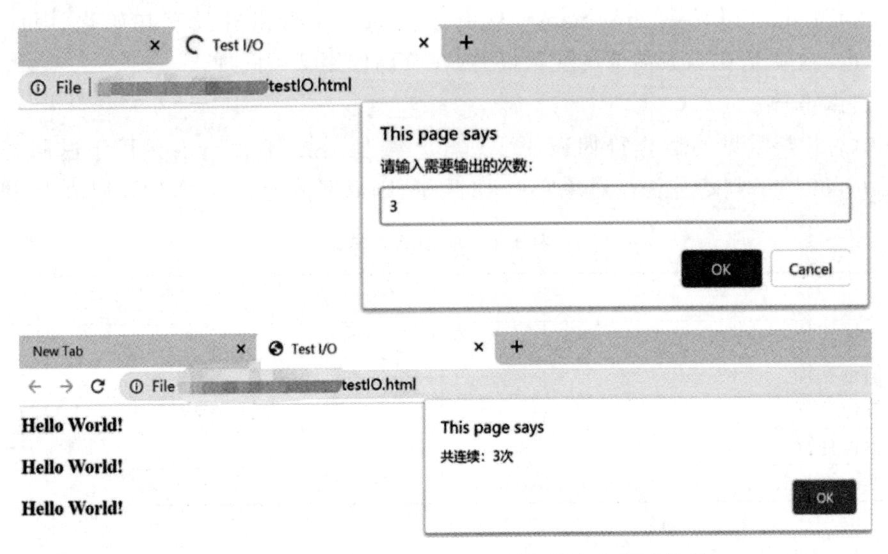

图 4.6　JS 输入输出在 Chrome 浏览器中的显示效果

while 循环语句则只有一个判断条件,条件满足时不断循环,条件不满足时则退出循环。

7. 注释

JS 中的注释包括以下两种形式:

(1) 单行注释:以"//"开头。任何位于"//"与行末之间的文本都不会被执行。

(2) 多行注释(注释块):多行注释以"/ * "开头,以" * /"结尾,任何位于"/ * "和" * /"之间的文本都不会被执行。

8. 函数

基本上所有的高级语言都支持函数,JS 也不例外。实际上,函数是 ECMAScript 的核心。在 JS 中函数的定义方式如下:

```
function functionName(arg0, arg1, ..., argN) {
  function body;
}
```

其中,function 是关键字,指出这是一个函数定义。functionName 是函数名称,遵循标识符规则。arg0, arg1,…, argN 是 N 个参数,参数个数不受限制,参数之间用逗号隔开。{ … }之间的代码是函数体,可以包含若干语句,甚至可以没有任何语句。

由于 JS 的函数也是一个对象,上述定义的函数实际上是一个函数对象,而函数名可以视为指向该函数的变量,因此,函数的定义还可以写成如下形式:

```
var functionName = function (arg0, arg1, ..., argN) {
    function body;
};
```

在这种方式下,function (arg0, arg1,..., argN) { ... }是一个匿名函数,它没有函数名,但是,这个匿名函数赋值给了变量 functionName,所以,通过变量 functionName 就可以调用该函数。

上述两种定义完全等价,需要注意第二种方式按照完整语法需要在函数体末尾加一个冒号,表示赋值语句结束。

函数可以通过其名字加上括号中的参数进行调用,函数的定义方法如下:

```
functionName(arg0, arg1, ..., argN);
```

这里新建一个 testFun.html 文件,代码如下:

```
<!-- testFun.html -->
<html>
<head>
<title>Test Function</title>
</head>
<body>
<script>
        function sum(a,b){
            return a + b;
        }
     alert(sum(2,3));
    </script>
</body>
</html>
```

在该文件中定义了函数 sum,实现两个参数的加法,并通过 alert 显示函数调用的结果。在 Chrome 浏览器中运行该程序,运行结果如图 4.7 所示。

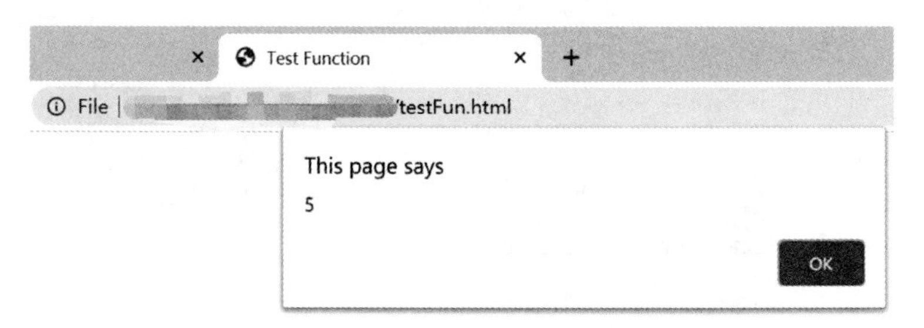

图 4.7 JS 中函数调用在 Chrome 执行效果

4.3.3 ES6 语法概述

到这里,相信读者已经掌握了上述 JS 中的核心语法,本章将继续讲解 JS 中的 ES6 语法。实际上,对于手机、平板、智慧屏和智能穿戴的开发,HarmonyOS 是支持 ES6 的,轻量级智能穿戴开发支持的 ES6 语法略有限，仅支持部分 ES6 语法。

ECMAScript 是 JS 实现的规范标准，随着 JS 的应用范围越来越广泛，JS 的规范标准也在不断更替以适应实际开发应用。ES6 全称为 ECMAScript 6.0，是 JS 的一个版本标准，发布于 2015 年 6 月。目前各大浏览器基本上支持 ES6 的新特性，其中，Chrome 和 Firefox 浏览器对 ES6 新特性最友好。此外，Node.js 同样支持 ES6 且其支持度更高，因此为了有更好的开发体验，这里对 ES6 语法进行一些基本的概念和用法介绍，想要深入学习的读者可以自行查阅学习资料。

1. let 和 const

ES6 新增加了两个重要的 JS 关键字：let 和 const。它们是 var 在块级声明的替代方案。块级声明用于声明在指定块的作用域之外无法访问的变量。块级作用域存在于函数内部和块中，即{}之间的区域。接下来将通过两个实例来说明 let 和 const 与 var 之间的区别。首先定义一个通过 var 声明变量的函数 funVar，将其嵌入 html 文件并调用 funVar(40)，运行效果如图 4.8 所示，代码如下：

```
function funVar(x){
    var y = x;
    if (x > 30){
        var y = 20;
        alert("Inner y = " + y);
    }
    alert("Outer y = " + y);
}
```

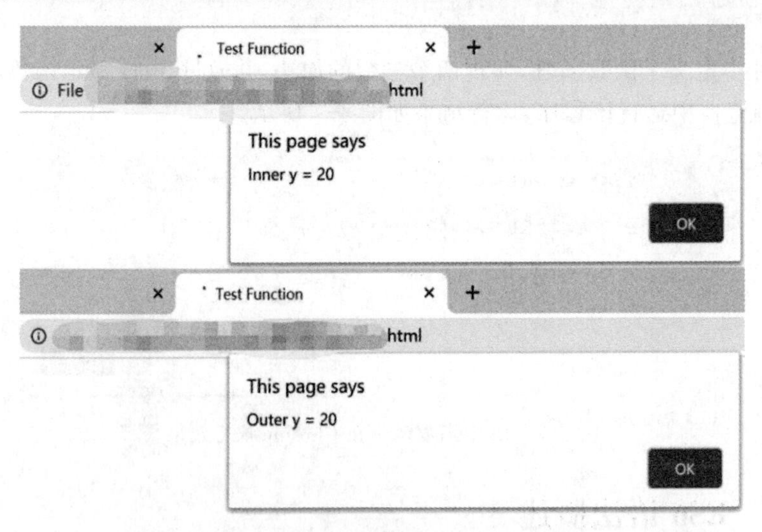

图 4.8 使用 var 进行块级变量声明的执行效果

从图 4.8 可以看出，使用 var 在内部的 if 代码块内进行变量声明并赋值后，在 if 代码块之外的变量也受到影响。

1）let 与 var 的区别

将上述代码中，if 代码块中的 var y＝x 改为 let y＝x，运行后所得效果如图 4.9 所示。

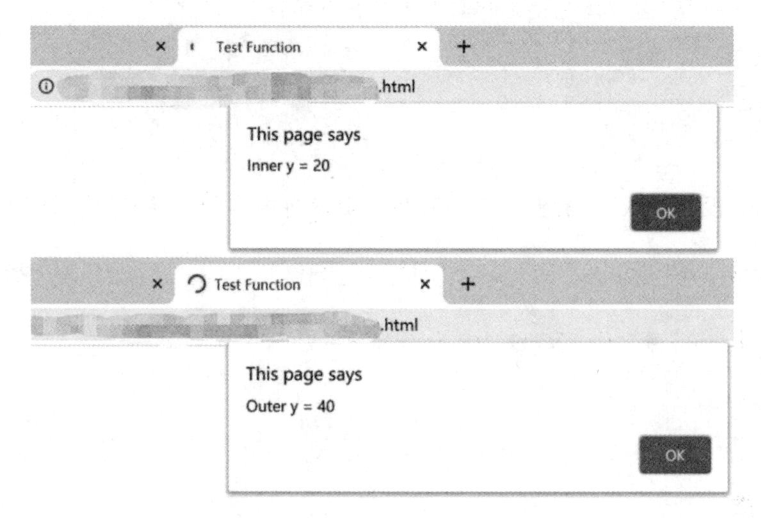

图 4.9 使用 let 进行块级变量声明的执行效果

从图 4.9 可以看出，不同于 var，使用 let 进行变量声明时只作用于内部代码块，对代码块之外的变量不起作用。

2）const 与 var 的区别

实际上，使用 var 对变量进行声明且赋值后，可以对该变量重新赋值，代码如下：

```
var name = "Bob";
name = Alice;
```

上述情况并不会报错，并且 name 最终的值为 Alice，而利用 const name ＝ "Bob"进行声明并赋值后，不能继续对 name 进行赋值，也不能将其重新声明和重新初始化，否则会报错。此外，const 一旦声明变量必须赋值，否则也会报错。感兴趣的读者可以自行进行实验。

2. Arrow Functions 箭头函数

在 ES6 中，箭头函数是一种新增特性，箭头函数是一种使用箭头（=>）定义函数的新语法。基本语法是“参数=>函数体”。箭头函数根据实际的使用场景有多种形式，根据使用需求，参数和函数体可以分别采取多种不同的形式。

（1）采用单一参数且函数功能是返回该参数的值，代码如下：

```
let name = value => value;
//实际上,上述箭头函数的一般函数形式为
let name = function (value){
      return value;
};
```

(2) 传入两个或两个以上的参数,要在参数的两侧添加一对小括号,代码如下:

```
let sum = (arg0, arg1 ,arg2) => arg0 + arg1 + arg2;
//实际上,上述箭头函数的一般函数形式为
let sum = function (arg0, arg1, arg2){
      return arg0 + arg1 + arg2;
};
```

(3) 当函数没有参数时,需要在声明时写一组没有内容的小括号,代码如下:

```
let getValue = () => 100;              //默认返回值为 100

//实际上,上述箭头函数的一般函数形式为
let getValue = function(){
     return 100;
}
```

3. class 类

ES6 中引入了 class(类)作为对象的模板。通过 class 关键字,可以定义类。class 的本质是 function。接下来看一个具体的实例,代码如下:

```
//Person.class
class Person{
    constructor(name){
        this.name = name;
    }
    sayname(){
        alert(this.name);
    }
}

let Student = new Person('Alice');
Student.sayname();
```

基于这个类创建的实例会自动执行 constructor 方法,可以把一些初始化的方法放在 constructor 里面。Student 这个实例可以使用 Person 类中的方法,运行效果如图 4.10 所示。

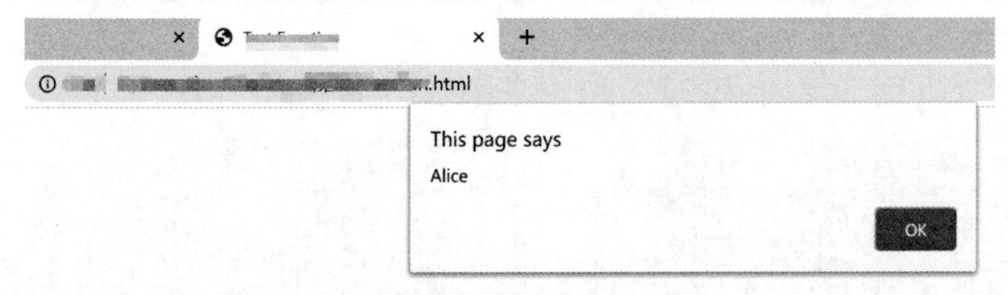

图 4.10 ES6 中的 class 类的用法演示

4. rest 参数

在 ES6 中引入了 rest 参数，它的形式是"...变量名"，用于获取函数的多余参数。rest 参数搭配的变量是一个数组，该变量将多余的参数放入数组中，一个实例代码如下：

```
function add(...values){
    let num = 0;
    for(let s of values){
        num += s;
    }
    return num;
}
console.log(add(1,2,3,4));
```

运行上述代码，所得到的运行结果如图 4.11 所示。可以看出，所得结果为 1+2+3+4=10，在接下来的解构赋值讲解部分，还会引入一小部分利用了 rest 的实例来加深对 rest 的认识。

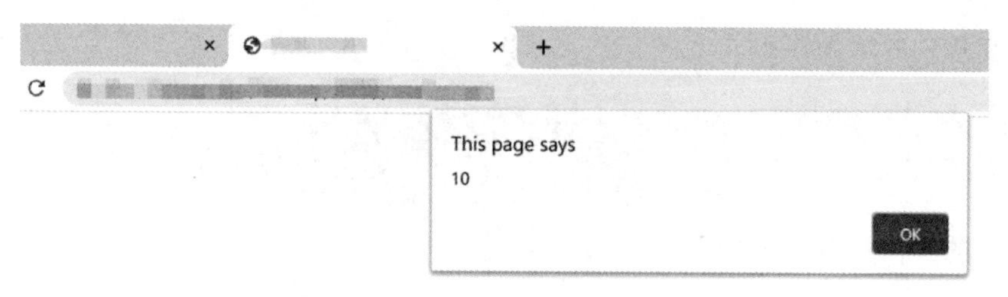

图 4.11 rest 参数使用实例

5. 解构赋值

ES6 允许按照一定模式，从数组和对象中提取值，以及对变量进行赋值，这被称为解构。解构赋值是对赋值运算符的扩展。针对数组或者对象进行模式匹配，然后对其中的变量进行赋值。

在数组的解构中，解构的目标若为可遍历对象，则皆可进行解构赋值。基本解构：可以从数组中提取值，按照对应位置，对变量赋值。只要等号两边的模式相同，左边的变量就会被赋予对应的值，代码如下：

```
let [a, b, c] = [1, 2, 3];              //基本解构：a = 1, b = 2, c = 3
let [a, [[b], c]] = [1, [[2], 3]];      //可嵌套：a = 1, b = 2, c = 3
let [a, , b] = [1, 2, 3];               //可忽略：a = 1, b = 3
let [a = 1, b] = [];                    //不完全解构：a = 1, b = undefined
let [a, ...b] = [1, 2, 3];              //剩余运算符：a = 1, b = [2, 3]
let [a, b, c, d, e] = 'hello';          //a = 'h', b = 'e', c = 'l', d = 'l', e = 'o'
```

注意，当解构模式的匹配结果是 undefined 时，会触发默认值作为返回结果，代码如下：

```
let [a = 2] = [];                //此时 a 匹配结果为 undefined,故返回默认值,即 a = 2
let [a = 3, b = a] = [];         //此时 a,b 均返回默认值: a = 3, b = 3
let [a = 3, b = a] = [1];        //此时 a 正常匹配 b,返回默认值: a = 1, b = 1
let [a = 3, b = a] = [1, 2];     //此时 a,b 均正常匹配: a = 1, b = 2
```

对对象的解构赋值,代码如下:

```
//name = 'aaa', city = 'bbb'
let { name, city } = { name: 'aaa', city: 'bbb' };
let { person : name } = { person: 'ddd' };                //name = 'ddd'
```

对象的解构与数组有一个重要的不同,数组的元素是按次序排列的,变量的取值由它的位置决定,而对象的属性没有次序,变量必须与属性同名,才能取到正确的值,代码如下:

```
let { city, name } = { name: 'aaa', city: 'bbb' };     //city = 'bbb', 'name' = 'aaa'
let { contry } = { name: 'aaa', city: 'bbb' };         //contry 为 undefined
```

可嵌套解构的示例代码如下:

```
let obj = {p: ['hello', {y: 'world'}] };
let {p: [x, { y }] } = obj;                          //x = 'hello', y = 'world'
```

可忽略的解构示例代码如下:

```
let obj = {p: ['hello', {y: 'world'}] };
let {p: [x, { }] } = obj;                           //x = 'hello'
```

不完全解构的示例代码如下:

```
let obj = {p: [{y: 'world'}] };
let {p: [{ y }, x ] } = obj;                        //x = undefined, y = 'world'
```

剩余运算符的示例代码如下:

```
//a = 10, b = 20, rest = {c: 30, d: 40}
let {a, b, ...rest} = {a: 10, b: 20, c: 30, d: 40};
```

解构默认值的示例代码如下:

```
let {a = 10, b = 5} = {a: 3};                       //a = 3, b = 5
let {a: aa = 10, b: bb = 5} = {a: 3};               //aa = 3, bb = 5
```

6. for of

for of 是 ES6 新增的循环方法。for of 不仅可以用于遍历数组,还可以用于遍历类数组对象和其他可迭代对象，但需要注意的是，它不会遍历自定义属性。通过 for of 进行数组遍历的具体示例代码如下：

```
var arr = ['alice','bob','cindy','david'];
for(var item of arr){
    console.log(item);
}
```

运行上述代码所得的结果如图 4.12 所示，可见遍历并输出了 arr 数组中的每个元素。

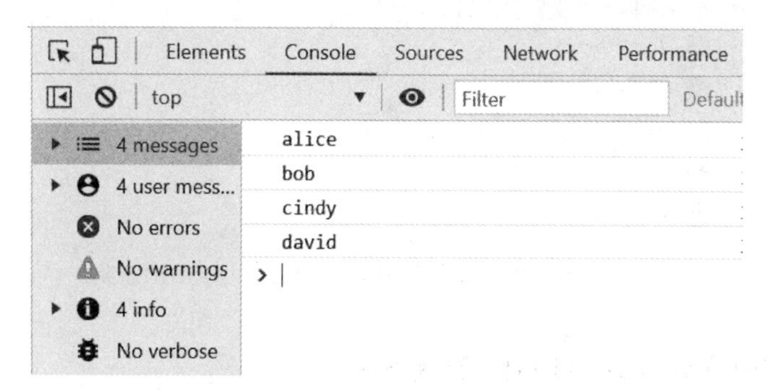

图 4.12 使用 for of 进行数组遍历

7. template strings 模板字符串

模板字符串是增强版的字符串，用反引号"`"标识，它可以被当作普通字符串使用，也可以用来定义多行字符串，或者在字符串中嵌入变量。基础语法的代码如下：

```
let message = `Hello World`;
console.log(message);
```

运行结果如图 4.13 所示，此时会打印输出 Hello World 字符串。

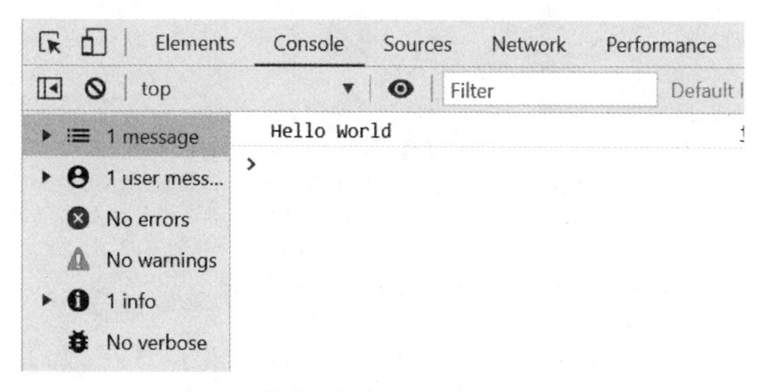

图 4.13 模板字符串的基础语法执行结果

在传统的 JS 中，如果使用单/双引号创建多行字符串，则字符串一定要在同一行。使用模板字符串则无须如此，代码如下：

```
let message = `Hello
World!`
console.log(message);
```

运行结果如图 4.14 所示，此时成功打印输出多行字符串。

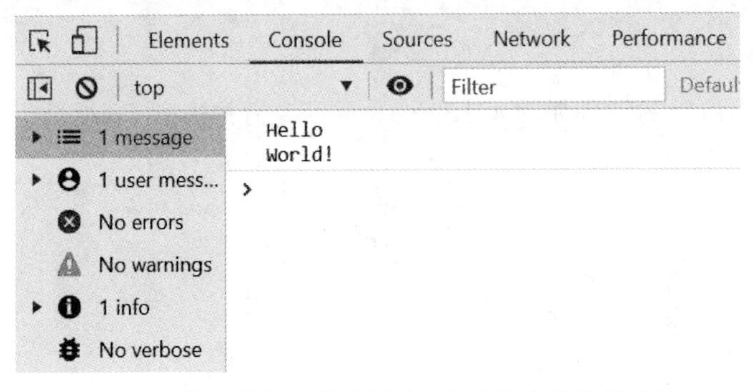

图 4.14　使用模板字符串创建多行字符串的结果展示

4.3.4　JavaScript、HML 及 CSS

经过以上语法知识的学习后，还必须了解 HML、CSS、JS 的相关概念及三者之间的关系。

首先，在 DevEco Studio 中创建一个新的 JS 项目，选择设备类型为 Phone，模板为 Empty Feature Ability(JS)。创建完成后可以看到，项目的目录结构及关键的目录说明如下：

```
MyJS ------------------------- 项目名称
  ...
  entry ----------------------- 项目入口
    src ----------------------- 代码部分
      main
        java ------------------- Java 代码
        js ------------------- JS 代码
          default ------------- 对应一个 JS Component
            i18n ------------- 存放多语言的 json 文件
              en-US.json ------ 定义了在英文模式下页面显示的变量内容
              zh-CN.json ------ 定义了在中文模式下页面显示的变量内容
            pages -------------- 存放多个页面
              index ------------ 前台页面文件夹
                index.css ------ 页面 css 样式
                index.hml ------ 页面 hml 结构
                index.js ------- 页面 js 代码
            app.js -------------- 用于全局 JavaScript 逻辑和应用生命周期管理
        resources --------------- 静态资源文件，如图片、视频等
        config.json ------------- 配置文件
```

这里简单介绍几个概念：

- JS Component：在 HarmonyOS 的 JS 工程中，可以存在多个 JS Component（例如 js 目录下的 default 文件夹就是一个 JS Component），一个 JS FA 对应一个 JS Component，可以独立编译、运行和调试。
- JS Page：Page 表示 JS FA 的一个前台页面，由 JS、HML 和 CSS 文件组成，是 Component 的最基本单元，构成了 JS FA 的每个界面。
- index. hml：在前文中提到过，HarmonyOS 应用开发中是没有 html 文件的，取而代之的是 hml 文件，其作用是定义 index 页面的布局，用于描述 index 页面中用到的组件，以及这些组件的层级关系。
- index. js：定义 index 页面的业务逻辑，例如数据绑定、事件处理等，用于描述页面中组件如何交互。
- index. css：css 的官方名字叫层叠样式表，定义了 index 页面的样式，描述页面中各组件的样子。
- index. hml、index. css 和 index. js 三者之间的关系：实际上，在 Web 开发中，html、css 及 js 文件共同构建了所有网页的展示和交互，其中，html 是主体，装载各种 DOM 元素，css 用来装饰 DOM 元素，js 控制 DOM 元素，三者之间是相辅相成的。同理，在 HarmonyOS 的应用开发中，hml 文件、css 文件和 js 文件之间的关系也是如此。

在 HarmonyOS 应用开发中，JS 文件与 HML 文件的关联方式有数据绑定和事件绑定。在数据绑定中，利用{{变量名}}获取存放于 js 文件中的变量值，从而实现数据绑定。以新建项目的 index. hml 文件和 index. js 文件中的代码为例，title 是 js 文件中定义的变量，在 hml 文件中通过{{title}}获取存放于 js 中的变量值从而实现数据绑定。js 文件中，变量 title 通过 onInit 方法被初始化为 zh-CN. json 文件中 world 字段的值，代码如下：

```
<! -- index.hml -->
<div class = "container">          //应用容器,使用 div 表示,class 名为 container
    <! -- 使用 text 标签显示内容,{{}}用获取变量的值 -->
    <text class = "title">
    {{ $t('strings.hello') }} {{title}}
</text>
</div>

//index.js
export default {                    //使用 export default 命令,为模块指定默认输出
    data: {                         //变量 title 存放在 data 中
title: ""
    },
    onInit() {
        this.title = this.$t('strings.world');
    }
}
```

除了数据绑定之外，在 hml 文件中还可以实现事件绑定。例如在 hml 标签中通过 onclick 绑定 changeTitle 事件，当 text 被单击时触发该事件，事件 changeTitle 的实现逻辑被定义在 js 文件中，代码如下：

```
<!-- index.hml -->
<!--绑定事件"changeTitle"-->
<text class = "title" onclick = "changeTitle">
    {{ $t('strings.hello') }} {{title}}
</text>

//index.js
export default {
    data: {
        title: ""
    },
    onInit() {
        this.title = this.$t('strings.world');
    },
    changeTitle() {                    //定义事件实现
        this.title = "鸿蒙"
    }
}
```

上述代码的执行结果如图 4.15 所示。当单击屏幕上的“您好 世界”字样时，触发 changeTitle 事件，从而屏幕字样改变为“您好 鸿蒙”，事件绑定成功。

图 4.15　事件绑定运行效果图

HarmonyOS 中的 CSS 与普通的 CSS 略有区别，有些属性用法不同，并且不同的标签对于 CSS 属性的支持程度也不一样，感兴趣的读者可以参考官方开发者文档。相同的是，在 HarmonyOS 中，同样利用 CSS 选择器来选择需要添加样式的元素，支持手机、平板、智慧屏和智能穿戴开发的选择器，如表 4.2 所示，但需要注意的是，支持轻量级智能穿戴开发的选择器与支持手机、平板、智慧屏和智能穿戴开发的选择器略有差异，仅支持表中前 3 种选择器。

表 4.2　鸿蒙系统中支持手机、平板、智慧屏和智能穿戴开发的 CSS 选择器

选择器	样例	样例描述
.class	.container	用于选择 class="container"的组件
#id	#titleId	用于选择 id="titleId"的组件
,	.title, .content	用于选择 class="title"和 class="content"的组件
tag	text	用于选择 text 组件
#id .class tag	#containerId .content text	非严格父子关系的后代选择器，选择具有 id="containerId"作为祖先元素，class="content"作为次级祖先元素的所有 text 组件。如需使用严格的父子关系，可以使用“>”代替空格，如：#containerId>.content

选择器的优先级计算规则与 w3c 规则保持一致，当多条选择器声明匹配到同一元素时，各类选择器优先级由高到低顺序为内联样式> id > class > tag。将项目文件 index.hml 和 index.css 中的代码进行修改，代码如下：

```
<!-- index.hml -->
<div id = "containerId" class = "container">
<text id = "titleId" class = "title">标题</text>
<div class = "content">
<text id = "contentId">内容</text>
</div>
</div>

/* index.css */
/* 对所有 div 组件设置样式:垂直显示 div 组件 */
div {
    flex-direction: column;
}
/* 对 class = "title"的组件设置样式:标题字体大小为 70px */
.title {
    font-size: 70px;
}
/* 对 id = "contentId"的组件设置样式: n 内容字体大小为 50px */
#contentId {
    font-size: 50px;
}
/* 对 class = "title"及 class = "content"的组件都设置 padding 为 100px */
.title, .content {
    padding: 100px;
```

```
}
/* 对 class = "container"的组件下的所有 text 设置样式:设置颜色为蓝色 */
.container text {
    color: #007dff;
}
/* 对 class = "container"的组件下的直接后代 text 设置样式: 设置颜色为红色 */
.container > text {
    color: #fa2a2d;
}
```

上述实例的运行结果如图 4.16 所示，需要说明的是，.container text 将“标题”和“内容”设置为蓝色，而.container > text 直接后代选择器将“标题”设置为红色。两者优先级相同，但直接后代选择器声明顺序靠后，将前者样式覆盖。

图 4.16　CSS 进行样式修改

此外，js 文件的主要作用是控制 DOM 元素，JS 获取 DOM 元素主要有两种方式，第一种通过 $refs 获取 DOM 元素，持有注册过 ref 属性的 DOM 元素或子组件实例的对象，代码如下：

```
<!-- index.hml -->
<div class = "container">
<image - animator class = "image - player" ref = "animator" images = "{{images}}" duration =
"1s" onclick = "handleClick"></image - animator>
</div>

//index.js
export default {
  data: {
    images: [
      { src: '/common/frame1.png' },
      { src: '/common/frame2.png' },
      { src: '/common/frame3.png' },
    ],
  },
  handleClick() {
    //获取 ref 属性为 animator 的 DOM 元素
    const animator = this.$refs.animator;
    const state = animator.getState();
    if (state === 'paused') {
      animator.resume();
    } else if (state === 'stopped') {
      animator.start();
    } else {
      animator.pause();
    }
  },
};
```

除此之外,还可以通过 $element 方法获取 DOM 元素,代码如下:

```
<!-- index.hml -->
<div class="container">
<image-animator class="image-player" id="animator" images="{{images}}" duration="1s"
onclick="handleClick"></image-animator>
</div>

//index.js
export default {
  data: {
    images: [
      { src: '/common/frame1.png' },
      { src: '/common/frame2.png' },
      { src: '/common/frame3.png' },
    ],
  },
  handleClick() {
    //获取 id 属性为 animator 的 DOM 元素
    const animator = this.$element('animator');
    const state = animator.getState();
    if (state === 'paused') {
      animator.resume();
    } else if (state === 'stopped') {
      animator.start();
    } else {
      animator.pause();
    }
  },
};
```

至此已经掌握了 HarmonyOS 中 JS 开发的比较关键的知识点,接下来,通过实例感受一下 HarmonyOS 中 JS 开发与 Java 开发的区别。

4.4 HarmonyOS 中的 JS 与 Java

4.1.3 节中对比了 Java 和 JS 的不同之处,那么具体到 HarmonyOS 的应用开发中,分别利用 Java 和 JS 创建一个播放网络视频的页面,通过对比实现流程来感受二者的差异。

4.4.1 Java 中的实现

如果想在 Java 中实现一个视频的播放控件,那么所需要做的主要工作如下。在开始之前,需要说明的是,由于在 Java 中实现一个自定义视频播放控件比较复杂,因此,我们在这里只列出了部分功能的实现代码及最后的实现效果。具体的实现过程读者可以参考第 9 章多媒体视频播放部分。

（1）声明一个 SurfaceProvider 类作为视频的外界承载控件，声明一个 Player 类对视频进行处理，代码如下：

```
SurfaceProvider surfaceProvider;
private Player player;
```

（2）对 surfaceProvider 进行初始化设置，代码如下：

```
//MainAbilitySlice.java
DirectionalLayout directionLayout = new DirectionalLayout(this);
directionLayout.setWidth(ComponentContainer.LayoutConfig.MATCH_PARENT);
directionLayout.setHeight(ComponentContainer.LayoutConfig.MATCH_PARENT);
directionLayout.setOrientation(Component.VERTICAL);
directionLayout.setPadding(32, 32, 32, 32);

surfaceProvider = new SurfaceProvider(this);
surfaceProvider.getSurfaceOps().get().addCallback(this);
surfaceProvider.setPadding(10, 80, 10 , 80);
surfaceProvider.setHeight(2200);
surfaceProvider.setWidth(1200);

directionLayout.addComponent(surfaceProvider);
```

（3）使用 Player 类，对 Player 类进行初始化，代码如下：

```
//MainAbilitySlice.java
player = new Player(this);
player.setPlayerCallback(new Player.IPlayerCallback() {
    @Override
    public void onPrepared() {}
    @Override
    public void onMessage(int i, int i1) {}
    @Override
    public void onError(int i, int i1) {}
    @Override
    public void onResolutionChanged(int i, int i1) {}
    @Override
    public void onPlayBackComplete() {}
    @Override
    public void onRewindToComplete() {}
    @Override
    public void onBufferingChange(int i) {}
    @Override
```

```
    public void onNewTimedMetaData(Player.MediaTimedMetaData mediaTimedMetaData) {}
    @Override
    public void onMediaTimeIncontinuity(Player.MediaTimeInfo mediaTimeInfo) {}
});
```

（4）调用 player，使用播放操作，代码如下：

```
Source source = new Source(localURL);
source.setRecorderVideoSource(Recorder.VideoSource.SURFACE);
player.setSource(source);

player.prepare();
player.play();
```

（5）声明实现 SurfaceOps.Callback 接口，并实现其方法，代码如下：

```
//example.class
public class example implements SurfaceOps.Callback{
        //……
    @Override
    public void surfaceCreated(SurfaceOps surfaceOps) {

        player.setSurfaceOps(surfaceProvider.getSurfaceOps().get());
        player.prepare();
    }
    @Override
    public void surfaceChanged(SurfaceOps surfaceOps, int i, int i1, int i2) {

    }
    @Override
    public void surfaceDestroyed(SurfaceOps surfaceOps) {
     }
}
```

（6）为 surfaceProvider 设置处理办法，代码如下：

```
surfaceProvider.getSurfaceOps().get().addCallback(this);
```

（7）添加权限（申请网络访问权限），在 config.json 页面中进行权限配置，具体的设置代码如下：

```
"reqPermissions": [
  {
    "name": "harmonyos.permission.INTERNET",
    "reason": "reason",
```

```
        "usedScene": {
          "ability": [
            "com.huawei.mytestapp"
          ],
          "when": "always"
        }
      }
    ]
```

经过以上一系列步骤，就完成了一个基于 Java 的视频播放页面，运行效果如图 4.17 所示。

图 4.17　基于 Java 开发的视频播放页面

4.4.2　JS 中的实现

基于 JS 实现一个视频的播放功能，则首先新建工程，设备类型为 Phone，模板为 Empty Feature Ability(JS)。在前面介绍了如何新建一个 JS 应用，此处不再赘述详细过程，主要开发步骤如下：

(1) 找到工程目录下的 pages 目录，并在其下新建一个文件夹，名字为 video，如图 4.18 所示。

(2) 在新建的 video 目录下新建 videos.hml 文件，如图 4.19 所示。

在 videos.hml 文件中编写界面，实现代码如下：

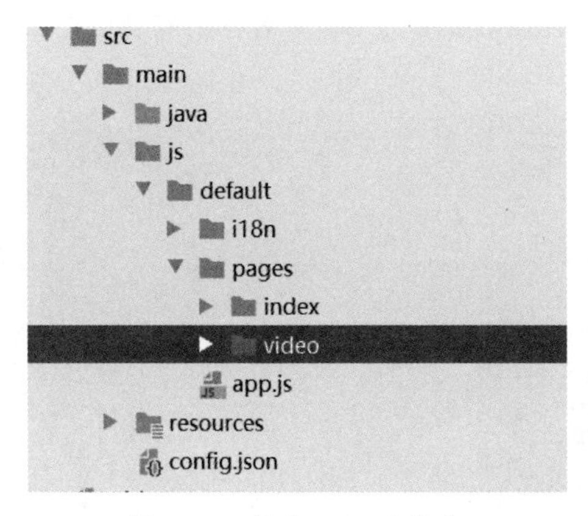

图 4.18 新建 video 文件夹

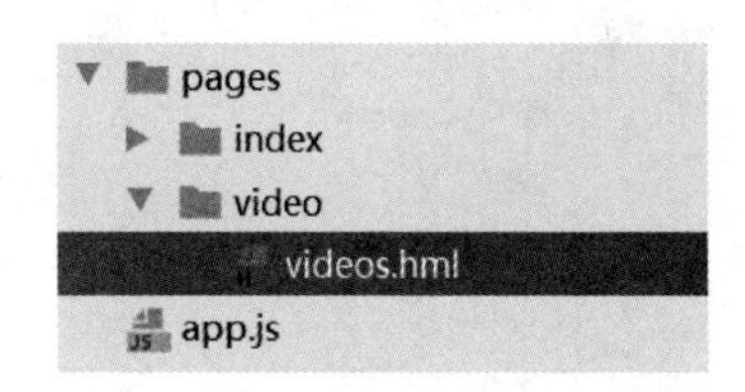

图 4.19 新建 videos. hml 文件

```
<! -- videos. hml -- >
< div class = "item - container">
< video class = " Video" autoplay = " true" src = " https://stream7. iqilu. com/10339/upload_
transcode/202002/18/20200218114723HDu3hhxqIT. mp4" >
</video >
</div >
```

其中,src 的内容为视频链接,可以将其替换为任意的网络 URL 视频链接。

(3) 为新建的 JS 页面注册页面信息,打开 config. json 配置文件,如图 4.20 所示。

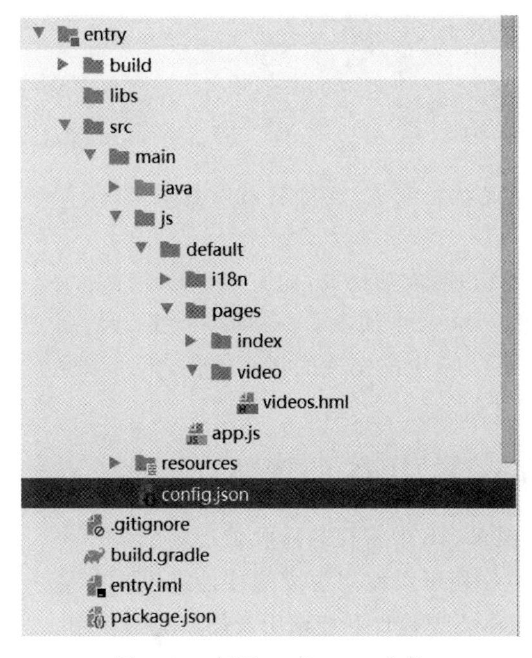

图 4.20 打开 config. json 文件

在其中找到 js 标签,并将标签内容进行修改,在 pages 的 tag 下对将刚刚新建的页面文件进行标记,代码如下:

```
"js": [
  {
"pages": [
      "pages/video/videos"
    ],
    "name": "default",
    "window": {
      "designWidth":720,
      "autoDesignWidth": false
    }
  }
]
```

(4) 为了让页面能够正常播放视频,还需要为 module 申请网络访问权限,否则无法从网络中获取视频流。权限配置同样是在 config.json 页面中进行设置。具体的设置代码如下:

```
"reqPermissions": [
  {
    "name": "harmonyos.permission.INTERNET",
    "reason": "reason",
    "usedScene": {
      "ability": [
        "com.huawei.jsplay"
      ],
      "when": "always"
    }
  }
]
```

至此,整个项目的设置就完成了,运行后可以看到,效果如图 4.21 所示。

从图 4.21 中可以看到视频成功播放,而实现视频播放功能的代码,也不过是在 videos.html 中加入了< video >标签,设置了视频路径。一个 div 调用了 JS 中已实现的视频控件,就可以完成整个播放逻辑,这就是 JS 的强大之处。

图 4.21　基于 JS 开发的视频播放页面

4.4.3　HarmonyOS 中 JS 的优缺点

JS 的优点是显而易见的,JS 中将视频播放处理集成为一个控件,只需调用 video TAG 便可以使用现成的控件进行视频播放的一系列操作,相比 Java 需要控件互相调用及播放类的使用,基本上 JS 所有的底层实现都被隐藏了起来,因此,JS 只需调用

video TAG,然后在 TAG 中声明视频链接就可以实现播放功能了。

而 JS 的缺点也恰恰是由其优点导致的,其简洁性和可使用性导致了可自定义性大大降低,例如图 4.21 中所示的视频播放控件的播放/暂停按钮、进度条拖动、全屏按键等均由控件本身定义好了,无法进行自定义修改,因此,在实际开发中需要根据实际需求选择合适的开发语言。

通过这一章对 JS 的相关概念、JS 的功能、JS 的语法等有了简单的了解,感兴趣的读者可以阅读相关的书籍进行更加深入的学习。同时也已经掌握了 HarmonyOS 中 JS 开发的基本流程,成功让开发界面“动”了起来。在接下来的章节中,将介绍 HarmonyOS 中的 JS UI 框架,从而对 HarmonyOS 的 JS 开发有一个更加深入的认识。

第5章 JS UI

5.1 关于 JS UI

5.1.1 JS UI 框架介绍

JS UI 框架是一种跨设备的高性能 UI 开发框架，支持声明式编程和跨设备多态 UI。它的基础能力主要体现在 3 个方面：

1）声明式编程

JS UI 框架采用类 HTML 和 CSS 声明式编程语言作为页面布局和页面样式的开发语言，让开发者避免编写 UI 状态切换的代码，页面业务逻辑则支持 ECMAScript 规范的 JS 语言。

2）跨设备

开发框架架构上支持 UI 跨设备显示能力，运行时自动映射到不同设备类型，开发者无感知，从而降低开发者多设备适配成本。

3）高性能

开发框架包含了许多核心的控件，如列表、图片和各类容器组件等，针对声明式语法进行了渲染流程的优化。

JS UI 整体架构如图 5.1 所示，包括应用层（Application）、前端框架层（Framework）、引擎层（Engine）和平台适配层（Porting Layer）。

1）Application

应用层表示开发者使用 JS UI 框架开发的 FA 应用，这里的 FA 应用特指 JS FA 应用。

2）Framework

前端框架层主要完成前端页面解析，以及提供 MVVM（Model-View-ViewModel）开发模式、页面路由机制和自定义组件等能力。

3）Engine

引擎层主要提供动画解析、DOM（Document Object Model）树构建、布局计算、渲染命令构建与绘制、事件管理等能力。

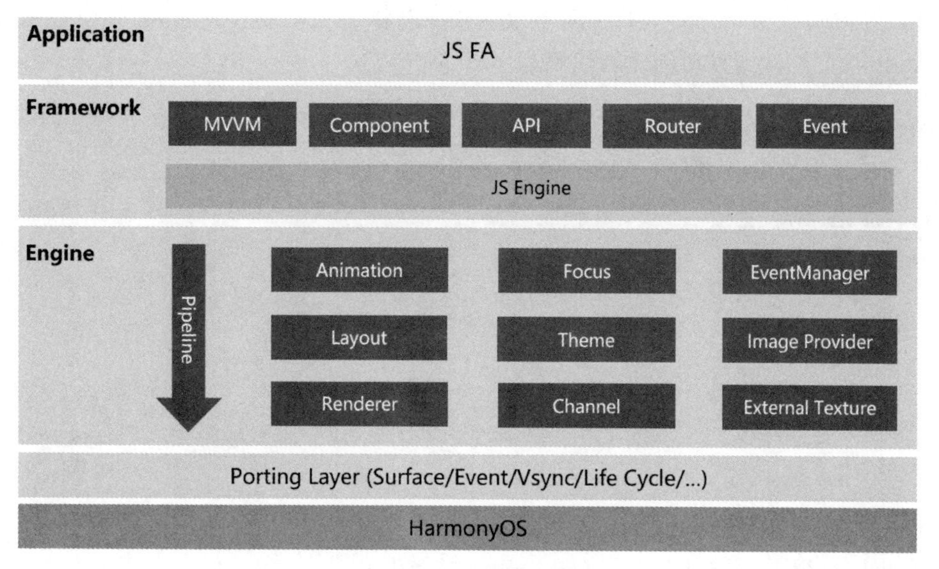

图 5.1 JS UI 整体架构图

4）Porting Layer

适配层主要完成对平台层进行抽象，提供抽象接口，可以对接到系统平台。例如：事件对接、渲染管线对接和系统生命周期对接等。

5.1.2 JS UI 主体介绍

JS UI 框架支持纯 JS、JS 和 Java 混合语言开发。JS FA 指基于 JS 或 JS 和 Java 混合开发的 FA。

新建一个工程，选择 Phone 设备下的 Empty Feature Ability (JS)模板，输入工程名称和包名。新建后的 entry 包结构如图 5.2 所示。

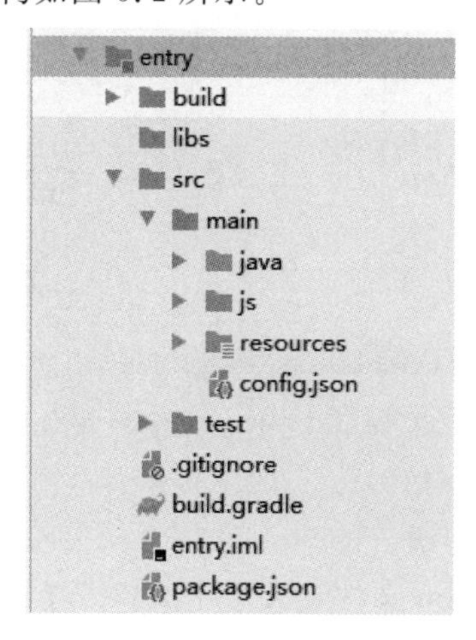

图 5.2 JS 工程目录

可以看出其中包含 java 和 js 两个文件夹，选择 entry→src→main→java→包名→MainAbility，可以看到自动创建的代码如下：

```
//MainAbility中的示例代码
public class MainAbility extends AceAbility {
    @Override
    public void onStart(Intent intent) {
        super.onStart(intent);
    }

    @Override
    public void onStop() {
        super.onStop();
    }
}
```

其中，JS FA 在 HarmonyOS 上运行时，必须需要基类 AceAbility，其继承自 Ability 类，所有应用运行入口类都应该从 AceAbility 类派生。

应用通过 AceAbility 类中 setInstanceName()接口设置该 Ability 的实例资源，并通过 AceAbility 窗口进行显示及全局应用生命周期管理，因此当加载 JS FA 时，需要通过 setInstanceName(String name)的参数 name 指明实例名称，实例名称与 config. json 文件中 profile. application. js. name 的值对应。若开发者未修改实例名，而使用了默认值 default，则无须调用此接口，如上述代码，新建工程时默认实例名为 default，因此不需要调用 setInstanceName()接口。若开发者修改了实例名，则需要在应用 Ability 实例的 onStart()中调用此接口，并将参数 name 设置为修改后的实例名称。多实例应用的 profile. application. js 字段中有多个实例项，使用时应选择相应的实例名称。如实例名称为 JSComponentName 时，必须在 super. onStart(Intent)前调用此接口，代码如下：

```
//MainAbility.java
public class MainAbility extends AceAbility {
    @Override
    public void onStart(Intent intent) {
setInstanceName("JSComponentName");     //参数为 config.json 配置文件中
module.js.name 的标签值
        super.onStart(intent);
    }
}
```

选择 entry→src→main→js，工程的 JS FA 开发目录如图 5.3 所示。

这里对每个文件进行具体介绍。

(1) i18n 文件夹用于存放多语言的 json 文件。en-US. json 文件定义了在英文模式下页面显示的变量内容，zh-CN. json 文件定义了中文模式下的页面内容。如 en-US. json 文件的代码如下：

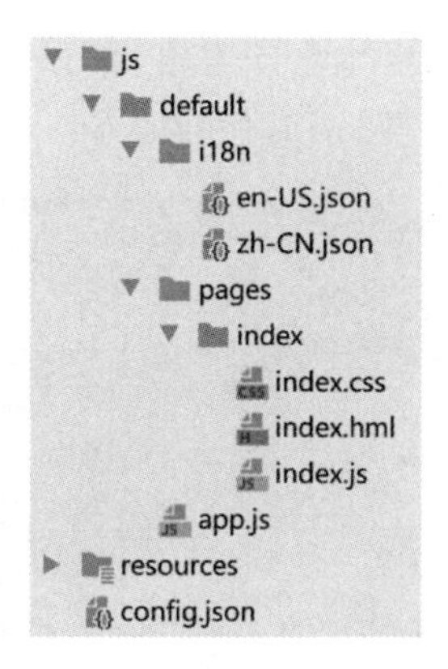

图 5.3 JS FA 开发目录

```
{
  "strings": {
    "hello": "Hello",
    "world": "World"
  },
  "Files": {
  }
}
```

(2) pages 文件夹用于存放多个页面，每个页面由 hml、css 和 js 文件组成。

index.hml 文件定义了 index 页面的布局、index 页面中用到的组件，以及这些组件的层级关系。以下面的代码为例，包含了一个 text 组件，内容为文本 Hello World，其中{{title}}采用了变量赋值的方式，在 index.js 文件中进行赋值，代码如下：

```
<div class = "container">
<text class = "title">
        {{ $t('strings.hello') }} {{title}}
</text>
</div>
```

index.css 文件定义了 index 页面的样式。如下面代码所示，该 css 文件定义了 index.hml 文件中 class=container 和 class=title 的容器或组件样式，代码如下：

```
.container {
    flex-direction: column;
    justify-content: center;
    align-items: center;
}

.title {
    font-size: 100px;
}
```

index.js 文件定义了 index 页面的业务逻辑，例如数据绑定、事件处理等。为 index.hml 文件中的变量 title 赋值字符串 World，代码如下：

```
export default {
    data: {
        title: ""
    },
    onInit() {
        this.title = this.$t('strings.world');
    }
}
```

5.2 开发第一个 JS FA 应用

5.2.1 页面布局说明

本节将逐步介绍如何开发一个 JS FA 应用。一个页面的基本元素包含标题区域、文本区域、图片区域等，每个基本元素内还可以包含多个子元素，开发者根据需求还可以添加按钮、开关、进度条等组件。在构建页面布局时，需要对每个基本元素思考以下几个问题：

(1) 该元素的尺寸和排列位置。

(2) 是否有重叠的元素。

(3) 是否需要设置对齐、内间距或者边界。

(4) 是否包含子元素及其排列位置。

(5) 是否需要容器组件及其类型。

在进行代码开发之前，首先要对整体页面布局进行分析，将页面分解为不同的部分，用容器组件来承载。将页面中的元素分解之后再对每个基本元素按顺序实现，可以减少多层嵌套造成的视觉混乱和逻辑混乱，提高代码的可读性，方便对页面进行后续的调整。应用的分解效果图如图 5.4 所示，其中，最上方的图片区可以通过滑动来观看不同的图片，中间的标题区可以进行收藏和取消收藏，最下方则为描述区，可以进行文字介绍。

图 5.4 JS FA 应用效果图

根据 JS FA 应用效果图，此页面一共分成 3 个部分：图片区、标题区、描述区。根据此分区，根节点的子节点应按列排列。

图片区和描述区分别使用 swiper 组件和 text 组件实现。标题区由两部分组成，以行来排列，其中

第一部分由两个 text 组件组成，分别为商品名称和商品标语，以列排列。第二部分由 image 组件和 text 组件组成，分别为代表收藏功能的星号和代表收藏次数的数字，以行排列，如图 5.5 所示。

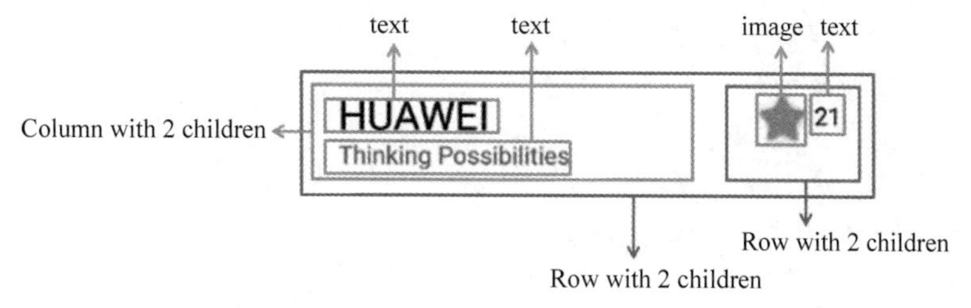

图 5.5 JS FA 标题区布局分析

5.2.2 构建布局

根据布局结构的分析，首先构建页面的基础布局。其中，实现图片区域通常用 image 组件实现，由于需要左右滑动图片，因此在 image 外层加入 swiper 滑动容器，swiper 容器提供了切换子组件显示的能力。图片资源放在 common 目录下，图片的路径要与图片实际所在的目录一致，需要注意，common 需要新建，级别须与 pages 目录平级。

标题区和描述区采用了最常用的基础组件 text，text 组件用于展示文本，文本内容需要写在标签内容区。要将页面的基本元素组装在一起，需要使用容器组件，如上述 swiper 容器。在页面布局中常用到 3 种容器组件，分别是 div、list 和 tabs。在页面结构相对简单时，可以直接用 div 作为容器，因为 div 作为单纯的布局容器使用起来更为方便，可以支持多种子组件。

在 index.hml 文件中实现页面基础布局，具体代码如下：

```
<! -- index.hml 文件代码示例 -->
<div class = "container">
<swiper class = "swiper - style">
<image src = "/common/Phone_00.jpg" class = "image - mode"></image>
<image src = "/common/Phone_01.jpg" class = "image - mode"></image>
<image src = "/common/Phone_02.jpg" class = "image - mode"></image>
<image src = "/common/Phone_03.jpg" class = "image - mode"></image>
</swiper>
<div class = "title - section">
<div class = "phone - title">
<text class = "phone - name">
            HUAWEI
</text>
<text class = "phone - definition">
            Thinking Possibilities
</text>
```

```
</div>
<div class = "favorite - image">
<image src = "{{unFavoriteImage}}" class = "image - size"onclick = "favorite"></image>
</div>
<div class = "favorite - count">
<text>{{number}}
</text>
</div>
</div>
<div class = "description - first - paragraph">
<text class = "description">{{descriptionFirstParagraph}}
</text>
</div>
<div class = "description - second - paragraph">
<text class = "description">{{descriptionSecondParagraph}}
</text>
</div>
</div>
```

在 index. hml 中，为每个组件和容器都定义了一个 class＝" *** "的样式，需要在 css 文件中依次对样式进行构建。如本例，在 index. css 文件中，需要设定的样式主要有：flex-direction 用于设置子组件容器的排列方式，paddind 用于设置内边距，font-size 用于设置字体大小，以及 swiper 组件的一些私有属性，如 indicator-color 用于设置导航点指示器的填充颜色，indicator-selected-color 用于设置导航点指示器选中的颜色，indicator-size 用于设置导航点指示器的直径大小等。具体代码如下：

```
/* index.css 文件代码示例 */
.container {
    flex - direction: column;
}
.swiper - style {
    height: 700px;
    indicator - color: #4682b4;
    indicator - selected - color: #ffffff;
    indicator - size: 20px;
}
.title - section {
    flex - direction: row;
    height: 150px;
}
.phone - title {
    align - items:flex - start;
    flex - direction: column;
    padding - left: 60px;
```

```
        padding-right: 160px;
        padding-top: 50px;
    }
    .phone-name {
        font-size: 50px;
        color: #000000;
    }
    .phone-definition {
        font-size: 30px;
        color: #7A787D;
    }
    .favorite-image {
        padding-left: 70px;
        padding-top: 50px;
    }
    .favorite-count {
        padding-top: 60px;
        padding-left: 10px;
    }
    .image-size {
        object-fit: contain;
        height: 60px;
        width: 60px;
    }
    .description-first-paragraph {
        padding-left: 60px;
        padding-top: 50px;
        padding-right: 60px;
    }
    .description-second-paragraph {
        padding-left: 60px;
        padding-top: 30px;
        padding-right: 60px;
    }
    .description {
        color: #7A787D;
    }
    .image-mode {
        object-fit: contain;
    }
```

在 index.hml 中，收藏的图片来源采用了 src="{{unFavoriteImage}}，text 标签也采用了{{descriptionFirstParagraph}}和{{descriptionSecondParag -raph}}的数据绑定形式，所以需要在 index.js 中对其进行赋值。index.js 代码如下：

```
//index.js 文件代码示例
export default {
    data: {
        unFavoriteImage: "/common/unfavorite.png",
        isFavorite: false,
        number: 20,
        descriptionFirstParagraph:"The breakthrough of visual boundaries, the exploration of photography and videography, the liberation of power and speed, and the innovation of interaction are now ready to be discovered. Embrace the future with new possibilities.",
        descriptionSecondParagraph: "Lighting up infinite possibilities. The quad camera of HUAWEI is embraced by the halo ring. It is a perfect fusion of reflection and refraction. Still Mate, but a new icon.",
    },
}
```

以上代码完成了基础的布局构建，接下来对页面中的收藏交互进行实现。

5.2.3 添加交互

添加交互通过在组件上关联事件实现，本节将介绍如何关联 click 事件，构建上述页面中的收藏功能，即单击星星图片，图片变成黄色，表示收藏，收藏数加 1，如图 5.6 所示。再次单击，星号恢复成原本颜色，表示取消收藏，收藏数减 1。

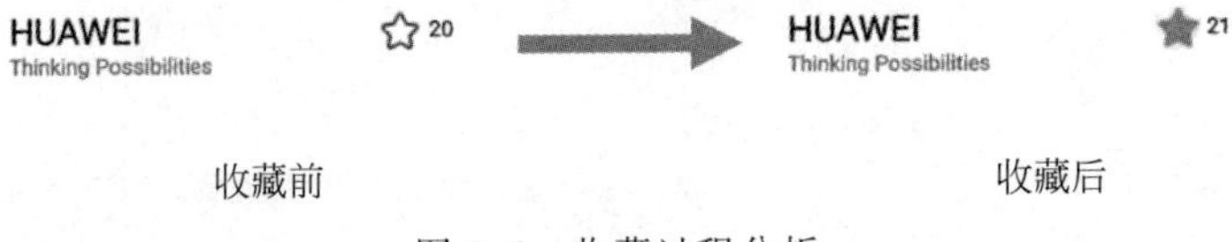

收藏前　　收藏后

图 5.6　收藏过程分析

收藏按钮通过一个 div 组件关联 click 事件实现。div 组件包含一个 image 组件和一个 text 组件。image 组件用于显示未收藏和收藏后的效果。click 事件函数会交替更新收藏和未收藏图片的路径。text 组件用于显示收藏数，收藏数也会在 click 事件的函数中同步更新。

index.js 文件用于构建页面逻辑，click 事件作为一个函数定义在 index.js 文件中，可以更改 isFavorite 的状态，从而更新显示的 image 组件和 text 组件。如果 isFavorite 为真，则更改收藏后的图片路径，并将点赞数加 1。该函数在 hml 文件中对应的 div 组件上生效。在 index.js 中加入代码如下：

```
//index.js 文件代码实现
export default {
    data: {
…
    },

    favorite() {
```

```
        var tempTotal;
        if (!this.isFavorite) {
            this.unFavoriteImage = "/common/favorite.png";
            tempTotal = this.number + 1;
        } else {
            this.unFavoriteImage = "/common/unfavorite.png";
            tempTotal = this.number - 1;
        }
        this.number = tempTotal;
        this.isFavorite = !this.isFavorite;
    }
}
```

运行程序,实现效果如图 5.7 所示。

顶部的 swiper 图片可进行左右滑动,且中间的标题区可进行收藏操作,收藏后效果如图 5.8 所示。

图 5.7 JS FA 运行效果图

图 5.8 JS FA 收藏运行效果图

5.3 常用组件

16min

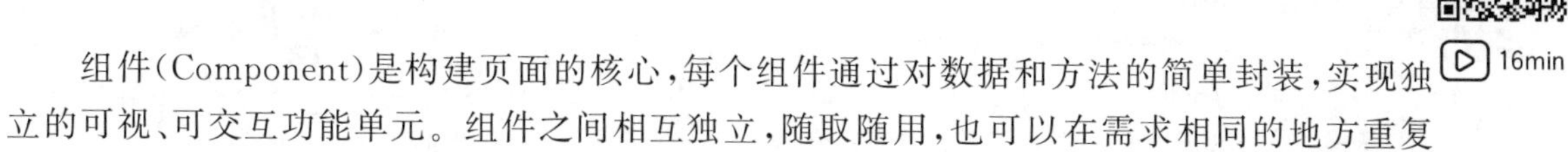

组件(Component)是构建页面的核心,每个组件通过对数据和方法的简单封装,实现独立的可视、可交互功能单元。组件之间相互独立,随取随用,也可以在需求相同的地方重复

使用。开发者还可以通过组件间合理的搭配定义满足业务需求的新组件，从而减少开发工作量。

根据组件的功能，可以将组件分为以下四大类，如表 5.1 所示。

表 5.1 组件的分类

组件类型	主要组件
基础组件	text、image、progress、rating、span、marquee、image-animator、divider、search、menu、chart
容器组件	div、list、list-item、stack、swiper、tabs、tab-bar、tab-content、popup、list-item-group、refresh、dialog
媒体组件	video
画布组件	canvas

合理使用控件可以编写出丰富多样的界面，下面介绍几种常用控件的使用方法。

5.3.1 基础组件

1. Text

实现标题和文本区域最常用的是基础组件 Text。Text 组件用于展示文本，可以设置不同的属性和样式，文本内容需要写在标签内容区。在页面中插入标题和文本区域的代码如下：

```
<!-- index.hml 实现在页面中插入标题和文本 -->
<div class="container">
<!-- 标题区域 -->
<text class="title_text">{{headTitle}}</text>
<!-- 第一段文字 -->
<div class="paragraph">
<text class="paragraph_text">{{paragraphFirst}}</text>
</div>
<!-- 第二段文字 -->
<div class="paragraph">
<text class="paragraph_text">{{paragraphSecond}}</text>
</div>
</div>

<!-- index.js -->
export default {
    data: {
        headTitle:"HUAWEI",
        paragraphFirst:"The breakthrough of visual boundaries, the exploration of photography and videography, the liberation of power and speed, and the innovation of interaction are now ready to be discovered. Embrace the future with new possibilities.",
```

```
        paragraphSecond: "Lighting up infinite possibilities. The quad camera of HUAWEI is
embraced by the halo ring. It is a perfect fusion of reflection and refraction. Still Mate, but a
new icon."
    },
}
<!-- index.css -->
.container{
    flex-direction: column;
}
.title_text{
    align-items:flex-start;
    flex-direction: column;
    padding-left: 60px;
    padding-right: 160px;
    padding-top: 50px;
    font-size: 50px;
}
.paragraph{
    padding-left: 60px;
    padding-top: 50px;
    padding-right: 60px;
}
```

示例效果图如图 5.9 所示。

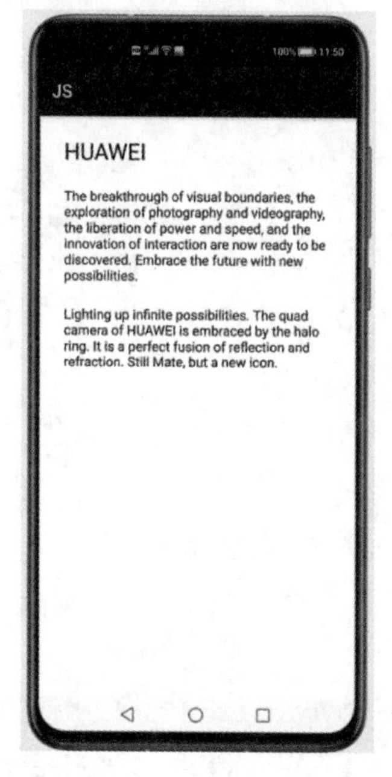

图 5.9 Text 组件示例效果图

2. Button

Button 是实现用户交互最常用的组件，可用于触发 JS 中的函数。可以使用 JS 提供的各种按钮(包括胶囊按钮、圆形按钮、文本按钮、弧形按钮、下载按钮)实现很多有趣的功能。例如，在 hml 文件中定义上述各种按钮，此外，在圆形按钮中加入了一张图标，将下载按钮的 onClick 属性绑定到了 js 文件中的 setProgress 函数中，将第二个胶囊按钮的 waiting 属性定义为 true，让它一直处于等待状态。实现代码如下：

```
<!-- index.hml 定义多种按钮 -->
<div class = "div - button">
<button class = "button" type = "capsule" value = "胶囊按钮"></button>
<button class = "button circle" type = "circle" icon = "common/logo.png"></button>
<button class = "button text" type = "text">文本按钮</button>
<button class = "button download" type = "download" id = "download - btn"
            onClick = "setProgress">{{downloadText}}</button>
<button class = "button" type = "capsule" waiting = "true">载入中...</button>
<button class = "button" type = "arc">弧形按钮</button>
</div>
```

上述下载按钮的 onClick 属性绑定到了 js 文件中的 setProgress 函数中，js 文件的代码如下：

```
//index.js
export default {
    data: {
        progress: 5,
        downloadText: "下载"
    },
    setProgress(e) {
        this.progress += 10;                    //每次单击增加 10% 的进度
        this.downloadText = this.progress + "%";
        if (this.progress >= 100) {
            this.downloadText = "完成";     //到达 100% 时显示完成
        }
    }
}
```

css 文件中定义了按钮的样式，具体实现代码如下：

```
/* index.css 定义按钮的样式 */
.div - button {
    flex - direction: column;
    align - items: center;
}
```

```
.button {
    margin-top: 15px;
}
.button:waiting {
    width: 280px;
}
.circle {
    background-color: #007dff;
    radius: 72px;
    icon-width: 72px;
    icon-height: 72px;
}
.text {
    text-color: red;
    font-size: 40px;
    font-weight: 900;
    font-family: sans-serif;
    font-style: normal;
}
.download {
    width: 280px;
    text-color: white;
    background-color: #007dff;
}
```

运行上述代码,最终的效果如图 5.10 所示。

图 5.10 Button 组件示例效果图

3. Menu

Menu 提供菜单组件，作为临时性弹出窗口，用于展示用户可执行的操作。当菜单中某个值被单击选中时，selected 事件将会被触发，同时返回 value 的值。在页面中插入菜单组件的示例代码如下：

```
<!-- index.hml 在页面中插入 Menu 组件 -->
<div class="container">
<text onclick="onTextClick" class="title-text">Show popup menu.</text>
<menu id="apiMenu" onselected="onMenuSelected">
<option value="Item 1">Item 1</option>
<option value="Item 2">Item 2</option>
<option value="Item 3">Item 3</option>
</menu>
</div>

/* index.js */
import prompt from '@system.prompt';
export default {
    onMenuSelected(e) {
        prompt.showToast({
            message: e.value
        })
    },
    onTextClick() {
        this.$element("apiMenu").show({x:280,y:120});
    }
}
```

运行上述代码，示例效果图如图 5.11 所示，可以看到实现了最简单的菜单效果。

图 5.11　Menu 组件示例效果图

5.3.2　List 组件

现如今，购物 App 已经成为生活中不可或缺的一部分，滑动浏览的商品列表是必不可少的基础组件，如图 5.12 所示。在本节将介绍 JS UI 中的 List 组件，用于呈现多行连续同类的数据。

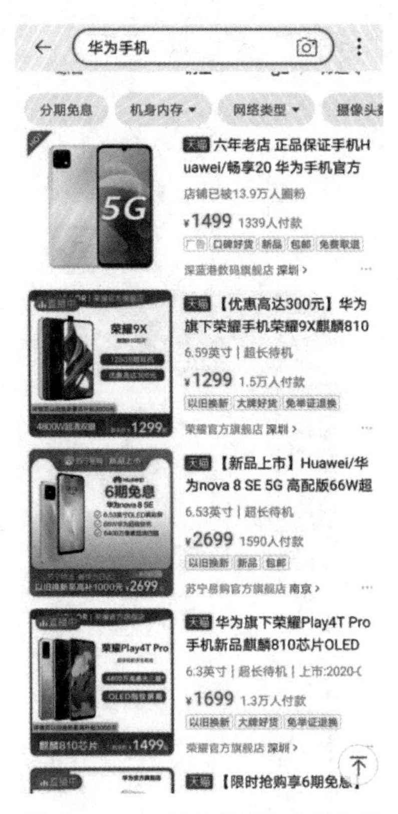

图 5.12　List 组件的实际应用实例

List 列表组件包含两类子组件，分别是 list-item 列表项和 list-item-group 列表项组。下面对两个子组件分别进行介绍。

1. list-item 列表项

这里先举一个简单的 list-item 例子，在 hml 文件中，编写代码如下：

```
<!-- list-item 示例 -->
<div class = "container">
<list>
<list-item><text>列表项文本 1 </text></list-item>
<list-item><text>列表项文本 2 </text></list-item>
<list-item><text>列表项文本 3 </text></list-item>
<list-item><text>列表项文本 4 </text></list-item>
```

```
<list-item><text>列表项文本 5</text></list-item>
<list-item><text>列表项文本 6</text></list-item>
<list-item><text>列表项文本 7</text></list-item>
<list-item><text>列表项文本 8</text></list-item>
<list-item><text>列表项文本 9</text></list-item>
<list-item><text>列表项文本 10</text></list-item>
<list-item><text>列表项文本 11</text></list-item>
<list-item><text>列表项文本 12</text></list-item>
    ...
<list-item><text>列表项文本 32</text></list-item>
</list>
</div>
```

运行上述代码,效果如图 5.13 所示。

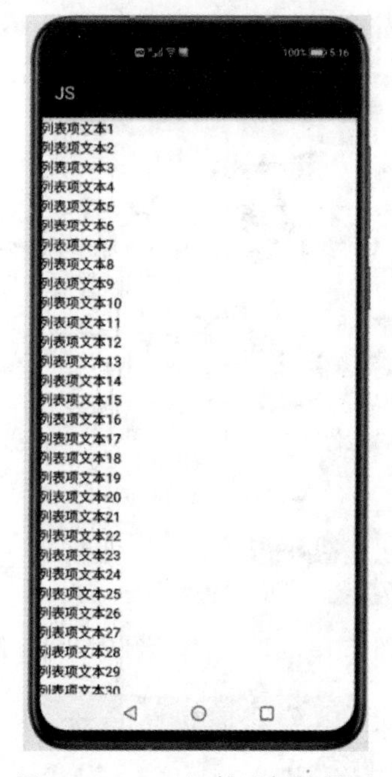

图 5.13　List 组件示例效果图

示例中枚举了 32 个列表项文本,一般情况下,都会在 js 文件中,通过数组的形式来存储列表项文本内容,代码如下:

```
<!-- index.hml -->
<div class="container">
<list>
<list-item><text>{{data}}</text></list-item>
</list>
</div>
```

```
/* index.js */
export default {
    data: {
        title: "",
        data : ["列表项文本 1","列表项文本 2","列表项文本 3","列表项文本 4"...,"列表项文
本 32"]
    }
}
```

事实上,32个列表项文本长度已经超过了屏幕长度,超出部分被自动隐藏起来,当手指滑动屏幕时显示剩余部分,这一过程一般称为scroller。在list-item属性设置中还有几个比较常用的属性:

(1) scrollbar用于设置侧边滑动栏的显示模式,默认值为off,即不显示,另外还可以设置为on(表示常驻显示)和auto(表示按需显示)。当设置为auto时,触摸屏幕时会显示滑动条且2s后自动消失。

(2) scrolleffect用于显示滑动效果,默认值为spring,是一个弹性物理动效,当滑动到边缘后可以根据初始速度或通过触摸事件继续滑动一段距离,然后松手后回弹。另外还可以设置为fade,实现渐隐动效,活动到边缘后展示一个波浪形的渐隐,根据速度和滑动距离的变化,渐隐也会发生一定的变化,也可以设置为no,即不设置滑动边缘效果。

(3) indexer用于展示侧边栏快速字母索引栏,一般需要配合item中的section属性一起使用。可能很多读者会对这个快速字母索引栏有些不解,不知为何物。其实这个索引栏类似于计算机文件中的按类型分类,可以将相同类型的文件分为一类,这里的索引栏也有异曲同工之妙,根据section中设置的值,将相同section的条目分为一类。indexer的参数可以为true,表示使用默认字母索引表,也可以为false,表示无索引,还可以自定义索引表,这里值得注意的是在自定义时“#”必须存在。实现自定义索引表的代码如下:

```
<!-- index.hml -->
<div class = "container">
<list scrollbar = "on" scrolleffect = "fade" indexer = "{{index}}">
<list-item section = "a"><text>列表项文本 1</text></list-item>
<list-item section = "b"><text>列表项文本 2</text></list-item>
<list-item section = "c"><text>列表项文本 3</text></list-item>
<list-item section = "d"><text>列表项文本 4</text></list-item>
<list-item section = "a"><text>列表项文本 1</text></list-item>
<list-item section = "b"><text>列表项文本 2</text></list-item>
<list-item section = "c"><text>列表项文本 3</text></list-item>
<list-item section = "d"><text>列表项文本 4</text></list-item>
    ...
</list>
</div>
```

上述代码中可以看到 indexer 需要配合 section 共同使用，indexer 设置的是 index 数组，由用户自定义，section 设置的是该 list-item 所属的索引列。index 数组定义在 js 文件中，代码如下：

```
/* index.js */
export default {
    data: {
        title: "",
        index : ["#","a","b","c"]
    }
}
```

上述代码的执行效果如图 5.14 所示。

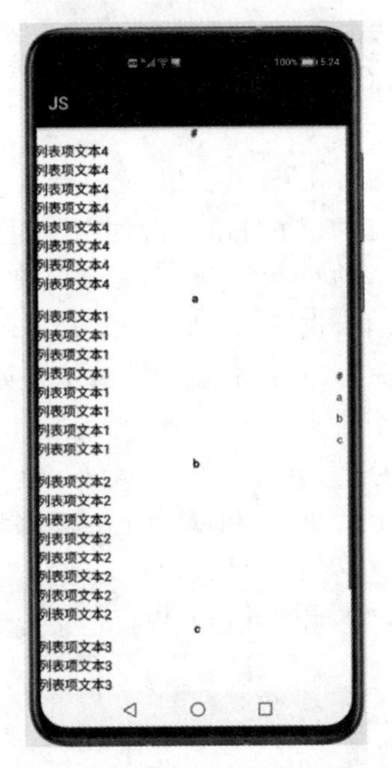

图 5.14 带索引的 List 组件示例效果图

此外，除了按照索引分类，indexer 还会在侧边栏添加索引值查找的功能，可以单击跳转到具体的索引列表。

(4) updateeffect 属性用于设置当 list 内部的 item 发生删除或新增时是否支持动效。当设置为 true 时，新增或删除 item 时播放过程动效。

2. list-item-group 列表项组

list-item-group 是一个具有折叠效果 list 列表，配合 list-item 使用，以 list-item-group 中的第一个 list-item 作为分类标准，以下都折叠进分类中，单击具有折叠和扩展的效果，代码如下：

```
<! -- index.hml -->
<div class="container">
<list>
<list-item-group>
<list-item class="item_style"><text class="big_size">水果</text></list-item>
<list-item><text class="font_size">苹果</text></list-item>
<list-item><text class="font_size">香蕉</text></list-item>
</list-item-group>
<list-item-group>
<list-item class="item_style"><text class="big_size">饮料</text></list-item>
<list-item><text class="font_size">可乐</text></list-item>
<list-item><text class="font_size">咖啡</text></list-item>
</list-item-group>
<list-item-group>
    <list-item class="item_style"><text class="big_size">零食</text></list-item>
<list-item><text class="font_size">薯片</text></list-item>
<list-item><text class="font_size">锅巴</text></list-item>
</list-item-group>
    </list>
</div>

/* index.css */
.item_style {
    background-color: pink;
}
.big_size {
    font-size: 50px;
}
```

运行上述代码，结果如图 5.15 所示，实现了一个具有折叠效果的列表。

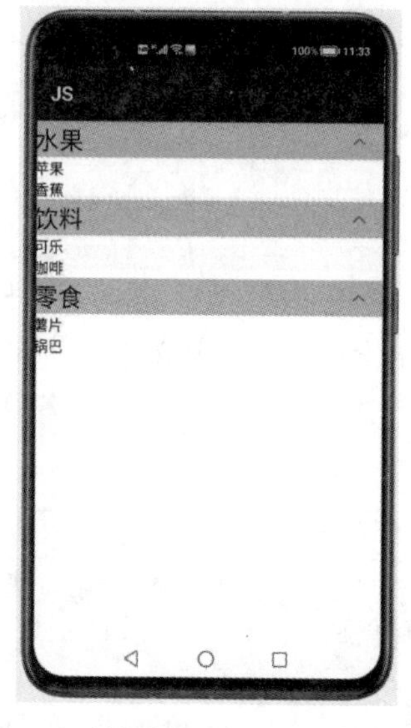

图 5.15 可折叠的 List 组件示例效果图

5.3.3 Tabs组件

有时需要在 App 的界面中滑动或者单击某个特定区域来显示不同的内容，类似于购物 App 中显示推送的功能，这时就需要用到 JS 提供的 Tabs 组件了。Tabs 组件仅包含一个 tab bar 和一个 tab content，其中，tab bar 可以用来显示一些简讯，而 tab content 可以用来显示主要的内容。

接下来开发一个观察华为手机三视图的小程序，在观察某个视图时界面能显示当前视图的名称及内容。其中，当前视图的名称可以用 tab bar 组件来显示，而视图内容可以用 tab content 来显示，hml 文件代码如下：

```
<! -- .hml 文件定义 Tabs 组件 -->
<div class = "container">
<tabs class = "tabs" vertical = "false" >
<tab - bar class = "tab - bar" mode = "fixed">
<text class = "tab - text"> black </text >
<text class = "tab - text"> orange </text >
<text class = "tab - text"> white </text >
</tab - bar >
<tab - content class = "tab - content" scrollable = "true">
<div class = "item - content" >
<image class = "item - image" src = "/common/1.png"></image >
</div >
<div class = "item - content" >
<image class = "item - image" src = "/common/2.png"></image >
</div >
<div class = "item - content" >
<image class = "item - image" src = "/common/2.jpg"></image >
</div >
</tab - content >
</tabs >
</div >
```

由代码可知，定义了一个 Tabs 组件，并且在 Tabs 组件内定义了一个 tab-bar 组件用于显示各视图的名称(black、orange、white)，还定义了一个 tab-content 组件用于显示 3 个视图。在 tab-bar 组件中定义了 3 个 Text 来显示视图名称，在 tab-content 中定义了 3 个 Image 来显示内容。

此外，还需要在 css 文件中设置具体的样式，css 文件的代码如下：

```
/* .css 文件代码设置样式 */
.container {
    flex - direction: column;          /* 调整主轴方向，将容器中的内容按列摆放 */
    justify - content: flex - start;   /* 将内容放在容器开头 */
```

```
    align-items: center;                    /* 内容对齐方式 */
}
.tabs {
    width: 100%;
}
.tab-bar {
    margin: 10px;
    height: 60px;
    border-color: #963C71;
    border-width: 1px;
}
.tab-text {
    width: 300px;
    text-align: center;
}
.tab-content {
    width: 100%;
    height: 80%;
    justify-content: center;                /* 将内容放在容器中央 */
}
.item-content {
    height: 100%;
    justify-content: center;
}
.item-image {
    object-fit: contain;                    /* 将内容缩放到全部显示 */
}
```

运行上述代码,效果如图 5.16 所示。

图 5.16 Tabs 组件示例效果图

可以看到定义的 tab-bar 在屏幕上方并显示了视图名称，通过单击 tab-bar 中各部分可以切换视图，同时通过拖动屏幕中内容的方法也可以切换视图（前提是在 tab-content 定义时将 scrollable 属性设置为 true）。

5.3.4 自定义组件

自定义组件是用户根据业务需求，将已有的组件组合起来，封装成新的组件，并作为新组件可以在工程中被多次调用，从而提高代码的可读性和可扩展性。自定义组件通过 element 引入宿主页面，使用方法的示例代码如下：

```
<element name = "comp" src = "../comp.hml"></element>
<div>
<comp prop1 = 'xxxx' @child1 = "bindParentVmMethod"></comp>
</div>
```

（1）name 属性指定自定义组件名称（非必填，若不填组件名称则为 hml 文件名）。src 属性指自定义组件 hml 文件路径，为了准确定位该组件位置，src 属性内容必须填写，需要注意，src 路径中"../"代表上一级目录索引。

（2）prop 属性用于组件之间的通信，可以通过< tag xxxx='value'>方式传递给组件，名称必须用小写。

（3）事件绑定：自定义组件中绑定子组件事件使用（on 或@）child1 语法，子组件中通过 this. $emit('child1', { params: '传递参数' })触发事件并进行传值，父组件执行 bindParentVmMe-thod 方法并接收子组件传递的参数。

下面尝试创建一个自定义组件，并在父组件中引入自定义组件的事件响应。首先在 page 文件夹中新建自定义组件目录 comp，同时创建自定义组件的基础 hml、css 和 js 文件，代码如下：

```
<!-- 新建 comp.hml 文件 -->
<div class = "item">
<text class = "title_style">{{title}}</text>
<text class = "text - style" onclick = "childClicked">单击这里查看隐藏文本</text>
<text class = "text - style" if = "{{showword}}"> hello world </text>
</div>

/* 新建 comp.css */
.item {
    width: 700px;
    flex - direction: column;
    height: 300px;
    align - items: center;
    margin - top: 100px;
}
```

```
.text - style {
    font - weight: 500;
    font - family: Courier;
    font - size: 40px;
}

/* 新建 comp.js */
export default {
    props: {
        title: {
            default: 'title'
        },
    },
    data: {
        showword: false,
    },
    childClicked () {
        this.$emit('eventType1', {text: "收到子组件参数"});
        this.showword = !this.showword;
    },
}
```

上述代码新创建了一个 comp 自定义组件，组件中设置了一个具有单击属性的文本内容，该单击效果会将自定义组件的数据传递给父组件，并将 show 值从初始化的 false 转换为 true，即显示 hello world 文本。

在父组件中通过 element 引入自定义组件，代码如下：

```
<!-- index.hml -->/
<element name = "comp" src = "../comp/comp.hml"></element>
<div class = "container">
<text>父组件: {{text}}</text>
<comp title = "自定义组件" @event - type1 = "textClicked"></comp>
</div>
/* index.js */
export default {
    data: {
        text: "开始"
    },
    textClicked (clicked) {
        this.text = clicked.detail.text;
    },
}
/* index.css */
.container {
    background - color: #f8f8ff;
```

```
    flex: 1;
    flex - direction: column;
    align - content: center;
}
```

运行上述代码,运行效果如图 5.17 所示。

图 5.17　自定义组件示例效果图

本示例中父组件通过添加自定义属性向子组件传递了名为 title 的参数,自定义子组件在 props 中接收,自定义组件可以通过 props 声明属性,父组件通过设置的属性向子组件传递参数,注意在命名 prop 时需要使用 camelCase 即驼峰式命名法,在外部父组件传递参数时需要使用 kebab-case 即用短横线分割命名,例如在上面示例代码中属性 eventType1 在父组件引用时需要转换为 event-type1。子组件也通过事件绑定向上传递了参数 text,父组件接收时通过 clicked. detail 获取数据。

父子组件之间的数据传输是单向的,一般只能从父组件传递给子组件,而子组件如果需要向上传递必须绑定事件,通过事件的 $emit 来传输。子组件获取来自父组件数据后,子组件不能直接修改父组件传递下来的值,可以通过将 props 传入的值用 data 接收后作为默认值,然后再对 data 的值进行修改。

如果需要观察组件中属性的变化,可以通过 $watch 方法增加属性变化回调,代码如下:

```
//comp.js
export default {
    props: ['title'],
    onInit() {
        this.$watch('title', 'onPropertyChange');
    },
    onPropertyChange(newV, oldV) {
        console.info('title 属性变化 ' + newV + '' + oldV);
    },
}
```

5.4 添加用户交互

5.4.1 手势事件

提到如何与 App 进行交互，首先想到的就是 Button 组件，其实 HarmonyOS 中绝大多数组件可以与用户进行交互，最常用的就是设置组件的 onclick 属性，代码如下：

```
<div class = "click - test" onclick = "Click" vertical = "false">
```

上述代码中，将 click-test 组件的 onclick 属性设置为 Click，这里的 Click 其实是在 js 文件中定义的一个函数，当此组件所包含的区域被单击时就会触发 Click 函数。这里依旧以 5.3.3 节中的 3 种不同颜色的手机为例，代码如下：

```
<! -- .hml 文件代码 -->
<div class = "container">
<div class = "click - test" onclick = "Click" vertical = "true">
< image class = "image" src = "{{Image}}"></image>
<text class = "image - name">{{Name}}</text>
</div>
</div>
```

在 hml 文件代码中定义了一个 div 组件，并将其 onclick 属性连接到了 js 文件中定义的 Click 函数，在 Click 函数中会根据 tmp 的值来决定每次单击后显示的视图及显示的视图名字，Click 函数在 js 文件中的实现代码如下：

```
//.js 文件代码
export default {
    data: {
        Image: "/common/1.png",                //最开始显示 black 视图
        Name: "black",
```

```
            tmp: 1,
        },
    Click() {
            if ( this.tmp == 0) {
                this.Image = "/common/1.png";                   //第 3 次单击跳到第一视图
                this.Name = "black";
            }
            else if ( this.tmp == 1 ) {
                this.Image = "/common/2.png";                   //第 1 次单击跳到第二视图
                this.Name = "orange";
            }
            else if ( this.tmp == 2 ) {
                this.Image = "/common/3.png";                   //第 2 次单击跳到第三视图
                this.Name = "white";
            }
            this.tmp = ( this.tmp + 1) % 3 ;
        }
    }
```

在 css 文件中对布局样式进行简单设计，代码如下：

```
.container {
    flex-direction: column;
    justify-content: flex-start;
    align-items: center;
}
.click-test {
    width: 100%;
}
.image {
    height: 80%;
    justify-content: center;
}
.image-name {
    color: #BCBCBC;
    width: 300px;
    text-align: center;
}
```

运行上述代码，当应用程序打开时，开始界面如图 5.18 所示。

当第 1 次单击界面容器之后，图片会进行跳转，从图 5.18 所示效果跳转至图 5.19 所示效果，如图 5.19 所示。

同理，当第 2 次单击之后效果如图 5.20 所示。第 3 次单击之后又会回到开始界面，即图 5.18 所示的效果，之后不断循环。

图 5.18 组件示例效果图

图 5.19 第 1 次手势交互后组件示例效果图

图 5.20 第 2 次手势交互后组件示例效果图

5.4.2 按键事件

按键事件是智慧屏上特有的手势事件，会在用户操作遥控器按键时触发。当用户单击一个遥控器按钮时，通常会触发两次 key 事件：先触发 action 为 0 即触发按下事件，再触发 action 为 1 即手指抬起事件。action 等于 2 的场景较少出现，一般为用户按下按键后不抬起即长按，此时 repeatCount 将返回次数。每个物理按键对应各自的按键 keycode 以实现不同的功能，代码如下：

```
<!-- index.hml -->
<div class = "card-box">
<div class = "content-box">
<text class = "content-text" onkey = "keyUp" onfocus = "focusUp"
onblur = "blurUp">{{up}}</text>
</div>
<div class = "content-box">
<text class = "content-text" onkey = "keyDown" onfocus = "focusDown"
onblur = "blurDown">{{down}}</text>
</div>
</div>
```

```
/* index.js */
export default {
    data: {
        up: 'up',
        down: 'down',
    },
    focusUp: function() {
        this.up = 'up focused';
    },
    blurUp: function() {
        this.up = 'up';
    },
    keyUp: function() {
        this.up = 'up keyed';
    },
    focusDown: function() {
        this.down = 'down focused';
    },
    blurDown: function() {
        this.down = 'down';
    },
    keyDown: function() {
        this.down = 'down keyed';
    },
}
```

按键事件通过获焦事件向下分发，因此上述示例中使用了 focus 事件和 blur 事件来明确当前焦点位置。当按上下键时，相应的 focused 状态将会响应。当失去焦点按键时，恢复到正常的 up 或 down 按键文本。按确认键后该键变为 keyed 状态。

5.4.3 页面路由

很多情况下，在开发 App 时不只使用一个页面，例如在购物 App 中有时需要从商品详情页面跳转到购物车页面，这时就需要在商品详情页面设置一个能跳转到购物车页面的入口，在浏览完购物车页面之后又需要回到之前的商品详情页面，此时就需要页面路由功能。

在页面路由中需要定义两个或两个以上页面，然后在各自页面的 js 文件中定义相应的路由函数来使用目标页面的 uri 跳转到目标页面。定义两个页面，两个页面的 hml 文件代码分别如下：

```
<!-- first.hml -->
<div class="container">
<div class="text-div">
<text class="title">
//这是第一个页面
```

```
</text>
</div>
<div class = "button - div">
<button type = "capsule" value = "跳转到第二页面" class = "button" onclick = "launch"></button>
</div>
</div>

<!-- second.hml -->
<div class = "container">
<div class = "text - div">
<text class = "title">
//这是第二个页面
</text>
</div>
<div class = "button - div">
<button type = "capsule" value = "回到第一个页面 by router.back" class = "button" onclick =
"launch"></button>
<button type = "capsule" value = "回到第一个页面 by router.push" class = "button" onclick =
"launch2"></button>
</div>
</div>
```

在第一个页面中定义了一个 button 并将其与 first.js 中的 launch 函数相关联，第二个页面中的两个 button 分别与 second.js 中的 launch 函数与 launch2 函数相关联，两个 js 文件中的代码如下：

```
//first.js
import router from '@system.router';
//指定的页面.在调用 router 方法之前,需要导入 router 模块
export default {
    launch: function () {
        router.push({
            uri: 'pages/second/second',                    //目标页面的路径
        })
    },
}

//second.js
import router from '@system.router'
export default {
    launch: function () {
        router.back();                                     //回到路由前的页面
    },
    launch2: function () {
```

```
            router.push({
                uri: 'pages/first/first',                          //目标页面的路径
            })
        }
    }
```

在第一个页面的 first.js 文件中，使用 router.push 来跳转到第二个页面，uri 代表目标页面的路径，然而回到之前的页面有两条途径，也就是在第二个页面中所使用的两种方法，router.back 和 router.push。其中，router.back 可以实现直接跳回原页面，所以不需要任何 uri，而 router.push 利用原页面的 uri 再次跳转到原页面。

下面是两个页面的 css 文件，代码如下：

```
/* first.css */
.container {
    flex-direction: column;
    justify-content: center;
    align-items: center;
}
.text-div{
    justify-content: center;
}
.button-div{
    justify-content: center;
}

/* second.css */
.container {
    flex-direction: column;
    justify-content: center;
    align-items: center;
}
.text-div{
    justify-content: center;
}
.button-div{
    justify-content: center;
}
```

为了实现页面跳转，需要将每个页面的 3 个文件(hml、css、js)都放入各自独立的文件夹，如图 5.21 所示。

之后还需要在 App 模块的 config.json 文件中注册已编写好的页面，代码如下：

```
"js": [
  {
    "pages": [
```

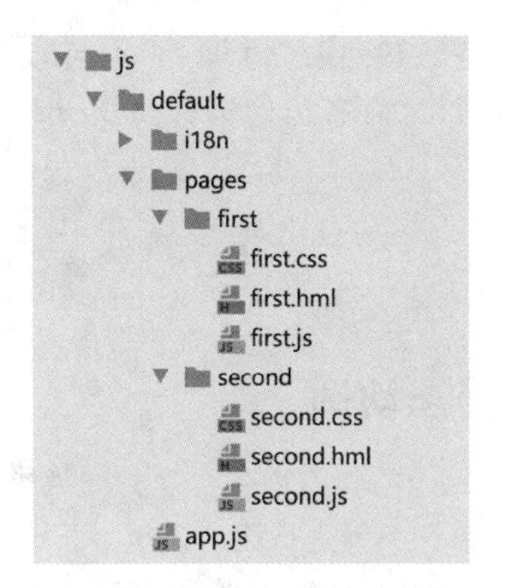

图 5.21 页面路由示例工程结构图

```
        "pages/first/first",
        "pages/second/second"
      ],
      ...
    }
  ]
```

pages 中排在第一位的页面将作为应用程序的默认页面，也就是打开应用程序后显示的第一个页面。运行上述代码，首先显示的第一个页面如图 5.22 所示。

单击“跳转到第二个页面”按钮将会跳转到第二个页面，如图 5.23 所示。

图 5.22 页面路由示例的第一个页面

图 5.23 页面路由示例的第二个页面

单击页面上的两个按钮都会跳转到第一个页面，不同的地方是左边的按钮使用返回方式回到原页面，因此更像返回操作，而右边的按钮使用 uri 跳转到原页面，因此更像前进操作。

5.5 动画

5.5.1 transform 静态动画

静态动画的核心是 transform 样式，其中包含 3 种变换类型，且每一次的样式设置只能支持一种类型的变化。下面对 3 种类型进行逐一讲解。

首先，translate 变换类型可以将组件沿水平或垂直方向移动一定的距离，创建 JS 项目工程，修改 index. hml 和 index. css 文件，示例是一个水平向右移动的示例代码，代码如下：

```
<! -- index.hml 实现水平向右移动 -- >
< div class = "container">
< text class = "translate"> hello </text >
</div >

/* index.css */
.container {
    flex - direction: column;
    align - items: center;
}
.translate {
    height: 300px;
    width: 400px;
    font - size: 100px;
    background - color: #008000;
    transform: translateX(300px);
}
```

上述代码实现了一个 text 组件水平向右移动的效果，translateX(300px)将 text 组件水平从基准线向右移动 300px，其中右为正值，左为负值。同理，若修改为 translateY()，则是以 *Y* 轴基准线为标准，下为正值，上为负值。运行结果如图 5.24 所示，灰线为基准线。

scale 样式可将组件沿横向或纵向，缩小或放大一定比例。下面对 text 组件进行横向放大，代码如下：

```
<! -- index.hml 实现横向放大 -- >
< div class = "container">
< text class = "scale"> hello </text >
</div >
```

```
/* index.css */
.container {
    flex-direction: column;
    align-items: center;
}
.scale {
    height: 300px;
    width: 400px;
    font-size: 100px;
    background-color: #008000;
    transform: scaleX(1.5);
}
```

其中,scaleX(1.5)表示将 text 文本横向放大 1.5 倍,运行上述代码可以直观看到效果,如图 5.25 所示。

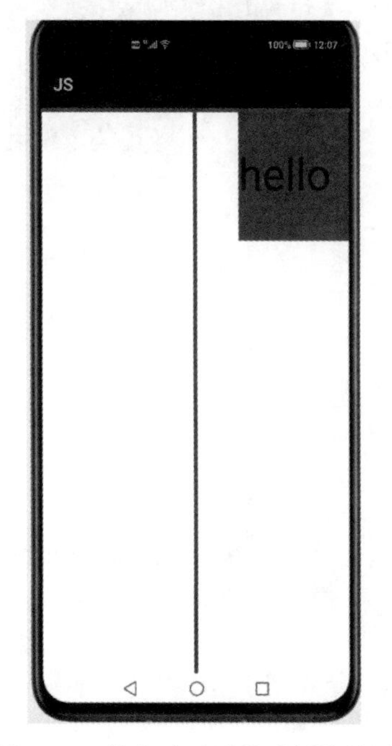

图 5.24 静态动画平移示例效果图

图 5.25 静态动画放大示例效果图

rotate 样式可以将组件沿横轴或纵轴或中心点,旋转一定的角度,如下示例将 text 文本绕 X 轴顺时针旋转 45°,代码如下:

```
<!-- index.hml 实现绕 X 轴顺时针旋转 -->
<div class = "container">
<text class = "rotate"> hello </text>
</div>
```

```
/* index.css */
.container {
    flex-direction: column;
    align-items: center;
}
.rotate {
    height: 300px;
    width: 400px;
    font-size: 100px;
    background-color: #008000;
    transform-origin: 200px 100px;
    transform: rotateX(45deg);
}
```

运行上述代码，结果如图 5.26 所示。

图 5.26 中 Z 轴垂直穿出屏幕。一般的 rotateX()和 rotateY()表示绕 X 和 Y 轴顺时针旋转，而 rotate()表示绕 Z 轴旋转。

与连续动画不同，静态动画只有开始状态和结束状态，而不能设置中间状态，如果需要设置中间的过渡状态和转换效果，则只能由连续动画实现。

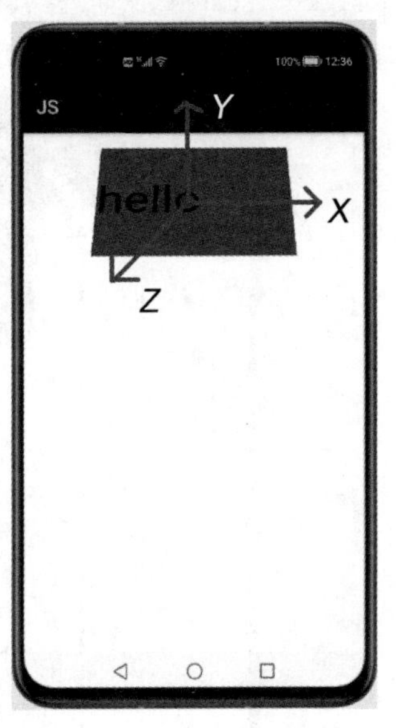

图 5.26 静态动画旋转示例效果图

5.5.2 animation 连续动画

前面讲过的静态动画只有开始状态和结束状态，而没有中间状态，因此静态动画看起来更像是图片之间的切换而不是真正意义上的动画。为了使动画变得连贯，HarmonyOS 中的 JS 引入了连续动画。连续动画最主要的功能贡献者就是 animation 样式，通过它可以定义动画的开始、结束状态及期间变化的速度。在引入连续动画之后，可以定义组件的宽、高、颜色和透明度等的变化速度和程度，利用这一点可以实现一些有趣的功能，例如进度条、渐变色块等。在本节中利用 animation 实现颜色、透明度和宽度变化的功能。首先新建 JS 工程，修改代码如下：

```
<!-- animation.hml 创建页面 -->
<div class="item-container">
<div class="group">
<text class="header">
<!-- 动画演示 -->
</text>
<div class="item {{colorParam}}">
```

```
<text class = "txt">
<!-- 颜色 -->
</text>
</div>
<div class = "item {{opacityParam}}">
<text class = "txt">
<!-- 透明度 -->
</text>
</div>
<input class = "button" type = "button" name = "" value = "开始" onclick = "show01"/>
</div>
</div>
```

在页面中定义了两个文本组件，分别用来演示颜色变化，以及透明度和宽度变化。input 组件的单击事件 show()在文件中的实现代码如下：

```
//animation.js
export default {
    data: {
        colorParam: '',
        opacityParam: '',
    },
    show: function () {
        this.colorParam = ''
        this.opacityParam = ''
        this.colorParam = 'color'
        this.opacityParam = 'opacity'
    },
}
```

其中，colorParam 和 opacityParam 分别与 hml 文件中的样式进行数据绑定，在 show()方法中，首先将这两个参数设置为默认格式，并在 css 文件中进行动画实现，css 文件代码如下：

```
/* animation.css */
.item-container {
    margin-bottom: 50px;
    margin-right: 60px;
    margin-left: 60px;
    flex-direction: column;
    align-items: flex-start;
}
.group {
    margin-bottom: 150px;
    flex-direction: column;
```

```
    align-items: flex-start;
}
.header {
    margin-bottom: 20px;
}
.item {
    background-color: #191FF7;
}
.txt {
    text-align: center;
    width: 200px;
    height: 100px;
}
.button {
    width: 200px;
    font-size: 30px;
    color: #ffffff;
    background-color: #09ba07;
}
.color {
    animation-name: Color-frames;                /* 动画由 Color-frames 定义 */
    animation-duration: 8000ms;
}
.opacity {
    animation-name: Opacity-frames;              /* 动画由 Opacity-frames */
    animation-duration: 8000ms;
}
@keyframes Color-frames {                        /* 颜色变换效果动画 */
    from {
        background-color: #191FF7;               /* 初始颜色 */
    }
    to {
        background-color: #09ba07;               /* 最终颜色 */
    }
} @keyframes Opacity-frames {                    /* 透明度、宽度变换效果动画 */
    from {
        width: 600px;                            /* 初始宽度 */
        opacity: 0.9;                            /* 初始透明度 */
    }
    to {
        width: 0px;                              /* 最终宽度 */
        opacity: 0.0;                            /* 最终透明度 */
    }
}
```

在.color 和.opacity 样式中，使用 animation-name 属性定义了各自的动画样式

(@keyframes)。其中@keyframes 可以自定义动画样式,例如代码中定义的 Color-frames 和 Opacity-frames,from 代表动画的开始状态,to 代表结束状态,期间的过渡动画由系统自动计算完成,当然也可以使用 animation-timing-function 属性来描述动画执行的速度曲线,使动画更加平滑。运行上述代码,初始页面如图 5.27 所示。

单击"开始"按钮后,会播放定义的动画(按时间顺序排列),如图 5.28 所示。

图 5.27　连续动画示例的初始状态

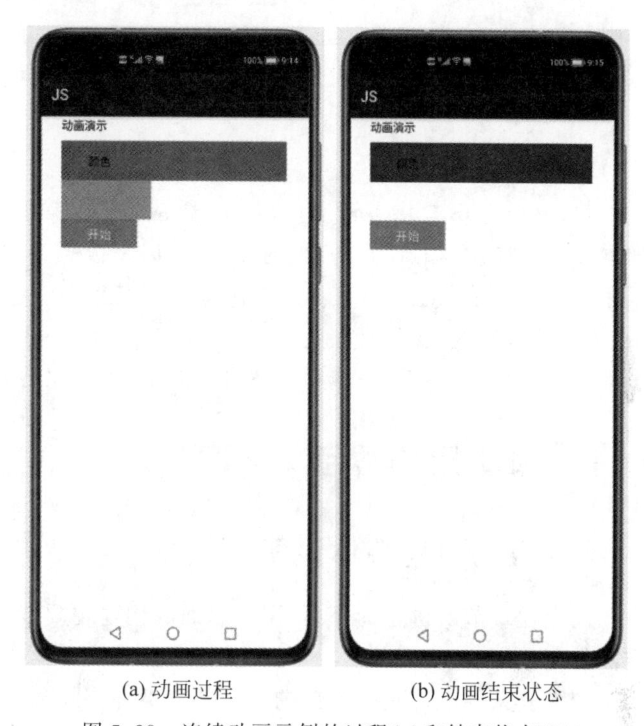

(a) 动画过程　　(b) 动画结束状态

图 5.28　连续动画示例的过程(a)和结束状态(b)

可以看到"颜色"块的背景色从蓝色最终变为绿色,"透明度"块的透明度从不透明变为全透明,并且宽度从最开始一直变为 0,类似一个反向的进度条。在这里还可以使用 animation 的其他属性来定义更多有趣的动画效果,读者可以自行尝试。

第 6 章

轻量级智能穿戴开发

13min

HarmonyOS 提供了轻量级智能穿戴应用开发，开发者可以在运动手表上开发跨设备协同工作的应用，如从与之匹配的智能手机、平板等各类设备上获取信息，便捷地显示在运动手表上，或通过对运动手表的操作来控制其他设备上的操作任务，为消费者带来更加灵活、智慧的分布式交互体验。

轻量级智能穿戴统一使用上述讲到的 JS 语言进行开发，这里带领大家体验一番。

6.1 构建用户界面

6.1.1 布局整体说明

同上述章节所讲述的那样，用户界面的构建主要以组件为基础，每个组件通过对数据和方法的简单封装，实现独立的可视化、可交互功能单元。在轻量级智能穿戴设备的开发中，个别较为复杂的组件是不支持的。大部分组件支持通用属性和通用样式，不同组件还支持其私有的属性和样式。根据组件的功能，将组件主要分为以下三大类：

- 基础组件：text、image、progress、marquee、chart 等。
- 容器组件：div、list、list-item、stack、swiper 等。
- 表单组件：input、slider、switch、picker-view 等。

目前，轻量级智能穿戴设备 Lite Wearable 开发均采用 JS 语言，支持 Empty Feature Ability 和 List Feature Ability 两种开发模板。同上述章节所讲，轻鸿蒙的页面构建方式与第 5 章 JS UI 基本一致，开发框架以 454px(这里，px 是逻辑像素，非物理像素)为基准宽度。

首先新创建一个轻量级智能穿戴设备的工程，选择 Lite Wearable 设备下的 Empty Feature Ability 模板，如图 6.1 所示。

创建完成后，会生成一个 Hello World 的工程，这里可以通过模拟器运行测试。单击右上角的运行按钮，在弹窗下方的 Available Huawei Lite Devices 中，选择华为轻量级穿戴设备 Huawei Lite Wearable Simulator，如图 6.2 所示，单击 OK 按钮即可启动模拟器。

运行模拟器后效果如图 6.3 所示，表盘中显示了 Hello World 字样。表盘可以完全模拟应用在手表上的运行效果，同时也支持使用鼠标单击或滑动等相应操作。

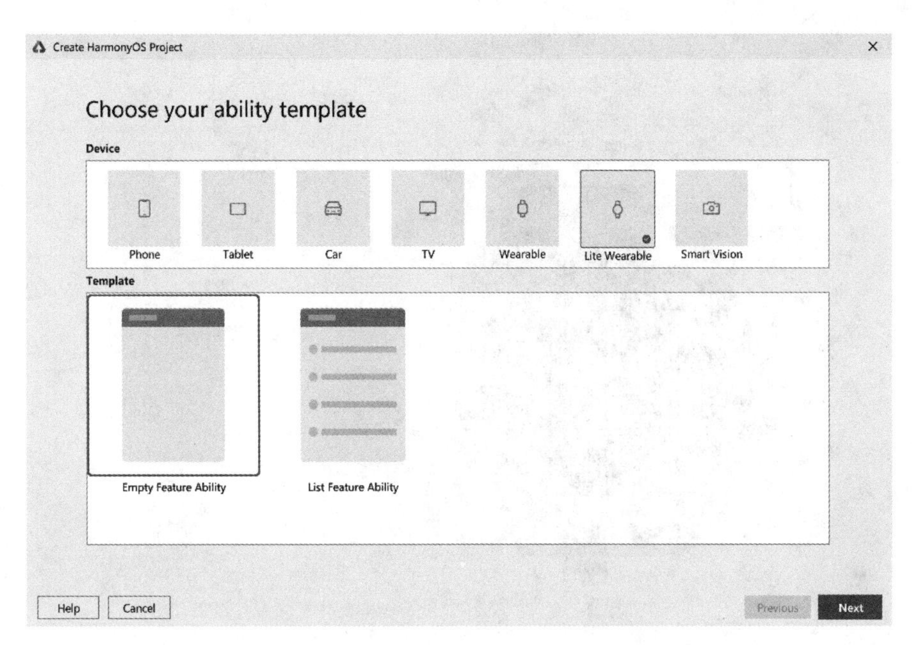

图 6.1 新建 Lite Wearable 设备工程

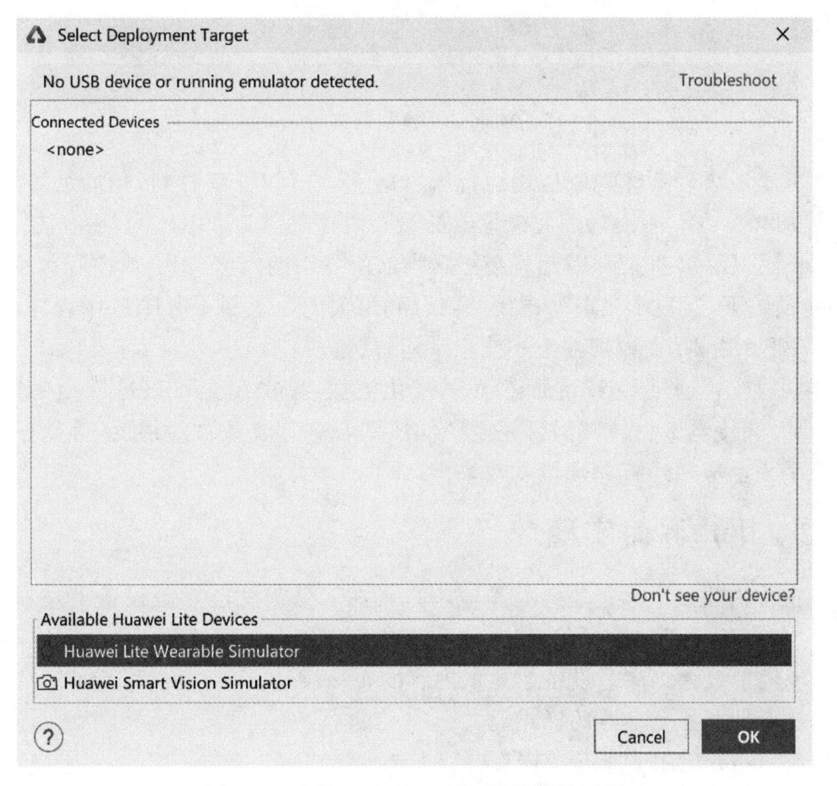

图 6.2 选择 Lite Wearable 模拟器并运行

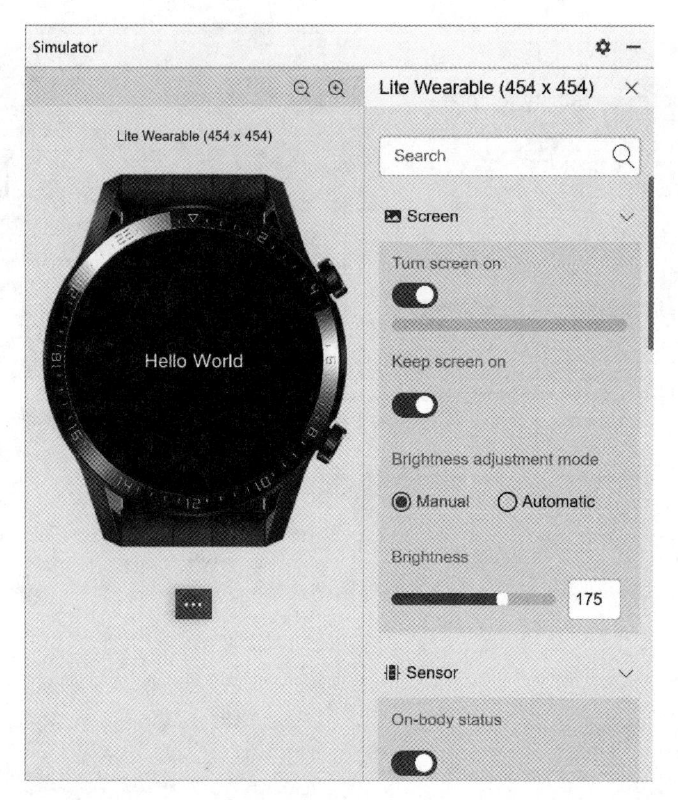

图 6.3 Lite Wearable 模拟器运行 Hello World

表盘下方显示了轻量级鸿蒙应用运行所占的 JS 内存，可以看到，轻量级智能手表的最大运行内存为 49144B(48KB)，因此模拟器的最大内存限制为 48KB。(注意，在 Debug 模式下，调试引擎会占用一定内存，因此只会显示当前所占用的内存，不会对内存大小进行限制。在非 Debug 模式下，会显示当前所占用的内存和最大内存限制 48KB，应用程序占用的内存不能大于 48KB。否则，在模拟器上不能正常运行。)

模拟器同时支持通过 GUI 界面注入场景化数据，单击设置可以进行相关配置，如“点亮/关闭屏幕”“亮度调节模式”“屏幕亮度”“气压”“心率”“步数”“经纬度”等数据，可以更加真实地模拟手表上应用运行的环境和效果。

6.1.2 用户界面实现

在本节中将构建一个出行服务打车应用。可在轻量级智能穿戴设备上左右滑动，显示“司机信息”“位置信息”和“车辆信息”。实现该打车应用 Demo，需要完成如下三步：

(1) 实现轻量级智能穿戴设备上显示内容的布局设计。

(2) 实现表盘上各个容器的样式定义。

(3) 实现 Demo 应用的交互逻辑设计。

实现后的效果图如图 6.4 所示。

首先对 Demo 的基础布局进行解析：

(1) 需要实现"司机信息""位置信息"和"车辆信息"的三屏左右滑动，需要使用 swiper 容器组件来控制。swiper 提供了切换子组件显示的能力。

(2) 在每一屏内(如司机信息)，需要展示 4 行信息，这 4 行信息需要定义 4 个容器进行存储。其中第一行包含 image 和 text 两个组件。

(3) 每条信息，需要从手机端动态地读取信息，所以需要设置对应的变量。

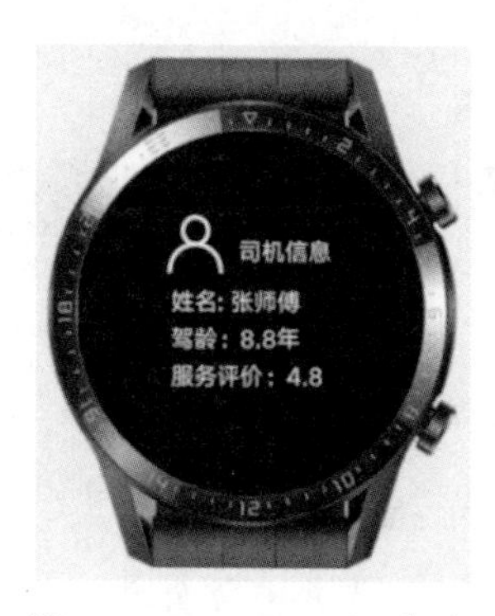

图 6.4 Lite Wearable 打车软件 Demo 效果图

在之前新建 Hello World 工程上进行打车应用 Demo 实现。首先选择 entry→src→main→js→default，新建 common 文件夹，用于存储 image 图片资源，然后选择 entry→src→main→js→default→pages→index，在 index.hml 中搭建页面的基础布局结构，实现代码如下：

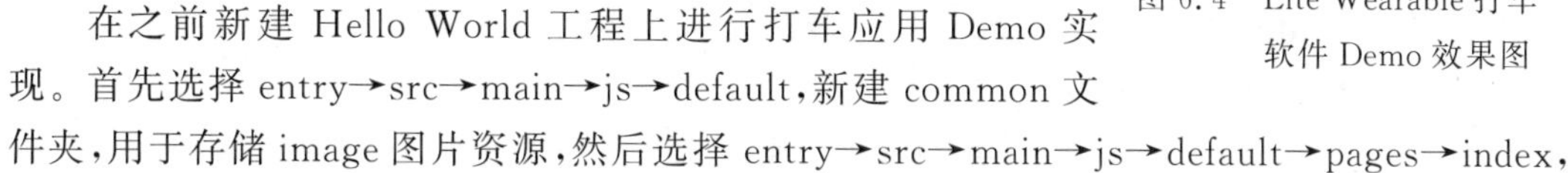

```
<! -- index.hml -- >
<! -- 关联 css 文件中的.container 样式代码块 -- >
< div class = "container">
<! -- 控制多屏之间的滑动 -- >
< swiper class = "swiper">
<! -- 定义"司机信息"页面内容 -- >
< div class = "bodyDriverInfo">
< div class = "titleAndImage">
< image class = "image" src = "common/driver.png"></image >
< text class = "title">司机信息</text >
</div >
< text class = "driverInfo">姓名: {{name}}</text >
< text class = "driverInfo">驾龄: {{age}}年</text >
< text class = "driverInfo">服务评价: {{score}}</text >
</div >
<! -- 定义"位置信息"页面内容 -- >
< div class = "bodyDriverInfo">
< div class = "titleAndImage">
< image class = "image" src = "common/location.png"></image >
< text class = "title">位置信息</text >
</div >
< text class = "locationInfo">{{position}}</text >
< text class = "locationInfo">距你{{dis}}米</text >
< text class = "locationInfo">{{status}}</text >
</div >
<! -- 定义"车辆信息"页面内容 -- >
< div class = "bodyDriverInfo">
< div class = "titleAndImage">
```

```
<image class = "image"src = "common/car.png"></image>
<text class = "title">车辆信息</text>
</div>
<text class = "carInfo">{{carType}}</text>
<text class = "carInfo">{{carColor}}</text>
<text class = "carInfo">{{carCard}}</text>
</div>
</swiper>
</div>
```

上述 hml 文件实现了页面的基础布局，其中通过 swiper 组件控制手表多屏之间的滑动，每个屏通过 div 容器进行存储，其中包括最上方标题的 div 容器和下方的 3 行 text 组件，所显示的数据均采用{{content}}的数据绑定形式。

每个布局中都必不可少样式的绘制，这里采用选择器样式的方式，即将所有的样式代码写到 css 文件中，然后通过 class、id 等方式和组件关联起来。

在上述 hml 布局文件中，每个容器都定义了一个如< div class＝"contain -er"></div >的 class 样式，该 class 用于定义每个容器内存放的内容的位置、元素的大小、字体、颜色、背景色等信息，因此，需要为 index. hml 中每个 class 定义具体的样式，打车应用 Demo 应用的样式文件 index. css 代码如下：

```
/* index.css */
/* 所有组件中 class = "container" 的组件都会使用该样式. */
.container {
    justify-content: center;
    align-items: center;
    width: 454px;
    height: 454px;
    background-color:black;
}
.swiper{
    width: 400px;
    height: 400px;
    background-color: black;
}
.bodyDriverInfo{
    flex-direction: column;
    align-items: center;
    width: 380px;
    height: 380px;
    background-color:black;
}
.titleAndImage{
    align-items: center;
```

```
    width: 350px;
    height: 100px;
    margin-left: 100px;
    margin-top: 60px;
    background-color:black;
}
.image{
    justify-content: center;
    width: 83px;
    height: 83px;
    margin-top: 10px;
    margin-left: 10px;
}
.title {
    font-size: 30px;
    margin-top: 30px;
    margin-left: 15px;
    width: 330px;
    height: 39px;
    background-color: black;
}
.driverInfo{
    font-size: 30px;
    margin-left: 120px;
    width: 330px;
    height: 39px;
    margin-top: 7px;
    background-color:black;
}
.locationInfo{
    font-size: 30px;
    text-align: center;
    margin-left: 20px;
    width: 330px;
    height: 39px;
    margin-top: 7px;
    background-color:black;
}
.carInfo{
    font-size: 30px;
    text-align: center;
    margin-left: 20px;
    width: 330px;
    height: 39px;
    margin-top: 7px;
    background-color:black;
}
```

其中，最外层的 container 的宽和高均设置为 454px，在轻鸿蒙设备端，以 454px 为基准宽度。接下来进行 Demo 中应用逻辑交互的实现。在 JS 文件中，需要实现打车 Demo 应用的“司机信息”“位置信息”和“车辆信息”的读取。index.js 中的具体代码如下：

```
/* index.js */
export default {
    onInit(){
        this.getInfo();
    },
//定义打车 Demo 信息变量
    data: {
        name:"",
        age:"",
        score:"",
        position:"",
        dis:"",
        status:"",
        carType:"",
        carColor:"",
        carCard:"",
    },
//实现打车数据的获取，本 Demo 以静态数据为例
    getInfo() {
        this.dis = 880;
        this.status = "预计 3 分钟后到达";
        this.position = "龙岗区居里夫人大道";
        this.name = "张师傅";
        this.age = 8.8;
        this.score = "4.8";
        this.carColor = "红色";
        this.carType = "兰博基尼 918";
        this.carCard = "粤 B 888888";
    },
}
```

其中，本 Demo 获取的打车数据采用了静态数据。在实际应用开发中，手表应从手机端获取数据，需要调用如下示例代码中的 FeatureAbility.subscribeMsg 接口，同时，手机端需要实现打车信息的推送功能。在 index.js 中，新增代码如下：

```
/* 实现运动手表从手机端动态获取数据信息.
同时手机端还需要实现数据的发送功能。 */
testMonitoSubMsg(){
var self = this;
FeatureAbility.subscribeMsg({
success: function(data){
```

```
var message = JSON.parse(data.message);
self.dis = message['dis'];
self.status = message['status'];
self.position = message['position'];
self.name = message['name'];
self.age = message['age'];
self.score = message['score'];
self.carType = message['carType'];
self.carColor = message['carColor'];
self.carCard = message['carCard'];
}
})
}
```

运行上述代码后，页面可左右滑动并可查看相应数据，效果如图 6.5 和图 6.6 所示。

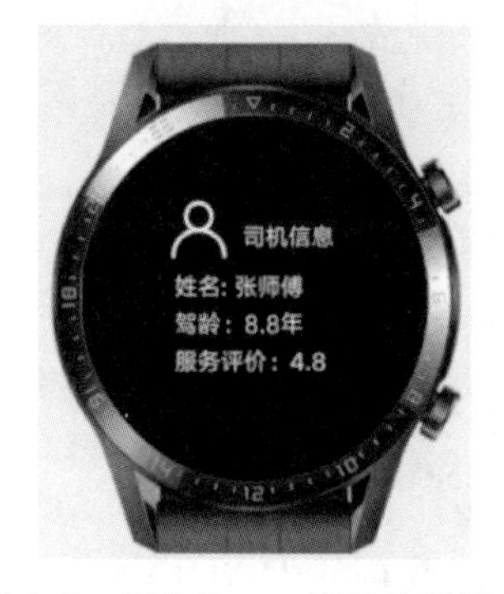

图 6.5　打车 Demo 运行效果图一

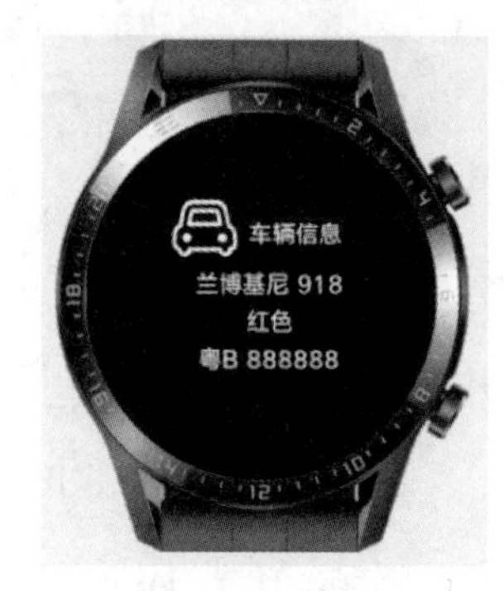

图 6.6　打车 Demo 运行效果图二

6.2　基本功能与系统能力

6.2.1　设备基本功能

在智能穿戴设备开发中，对于设备的一些基本功能，如应用配置、定时器、日志打印与页面路由等，HarmonyOS 均为开发者提供了相应的功能模块和便捷的接口。

例如，在运动手表的应用开发中，每个应用都必须有自己的应用名称，并在当前的应用配置文件中声明，开发者通常需要获取这些信息，包括应用名称、版本名称及版本号。这里需要用到 HarmonyOS 为我们提供的应用上下文的基本功能。首先，需要引用所需基本功能的模块。如上述所讲的应用上下文功能，HarmonyOS 提供了 system. app 功能模块，可在 index. js 中进行引用，代码如下：

```
import app from '@system.app';
```

通过调用 app. getInfo()接口，获取应用配置文件中声明的配置信息，通过 console. log()打印 Debug 级别的文本日志信息。其中，运动手表为开发者提供的日志打印基本功能不需要导入模块，通过 console. debug|log|info|warn|error()即可打印所需文本信息。在 index. js 中加入下列代码，并运行右上角 Debug 按钮，代码如下：

```
/* index.js */
import app from '@system.app';        //导入应用上下文模块
export default {
    onInit(){
        this.getAppInfo();
    },
    …
    getAppInfo:function(){
        var info = app.getInfo();
        console.log(JSON.stringify(info));
    },
}
```

运行后，在下方的 Debug 窗口中可以看到，在应用启动后，窗口打印出了当前应用的配置信息，如图 6.7 所示。

```
[Info] Application onCreate
[Debug] {"appName":"Demo","versionName":"1.0","versionCode":1}
```

图 6.7　应用上下文功能 Debug 输出结果

其中，当前应用的名称为 Demo，版本名称为 1.0，版本号为 1。打开应用配置文件 config. json，可以看到配置文件中的信息与 app. getInfo()获取的信息一致。

```
{
  "app": {
    "bundleName": "com.huawei.litewearable",
    "vendor": "huawei",
    "version": {
      "code": 1,
      "name": "1.0"
    },
…
```

还可以获取应用当前的语言和地区，基本方法与上述相同。这里需要导入应用配置模块 system. configuration，代码如下：

```
import configuration from '@system.configuration';
```

在 index. js 中通过调用 configuration. getLocale()获取应用的语言和地区，默认情况下与系统的语言和地区同步，代码如下：

```
/ * index.js * /
import configuration from '@system.configuration';          //导入应用配置模块
    export default {
        onInit(){
        this.getConfigure();
    },
...
    getConfigure:function(){
        const localeInfo = configuration.getLocale();
        console.info("当前语言为" + localeInfo.language);
        console.info("当前地区：" + localeInfo.countryOrRegion);
        console.info("文字布局方式为" + localeInfo.dir);
    },
}
```

运行 Debug 后输出结果如图 6.8 所示，其中当前语言为 zh 即中文，当前地区为 CN 即中国，表盘的文字布局方式为 ltr 即从左向右。

```
[Info] Application onCreate
[Info] 当前语言为：zh
[Info] 当前地区：CN
[Info] 文字布局方式为：ltr
```

图 6.8　应用配置功能 Debug 输出结果

在运动手表开发中，通常需要用到与定时器相关的功能，例如闹钟时隔几分钟后唤醒，或倒计时等场景。这里可以体验一下定时器的基本用法。定时器功能无须进行模块导入，可以通过 setTimeout()方法设置一个定时器，定时器在到达对应时间后，会执行相关函数。如在 index.js 文件中，加入代码如下：

```
/ * index.js * /
export default {
    onInit(){
        ...
    },
    onShow() {
console.log('onShow()方法执行');
        var timeoutID = setTimeout(function() {
            console.log('延迟 5s');
        }, 5000);
    },
...
}
```

当页面显示上述代码的时候，会调用 onShow()方法，打印“onShow()方法执行”，并在 onShow()方法中调用了 setTimeout()，设置延时时间为 5s(代码中所用时间单位为 ms)，在等待时间为 5s 后，会执行 setTimeout()中的执行函数，即打印“延迟 5s”。运行 Debug 后输出窗口结果如图 6.9 所示,其中两句输出文本的间隔时间为 5s。

```
[Info] Application onCreate
[Debug] onShow()方法执行
[Debug] 延迟5s
```

图 6.9 定时器功能 Debug 输出结果

setTimeout()在到达设定的延迟时间后,会调用一次执行函数,而 setInterval()会重复调用执行函数,每次调用之间都间隔所设定的固定时间延迟。下面设定一个 10s 的倒计时,并通过 Console.log()打印倒计时结果。在 index.js 文件中更改代码如下:

```
/* index.js */
var second = 10;
export default {
    onInit(){
…
    },
    onShow() {
        console.log('onShow()方法执行');
        var intervalID = setInterval(function() {
            console.log('倒计时' + second-- + 's');
            if(second == 0){
                clearTimeout(intervalID);
                console.log('倒计时清除');
            }
        }, 1000);
    },
…
}
```

其中设置倒计时 second 为 10s,在定时器 setInterval()中,每间隔 1s 会执行内部方法,输出当前倒计时时间并将倒计时−1,直至倒计时为 0s,执行 clearInterval()清除设置的重复定时器 intervalID。运行 Debug 后输出结果如图 6.10 所示。

```
[Info] Application onCreate
[Debug] onShow()方法执行
[Debug] 倒计时10s
[Debug] 倒计时9s
[Debug] 倒计时8s
[Debug] 倒计时7s
[Debug] 倒计时6s
[Debug] 倒计时5s
[Debug] 倒计时4s
[Debug] 倒计时3s
[Debug] 倒计时2s
[Debug] 倒计时1s
[Debug] 倒计时清除
```

图 6.10 倒计时 Debug 输出结果

上述是 HarmonyOS 为轻量级智能穿戴设备提供的一些基本能力的简单用法,下面再学习一些系统能力,这些能力可以为开发者在应用开发中提供丰富便捷的接口,以简化开发过程。

6.2.2 系统能力

HarmonyOS 还提供了丰富的接口,用来调用系统能力,如震动、传感器、地理位置、设备信息、屏幕亮度等,方便开发者进行功能开发。通过导入对应功能模块并调用相应接口,就可对系统能力进行获取和配置。

如在 6.1 节的打车 Demo 中，通常需要手表屏幕保持常亮来实时查看打车信息。要想获取或设置屏幕亮度，首先需要导入屏幕亮度功能模块 system. brightness，代码如下：

```
import brightness from '@system.brightness';
```

导入后即可对设备屏幕亮度进行获取和设置，通过 brightness. getValue()方法可以获取设备当前的屏幕亮度值，通过 brightness. setValue()方法可以设置设备当前的屏幕亮度值等。这里通过 Harmony 提供的 brightness. setKeepScreenOn()设置屏幕是否处于常亮状态。在 index. js 文件中增加代码如下：

```
/* index.js */
import brightness from'@system.brightness';            //导入屏幕亮度设置模块
export default {
    onInit(){
        this.getInfo();
        this.testSetBrightnessKeepScreenOn();
    },

…

//实现 GT 表屏幕常亮
    testSetBrightnessKeepScreenOn:function() {
        var self = this;
        brightness.setKeepScreenOn({
            keepScreenOn:true,
            success:function() {
            },
            fail:function(data, code) {
            }
        });
    },
}
```

重新运行代码，会发现屏幕始终保持在常亮状态。

除此之外，还可以通过系统能力获取设备信息。同上，需要先在 index. js 文件中导入设备信息模块，代码如下：

```
import device from '@system.device';
```

通过 device. getInfo()接口获取当前设备的各类信息，包括设备的品牌、生产商、型号等信息，系统语言、地区等信息，以及屏幕宽和高、形状等信息。在 index. js 中新建 getDeviceInfo 方法并调用 device. getInfo()接口，若返回 success，即表示接口调用成功，可通过 console. log()的方式打印获取的信息。实现代码如下：

```
/* index.js */
import device from '@system.device';                          //导入设备信息模块
export default {
    onInit(){
        …
        this.getDeviceInfo();
    },
…
    getDeviceInfo:function() {
        device.getInfo({
            success: function(data) {
                console.log('成功获取设备名称为:' + data.brand);
                console.log('成功获取设备型号为:' + data.model);
                console.log('当前可使用的窗口宽度为:' + data.windowWidth);
            },
            fail: function(data, code) {
                console.log('fail get device info code:'+ code + ', data: ' + data);
            },
        });
    }
}
```

运行 Debug 后从 Console 输出窗口中可以看到，在页面进入 onInit()之后，成功打印出所需的设备名称、型号和可使用的窗口宽度信息，如图 6.11 所示。

```
[Info] Application onCreate
[Debug] 成功获取设备名称为:HUAWEI
[Debug] 成功获取设备型号为:VID-B19
[Debug] 当前可使用的窗口宽度为:454
```

图 6.11 设备信息能力 Debug 输出结果

HarmonyOS 还提供了很多其他的轻量级智能穿戴设备的系统能力，可以查看官方文档，通过调用接口便捷地调用系统能力。

6.2.3 应用生命周期

在应用开发中，需要掌握应用页面的生命周期才能写出合理流畅的应用，从而拥有较好的用户体验。

在 HarmonyOS 的运动手表开发中，应用生命周期主要有两个：

- onCreate：在应用创建时调用。
- onDestroy：在应用被销毁时触发。

在一个应用中，一般会包含多个页面，每个页面都有自身的声明周期，一个页面一般包括 5 个声明周期：

- onInit：在页面创建时被调用，表示页面的数据已经准备好了，可以使用 js 文件中的 data 数据。
- onReady：表示页面已经编译完成，可以将界面显示给用户。
- onShow：在页面显示时被调用。这里需要注意，JS UI 只支持应用同时运行并展示一个页面，当打开一个新页面时，上一个页面就自动被销毁了。

- onHide：在页面消失时被调用。
- onDestroy：在页面销毁时被调用。

整个应用和各个页面的声明周期示意图如图 6.12 所示。

这里通过一个简单的实例，直观感受应用和页面的生命周期。新建一个项目 AppLife，依然选择 Lite Wearable 设备下的 Empty Feature Ability 模板，创建完成后，在 pages 目录下新建一个页面 details，创建后工程的 pages 目录如图 6.13 所示。

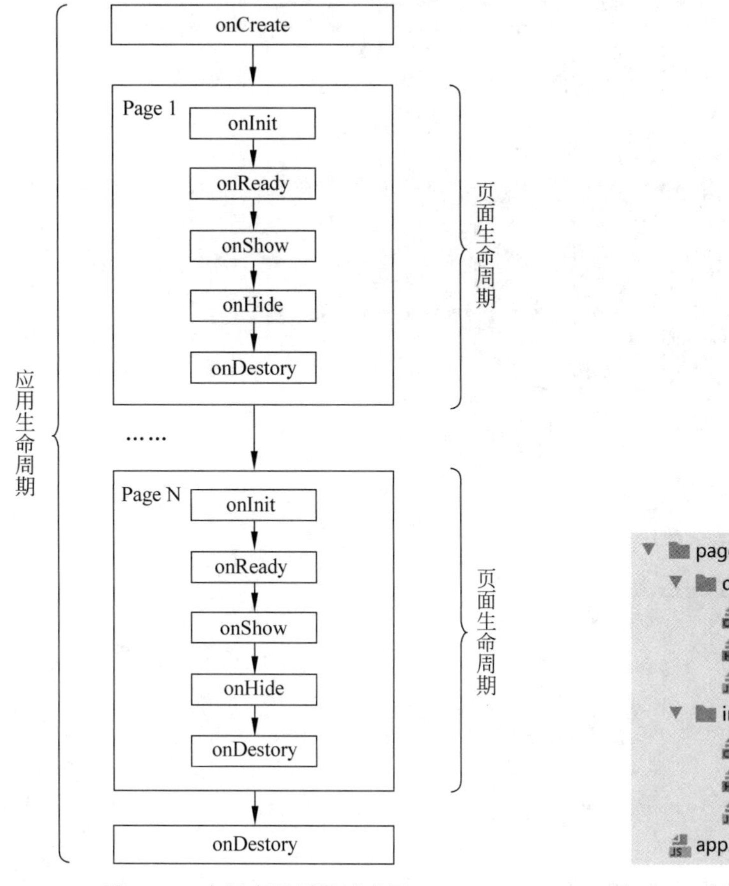

图 6.12　应用声明周期示意图

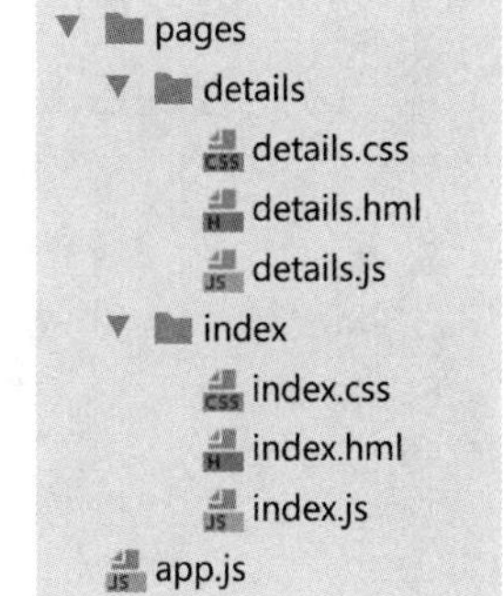

图 6.13　新建工程 pages 目录

在 config.json 文件中增加新建页面的注册，代码如下：

```
{
  …
  "module": {
    …
    "js": [
```

```
        {
          "pages": [
            "pages/index/index",
"pages/details/details"
          ],
          "name": "default"
        }
      ]
    }
}
```

这里简单实现两个页面之间的路由。在 index.hml 中修改代码如下：

```
<! -- index.hml -->
<div class = "container">
<text class = "title">
<! -- 这是主页面 -->
</text>
<input class = "btn" type = "button" value = "跳转到下一页" onclick = "clickAction"></input>
</div>
```

在 index.css 中修改页面布局代码如下：

```
/* index.css */
.container {
    display: flex;
    flex-direction: column;
    justify-content: center;
    align-items: center;
    left: 0px;
    top: 0px;
    width: 454px;
    height: 454px;
}
.title {
    font-size: 30px;
    text-align: center;
    width: 454px;
    height: 100px;
    margin: 10px;
}
.btn {
    width: 200px;
    height: 50px;
}
```

在 index.js 中修改代码如下，其中依次在 index 主页面的每个生命周期事件中通过 console.log()打印了一句话，以通过查看日志打印的方式理解生命周期。index.js 代码如下：

```
/* index.js */
import router from '@system.router'

export default {
    clickAction(){
        router.replace({
            uri: 'pages/details/details'
        });
    },
    onInit() {
        console.log("主页面 onInit()被调用");
    },
    onReady(){
        console.log("主页面 onReady()被调用");
    },
    onShow() {
        console.log("主页面 onShow()被调用");
    },
    onHide(){
        console.log("主页面 onHide()被调用");
    },
    onDestroy() {
        console.log("主页面 onDestroy()被调用");
    },
}
```

随后在新建的子页面 details.hml 中修改代码如下：

```
<!-- details.hml -->
<div class="container">
<text class="title">
<!-- 第二个页面 -->
</text>
<input class="btn" type="button" value="返回" onclick="clickAction"></input>
</div>
```

details.css 代码与 index.css 代码相同即可。修改 details.js 代码，同理在每个生命周期中打印一条日志信息，details.js 代码如下：

```
/* details.js */
import router from '@system.router'

export default {
```

```
        clickAction(){
            router.replace({
                uri: 'pages/index/index'
            });
        },
        onInit() {
            console.log("第二个页面 onInit()被调用");
        },
        onReady(){
            console.log("第二个页面 onReady()被调用");
        },
        onShow() {
            console.log("第二个页面 onShow()被调用");
        },
        onHide(){
            console.log("第二个页面 onHide()被调用");
        },
        onDestroy() {
            console.log("第二个页面 onDestroy()被调用");
        },
    }
```

最后，打开 app.js 文件，在新建项目时已自动生成了应用生命周期事件，这里修改为 Debug 下的日志打印方式，代码如下：

```
export default {
    onCreate() {
        console.log("应用创建 onCreate()");
    },
    onDestroy() {
        console.log("应用销毁 onDestroy()");
    }
};
```

现在运行 Debug 程序，效果如图 6.14 所示。

同时在下方 Debug 窗口中查看 console.log()日志打印信息，如图 6.15 所示。

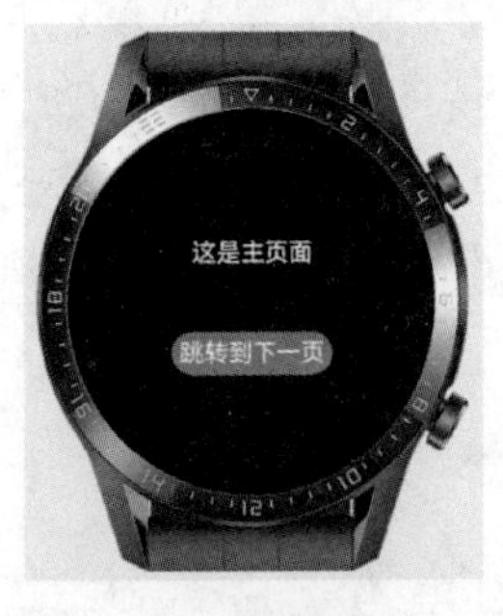

图 6.14 index 主页面

```
[Debug] 应用创建onCreate()
[Debug] 主页面onInit()被调用
[Debug] 主页面onReady()被调用
[Debug] 主页面onShow()被调用
```

图 6.15 启动应用时日志打印

可以看到，当应用程序第一次被创建并展示第一个页面时，首先会执行应用的 onCreate()方法，随后在启动并显示主页面时，依次会执行主页面的 onInit()、onReady()和 onShow()方法。这时单击模拟器表盘中的"跳转到下一页"，启动 details 页面，如图 6.16 所示。

图 6.16　details 页面

这时日志打印信息如图 6.17 所示。

可以看到，当应用从一个页面 A 跳转到另一个页面 B 时，首先会调用页面 A 的 onHide()和 onDestroy()函数，当页面 A 销毁后，会依次调用页面 B 的 onInit()、onReady()和 onShow()函数来初始化和显示页面 B。因为在运动手表中，没有后台页面，所以在某一时刻只能运行并显示一个页面，且这时应用的 onCreate()方法不会被执行，因为只有页面被销毁，而应用并未被销毁。

同理，单击 details 页面的返回时，日志打印结果如图 6.18 所示，会先销毁当前 details 页面，并创建和显示 index 主页面。

```
[Debug] 主页面onHide()被调用
[Debug] 主页面onDestroy()被调用
[Debug] 第二个页面onInit()被调用
[Debug] 第二个页面onReady()被调用
[Debug] 第二个页面onShow()被调用
```

图 6.17　跳转 details 页面时日志打印

```
[Debug] 第二个页面onHide()被调用
[Debug] 第二个页面onDestroy()被调用
[Debug] 主页面onInit()被调用
[Debug] 主页面onReady()被调用
[Debug] 主页面onShow()被调用
```

图 6.18　返回 index 页面时日志打印

6.3　手表应用推送至真机

首先需要将开发的手表侧 demo 编译、构建并生成 HAP 包，后缀为.hap。生成的 HAP 包可以独立部署和运行在搭载 HarmonyOS 的设备上。在主菜单栏中选择 File→Project Structure，在 Modules→entry(模块名称)→Signing Configs → debug 窗口中，配置指定模块的调试签名信息，如图 6.19 所示。

在主菜单栏中选择 Build→Build App(s)/Hap(s)→Build Debug Hap(s)，生成 Debug HAP 包。随后将该 HAP 包推送至运动手表即可。

运动手表的 HarmonyOS 应用安装，依赖华为手机上的运动健康和应用调测助手 App 辅助进行。首先需要将"运动健康 App"升级至最新版本，并且从华为应用市场安装"应用调测助手 App"。使用 USB 连接线将手机和计算机进行连接，确保连接状态是正常的，并在手机上选择传输文件连接方式。在工程目录中的 Build→outputs→hap 中选择生成的 HAP，通过手工复制的方式将 HAP 复制至手机中的"/sdcard/haps/"目录。若在手机存储根目录下没有 haps 文件夹，则手工创建 haps 文件夹，再复制 HAP 包到该文件夹下。

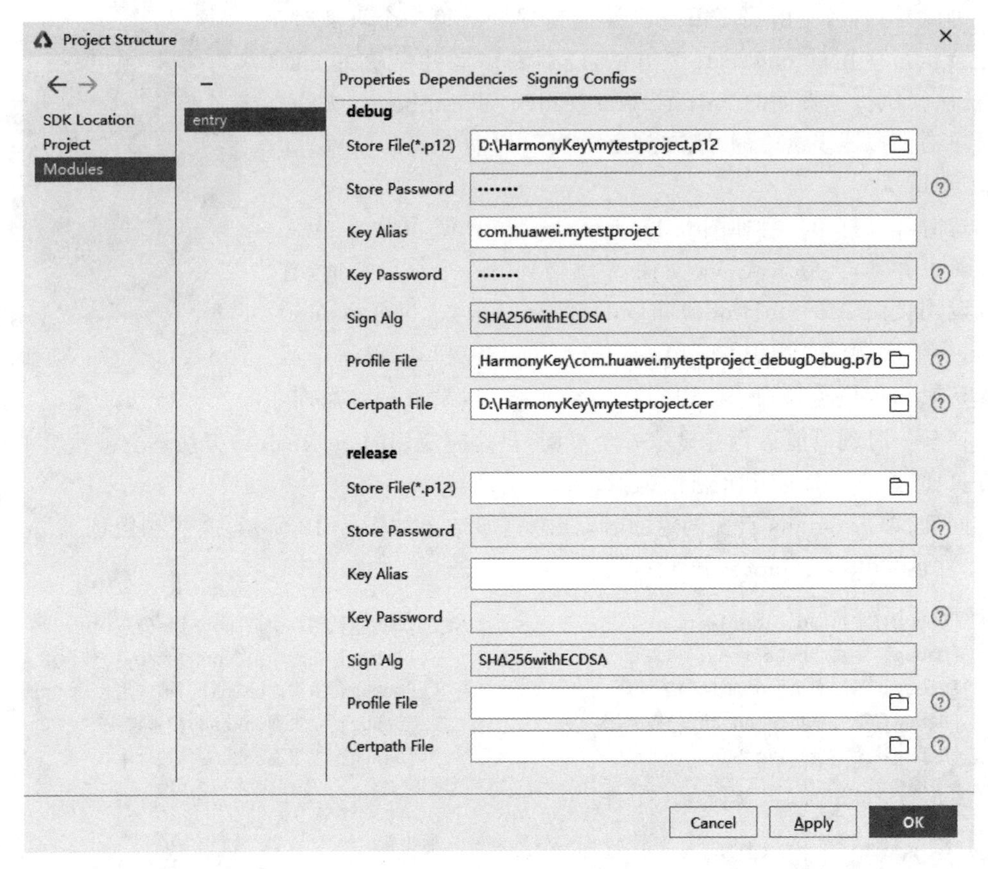

图 6.19　配置签名信息

随后通过蓝牙，将运动手表与华为手机进行连接。进入运动健康 App，在“设备”页签中，单击添加设备按钮，如图 6.20 所示。

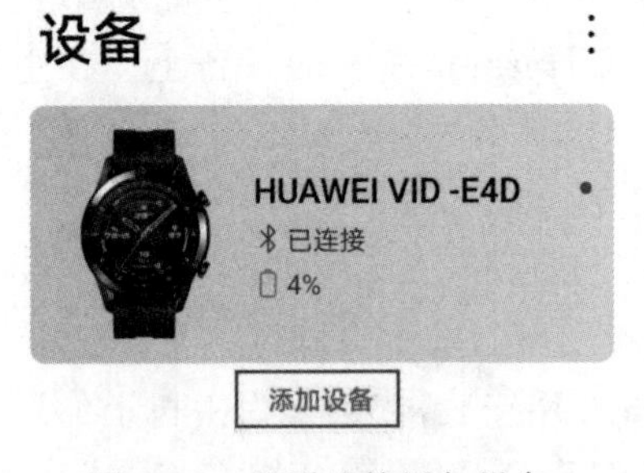

图 6.20　蓝牙连接添加设备

进入手表列表中，选择对应的运动手表型号。单击开始配对，按照页面指引完成手表与华为手机之间的连接，直至显示配对成功。此时打开应用调测助手 App，页面会显示已经与华为手机连接的运动手表设备。(如果运动手表与华为手机未连接，需单击应用调测助手

App 界面的连接设备按钮,手机会自动打开运动健康 App 添加设备)。单击应用调测助手 App 界面中的安装手表应用按钮,选择需要安装的 HarmonyOS 安装包进行安装,如图 6.21 所示。

图 6.21 应用调测助手 App 应用管理

安装完成后,打开运动手表应用程序列表,会发现已安装的应用图标,说明可成功运行该 HarmonyOS 应用。

第 7 章 Ability

14min

7.1 关于 Ability

Ability 即能力,代表了 HarmonyOS 工程中最重要最核心的功能。一个 HarmonyOS 应用可以包含多个 Ability,Ability 可以分为 Feature Ability(简称 FA)和 Particle Ability(简称 PA)两种类型,每种类型为开发者提供了不同的模板,以便实现不同的业务功能。FA 能力只支持 Page Ability,代表了 UI 的功能,PA 支持 Service Ability 和 Data Ability 两种模板,Service 模板用于提供后台运行任务的能力,Data 模板用于对外部提供统一的数据访问抽象。后面的章节将分别讲解这 3 种 Ability 的用法。

在使用 Ability 时必须在配置文件 config. json 中注册该 Ability,设置相应的属性,该文件存储在每个应用程序的 Java 代码的根目录中,具体结构如图 7.1 所示。

```
{
  "module": {
    "package": [...],
    "name": [...],
    "reqCapabilities": [...],
    "deviceType": [...],
    "distro":[...] ,
    "abilities": [
      {
        "name": [...],
        "description": [...],
        "icon": [...],
        "label": [...],
        "launchType": [...],
        "orientation": [...],
        "visible": [...],
        "skills": [...],
        "type": "page",
        "formEnabled": [...]
      }
    ],
    "reqPermissions": [...]
  }
```

图 7.1 Ability 注册配置示例

7.2 Page Ability

7.2.1 概述

在 HarmonyOS 手机上完成一个应用，首先必不可少的是 UI 页面，UI 就是程序的窗户，UI 页面不仅仅展示了程序运行的结果，一切用户与程序的交互也都通过 UI 来完成，接下来学习一下 UI 到底是如何生成的。FA 旨在与用户进行交互，Page Ability 是 FA 唯一支持的模板，也就是说 Page Ability 承载了 UI 的功能。一个 Page 实例可以包含一组相关页面，每个页面用一个 AbilitySlice 实例表示。当一个 Page 由多个 AbilitySlice 共同构成时，这些 AbilitySlice 页面提供的业务能力应具有高度相关性。例如，新闻浏览功能可以通过一个 Page 实现，其中包含了两个 AbilitySlice：一个 AbilitySlice 用于展示新闻列表，另一个 AbilitySlice 用于展示新闻详情。Page 和 AbilitySlice 的关系如图 7.2 所示。

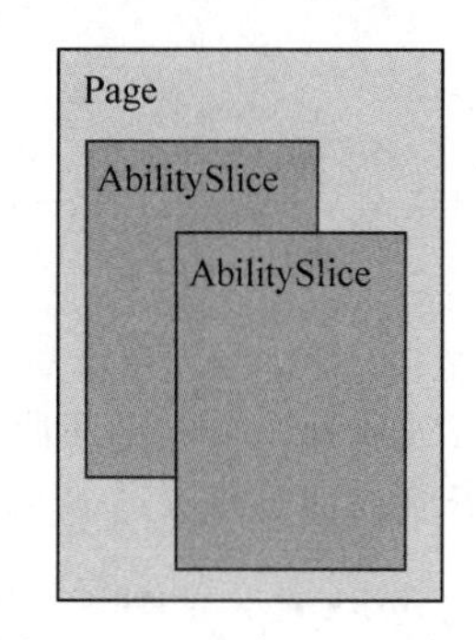

图 7.2 Page 与 AbilitySlice

相比于桌面场景，移动场景下应用之间的交互更为频繁。与单个应用专注于某个方面的能力开发情景类似，在需要其他能力辅助时，会调用其他应用提供的能力，例如，外卖应用提供了联系商家的业务功能入口，当用户在使用该功能时，会跳转到通话应用的拨号页面，HarmonyOS 支持不同 Page 之间的跳转，并可以指定跳转到目标 Page 中某个具体的 AbilitySlice。

7.2.2 路由配置

虽然一个 Page 可以包含多个 AbilitySlice，但是 Page 进入前台时页面默认只展示一个 AbilitySlice。默认展示的 AbilitySlice 通过 setMainRoute()方法来指定。接下来新建一个项目，看一看 AbilitySlice 是如何完成 UI 显示功能的。

首先选择 IDE 左边的 File→New→New Project，如图 7.3 所示。

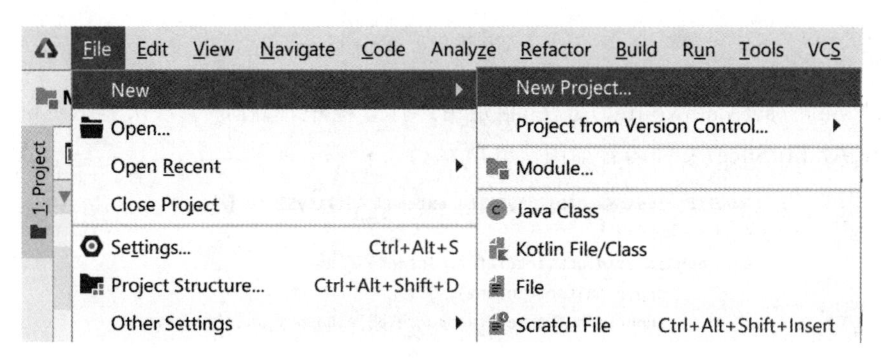

图 7.3 新建 Project 导向

在所弹出的页面选择创建一个空的 FA，如图 7.4 所示。

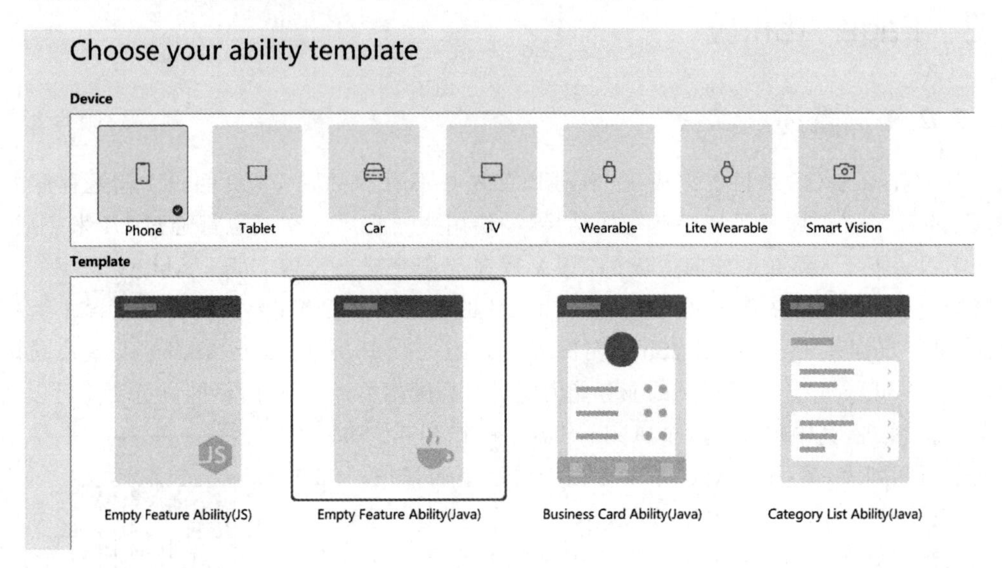

图 7.4　选择一个空白 FA 模板

完成新项目的创建后，单击左侧工程目录中的 MainAbility（该 Ability 为 Page Ability），可以看到代码如图 7.5 所示。

```
public class MainAbility extends Ability {
    @Override
    public void onStart(Intent intent) {
        super.onStart(intent);
        super.setMainRoute(MainAbilitySlice.class.getName());
    }
}
```

图 7.5　MainAbility 文件内容

其中，图中所选中的代码如下：

```
super.setMainRoute(MainAbilitySlice.class.getName());
```

通过 super.setMainRoute()方法，指定用户打开程序后，程序会默认展示 AbilitySlice，继续打开 AbilitySlice，文件内容如图 7.6 所示。

```
public class MainAbilitySlice extends AbilitySlice {
    @Override
    public void onStart(Intent intent) {
        super.onStart(intent);
        super.setUIContent(ResourceTable.Layout_ability_main);
    }
```

图 7.6　MainAbilitySlice 文件内容

AbilitySlice.java 中只有一行逻辑代码，在 Java UI 章节中讲解过，这行代码将 Layout_ability_main.xml 布局文件应用到这个 Slice 中，作为这个 Slice 的根布局，代码如下：

```
super.setUIContent(ResourceTable.Layout_ability_main);
```

打开 Layout_ability_main.xml 布局文件，文件内容如图 7.7 所示。

```
<?xml version="1.0" encoding="utf-8"?>
<DirectionalLayout
    xmlns:ohos="http://schemas.huawei.com/res/ohos"
    ohos:height="match_parent"
    ohos:width="match_parent"
    ohos:orientation="vertical">

    <Text
        ohos:id="$+id:text_helloworld"
        ohos:height="match_content"
        ohos:width="match_content"
        ohos:background_element="$graphic:background_ability_main"
        ohos:layout_alignment="horizontal_center"
        ohos:text="Hello World"
        ohos:text_size="50"
    />

</DirectionalLayout>
```

图 7.7 Layout_ability_main.xml 文件内容

从图 7.7 中可以看到根布局为 DirectionalLayout，在其中放置了一个 Text 组件，用于展示 Hello World。运行程序，所得效果如图 7.8 所示。

看完默认展示的界面后，还可以添加其他的 AbilitySlice，通过 addActionRoute()方法为此 AbilitySlice 配置一条路由规则。此时，当其他 Page 实例期望导航到此 AbilitySlice 时，可以在 Intent 中指定 Action，本章 7.2.4 节将进行详细介绍。

7.2.3 Page 与 AbilitySlice 的生命周期

实际上，Page 实例生命周期内的状态随着系统管理或用户操作等行为而不断转换。为了使 Page 正确应对类似于资源释放等的一系列状态变化，Ability 类提供了回调机制。这种回调机制能让 Page 及时感知系统管理或用户操作等外界变化，有助于提升应用的性能及稳健性。下面将具体讲解 Page 实例生命周期中的不同状态，以及不同状态之间的转换方式。

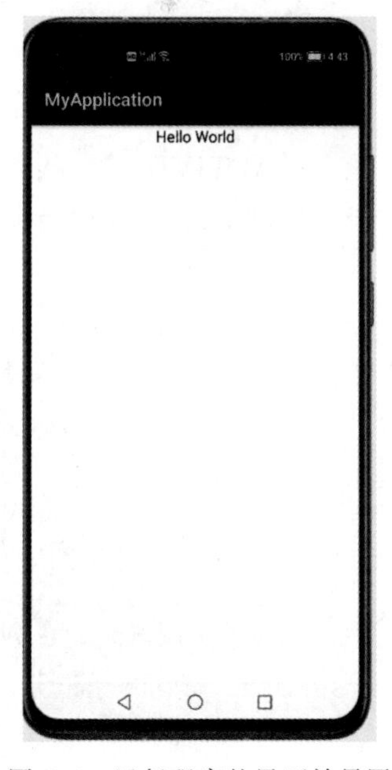

图 7.8 运行程序的界面效果图

Page 的生命周期如图 7.9 所示。

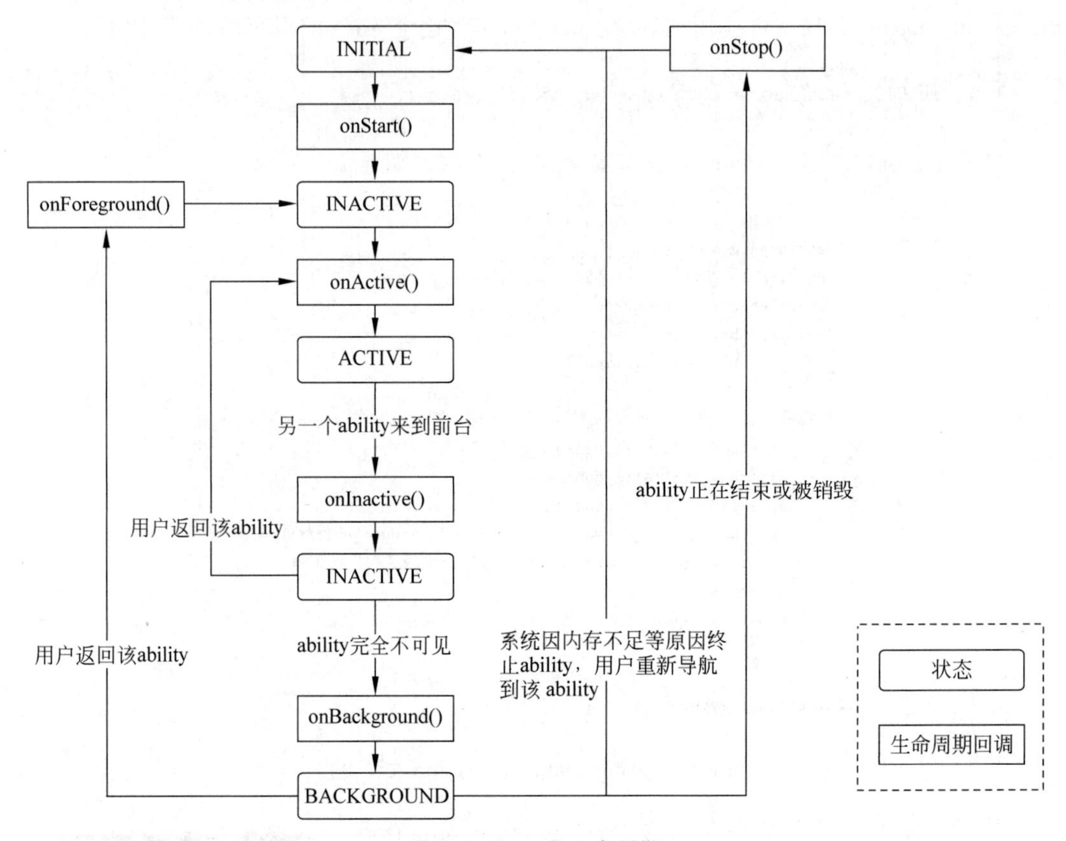

图 7.9 Page 的生命周期

由图 7.9 可见，Page 实例生命周期内的状态包括 INITIAL（初始状态）、INACTIVE（可见状态）、ACTIVE（前台状态）及 BACKGROUND（后台状态）。Ability 类提供了一系列回调机制协助 Page 实例进行状态转换，回调方法及作用如下：

1）onStart()

当系统首次创建 Page 实例时，触发 onStart()回调，Page 由此进入 INACTIVE 状态。需要注意，onStart()回调在 Page 实例生命周期过程中仅触发一次。此外，由于需要在此配置默认展示 AbilitySlice，因此开发者必须重写该方法。将 MainAbilitySlice 设置为默认的展示页面，代码如下：

```
@Override
public void onStart(Intent intent) {
    super.onStart(intent);
    setMainRoute(MainAbilitySlice.class.getName());
}
```

2) onActive()

Page 进入 INACTIVE 状态后来到前台,系统调用 onActive()使其进入 ACTIVE 状态,从而实现与用户进行交互。然而,当用户单击返回键或发生其他某类事件时,会导致 Page 失去焦点(对用户不可操作,可能可见,也可能不可见)从而触发 onInactive()回调使其回到 INACTIVE 状态,否则 Page 将持续保持在 ACTIVE 状态。

3) onInactive()

当 Page 失去焦点时,系统会调用 onInactive()方法使其进入 INACTIVE 状态。此后,Page 也可以触发 onActive()回调回到 ACTIVE 状态,因此,开发者通常需要成对实现 onActive()和 onInactive(),并在 onActive()中获取在 onInactive()中被释放的资源。此外,开发者可以在此回调中实现 Page 失去焦点时应表现的恰当行为。常见的场景有从屏幕顶部拉出 Notification 或者被非全屏的 Page 不完全遮挡等。

4) onBackground()

如果 Page 不再对用户可见,系统将调用 onBackground()回调,Page 由此进入 BACKGROUND 状态。需要注意的是,开发者应该在此方法中释放 Page 不可见时的无用资源,或执行较为耗时的状态保存操作。

5) onForeground()

Page 处于 BACKGROUND 状态时仍然驻留在内存中,当进行一些操作触发 Page 回到前台(例如用户重新导航到该 Page)时系统首先调用 onForeground()方法通知开发者 Page 即将回到前台,而后 Page 才回到 INACTIVE 状态。需要说明的是,onBackground()中释放掉的资源需要在 onForeground()回调中重新申请。最后 Page 进一步回到 ACTIVE 状态,系统会通过 onActive()回调通知开发者。

6) onStop()

系统调用 onStop()回调销毁 Page,并通知用户释放系统资源。Page 被销毁的原因可能包括以下几点:

(1) 用户使用系统管理功能关闭指定 Page,例如使用任务管理器关闭 Page。

(2) 用户的某些行为触发 Page 的 terminateAbility()方法调用,例如用户使用了应用的退出功能。

(3) 配置变更引发系统暂时销毁 Page 并重建。

(4) 系统进行资源管理时自动销毁处于 BACKGROUND 状态的 Page。

接下来介绍 AbilitySlice 的生命周期。作为 Page 的组成单元,AbilitySlice 的生命周期依托于其所属 Page 的生命周期,它们具有相同的生命周期状态和同名的回调。当 Page 生命周期发生变化时,它的 AbilitySlice 也会发生相应的生命周期变化。此外,AbilitySlice 还具有独立于 Page 的生命周期变化,以保证在同一 Page 中的不同 AbilitySlice 之间导航时,Page 的生命周期状态不会改变。

AbilitySlice 的生命周期回调与 Page 的相应回调类似,因此不再赘述。由于 AbilitySlice 承载具体的页面,开发者必须重写 AbilitySlice 的 onStart()回调,并在此方法中通过 setUIContent()方法设置页面，代码如下：

```
public class MainAbilitySlice extends AbilitySlice {

    private PositionLayout myLayout = new PositionLayout(this);

    @Override
    protected void onStart(Intent intent) {
        super.onStart(intent);
        setUIContent(myLayout);
    }
}
```

AbilitySlice 实例创建和管理通常由应用负责，系统仅在特定情况下创建 AbilitySlice 实例。例如，通过导航启动某个 AbilitySlice 时，由系统负责实例化，但是在同一个 Page 中不同的 AbilitySlice 间导航时则由应用负责实例化。

在了解完两者的生命周期之后，它们之间的关联又是什么呢？当 AbilitySlice 处于前台且具有焦点时，其生命周期状态随着所属 Page 的生命周期状态的变化而变化。当一个 Page 拥有多个 AbilitySlice 时，例如 MainAbility 下有 FooAbilitySlice 和 BarAbilitySlice，当前 FooAbilitySlice 处于前台并获得焦点，并即将导航到 BarAbilitySlice，在此期间的生命周期状态变化顺序如下：

(1) FooAbilitySlice 从 ACTIVE 状态变为 INACTIVE 状态。

(2) BarAbilitySlice 则从 INITIAL 状态首先变为 INACTIVE 状态，然后变为 ACTIVE 状态(假定此前 BarAbilitySlice 未曾启动)。

(3) FooAbilitySlice 从 INACTIVE 状态变为 BACKGROUND 状态。

对应两个 Slice 的生命周期方法回调顺序为 FooAbilitySlice.onInactive()→BarAbilitySlice.onStart()→BarAbilitySlice.onActive()→FooAbilitySlice.onBackground()。

需要注意的是，在整个流程中，MainAbility 始终处于 ACTIVE 状态，但是，当 Page 被系统销毁时，其所有已实例化的 AbilitySlice 将联动销毁，而不仅是处于前台的 AbilitySlice。

接下来通过实例来验证上述生命周期。

首先新建两个 Slice：FooAbilitySlice 和 BarAbilitySlice，然后在两个 Slice 中分别初始化 HiLogLabel，然后在上述生命周期中打上 Hilog，这里只展示其中一个 Slice 的写法，另一个只需更改名字，FooAbilitySlice 的实现代码如下：

```
//FooAbilitySlice 的实现代码
public class FooAbilitySlice extends AbilitySlice {
private HiLogLabel hiLogLabel = new HiLogLabel(HiLog.LOG_APP, 0, "Slicetest");
    @Override
```

```
    public void onStart(Intent intent) {
        super.onStart(intent);
        super.setUIContent(ResourceTable.Layout_ability_main);
HiLog.info(hiLogLabel, " FooAbilitySlice onStart 运行中");
    }
    @Override
    public void onActive() {
        super.onActive();
HiLog.info(hiLogLabel, " FooAbilitySlice onActive 运行中");
    }
    @Override
    public void onInactive() {
        super.onInactive();
HiLog.info(hiLogLabel, "FooAbilitySlice onInactive 运行中");
    }
    @Override
    public void onBackground() {
        super.onBackground();
HiLog.info(hiLogLabel, "FooAbilitySlice onBackground 运行中");
    }
    @Override
    public void onForeground(Intent intent) {
        super.onForeground(intent);
HiLog.info(hiLogLabel, "FooAbilitySlice onForeground 运行中");

    }
    @Override
    public void onStop() {
        super.onStop();
HiLog.info(hiLogLabel, "FooAbilitySlice onStop 运行中");
    }
}
```

然后需要实现页面跳转功能，在 onStart()中可以初始化一个 Button 实现此功能，通过把 Button 写在 XML 文件中实现，代码如下：

```
//初始化 Button 实现页面跳转
<?xml version = "1.0" encoding = "utf - 8"?>
< DirectionalLayout
    xmlns:ohos = "http://schemas.huawei.com/res/ohos"
    ohos:height = "match_parent"
    ohos:width = "match_parent"
    ohos:orientation = "vertical">

< Text
```

```
            ohos:id = "$ + id:text_helloworld"
            ohos:height = "match_content"
            ohos:width = "match_content"
            ohos:background_element = "$ graphic:background_ability_main"
            ohos:layout_alignment = "horizontal_center"
            ohos:text = "FooAbilitySlice"
            ohos:text_size = "150"
        />
    <Button
            ohos:id = "$ + id:change"
            ohos:height = "match_content"
            ohos:width = "match_content"
            ohos:text = "跳转"
            ohos:text_size = "150"
            ohos:background_element = "$ graphic:Button"
            ohos:layout_alignment = "center"
            />
    </DirectionalLayout>
```

在 onStart()函数中获得 XML 文件中的 Button，实现跳转功能(BarAbilitySlice 将在 7.2.4 节中详细介绍)：

```
    @Override
        public void onStart(Intent intent) {
            super.onStart(intent);
            DirectionalLayout directionalLayout  =  (DirectionalLayout) LayoutScatter.getInstance
    (this).parse(ResourceTable.Layout_ability_main,null,false);
    Button button = (Button) directionalLayout.findComponentById(ResourceTable.Id_change);
            button.setClickedListener(new Component.ClickedListener() {
                @Override
                public void onClick(Component component) {
    present(new BarAbilitySlice(),new Intent());
                }
            });
            super.setUIContent(directionalLayout);
    HiLog.info(hiLogLabel, " FooAbilitySlice onStart 运行中");
        }
```

运行后可以看到效果如图 7.10 所示。

打开 HiLog，选择 Info 模式，在查询栏中输入 Slicetest，可以看到生命周期信息如图 7.11 所示。

由图 7.11 可知，在 FooAbilitySlice 开始执行时 onStart()被触发，当 FooAbility 前台显示时 onActive()被触发。

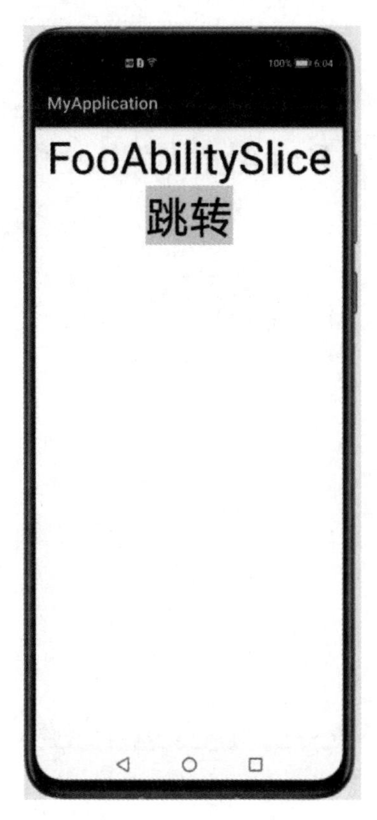

图 7.10 运行效果图

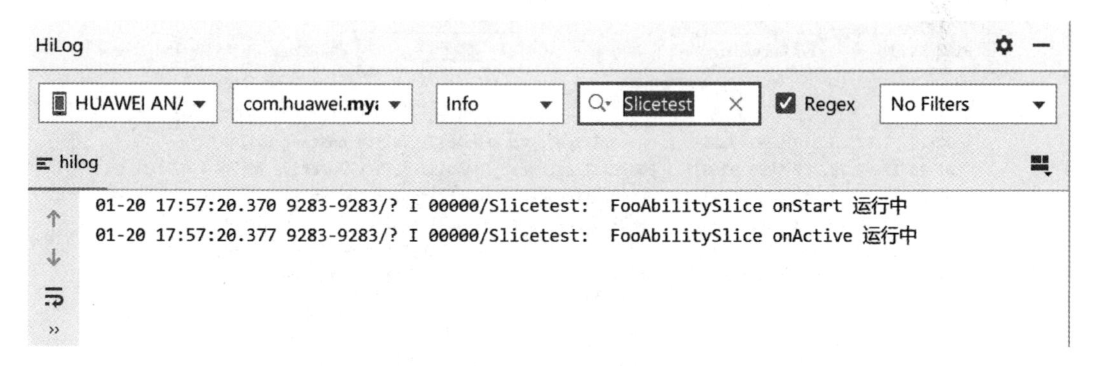

图 7.11 生命周期展示图

然后单击“跳转”按钮，运行效果如图 7.12 所示。

此时跳转到 BarAbilitySlice 页面，继续查看 HiLog，如图 7.13 所示。

可以看到 FooAbilitySlice 从 ACTIVE 状态变为 INACTIVE 状态；BarAbilitySlice 则从 INITIAL 状态首先变为 INACTIVE 状态，然后变为 ACTIVE 状态；FooAbilitySlice 从 INACTIVE 状态变为 BACKGROUND 状态。

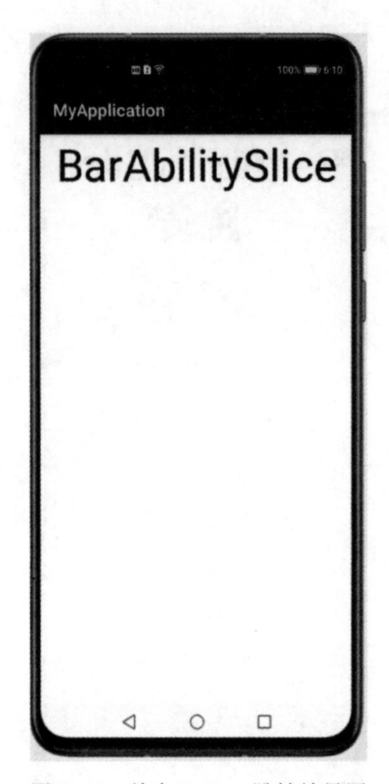

图 7.12　单击 Button 跳转效果图

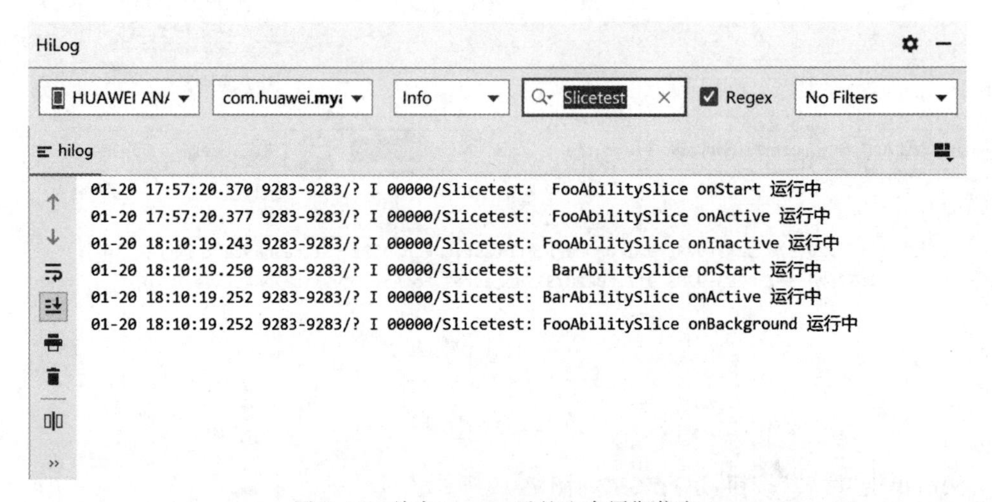

图 7.13　单击 Button 后的生命周期信息

接下来把页面暂时切换到菜单状态，让页面暂时处于后台状态，效果如图 7.14 所示。

查看 HiLog，可以发现在切换后台状态时，OnInactive()先被激活，然后 OnBackground()被激活，HiLog 信息如图 7.15 所示。

图 7.14 页面切换到菜单状态

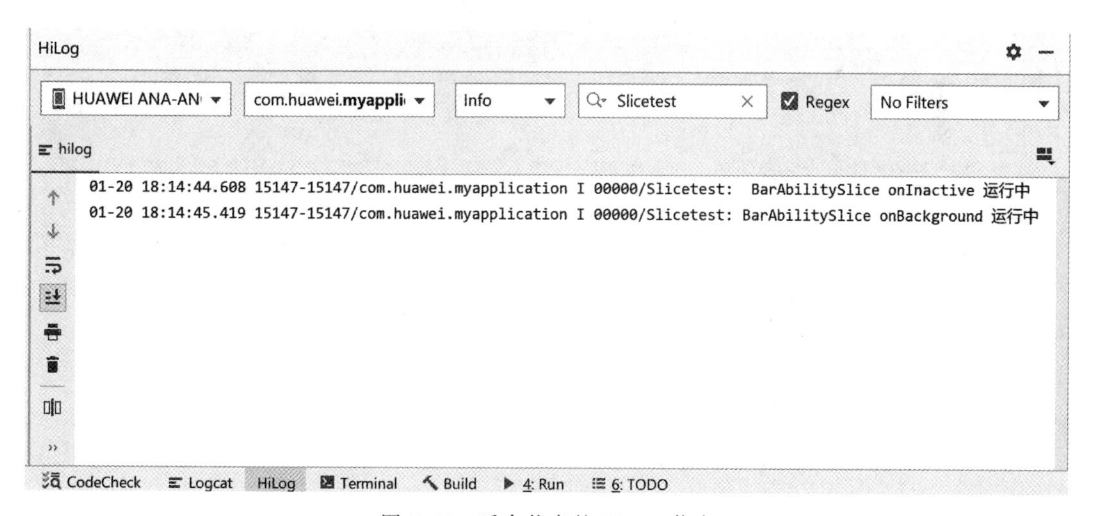

图 7.15 后台状态的 HiLog 信息

相应地，把手机再切换回 BarAbilitySlice 的主程序页面，效果如图 7.16 所示。

可以清楚地看到程序又从后台状态转换为前台状态，然后激活，展现在使用者面前，如图 7.17 所示。

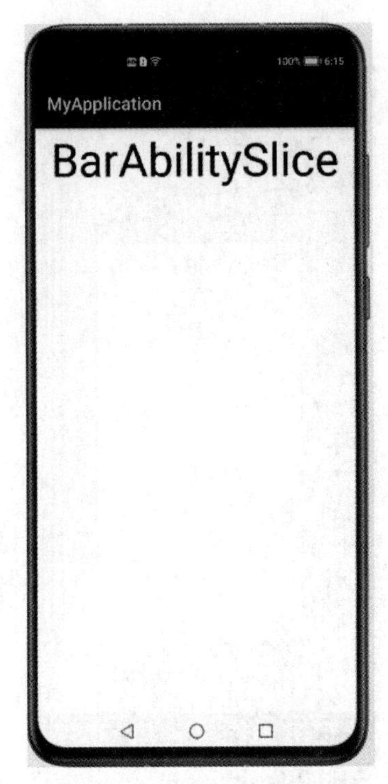

图 7.16　切换回 BarAbilitySlice 的主程序页面

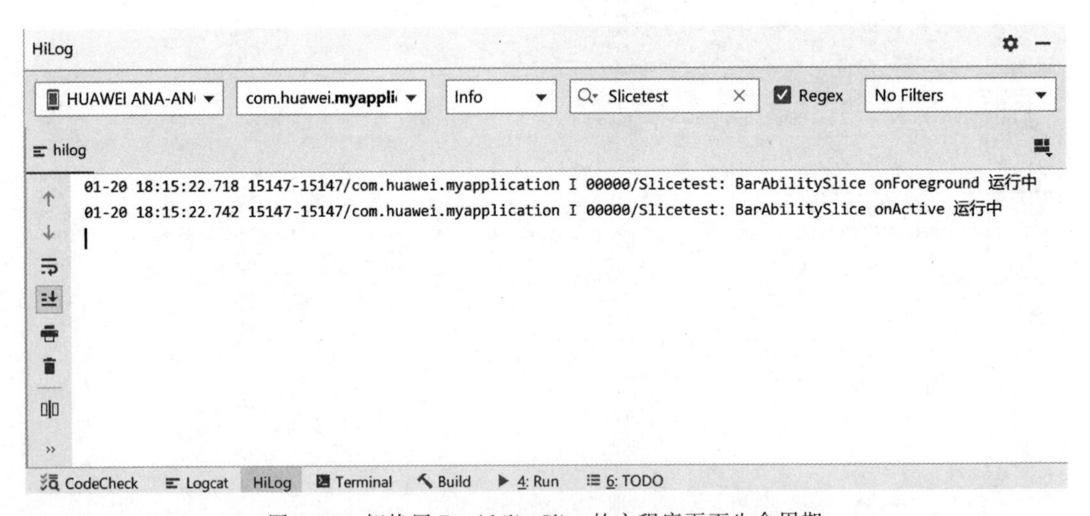

图 7.17　切换回 BarAbilitySlice 的主程序页面生命周期

最后，在后台中把这个程序上滑，关闭页面后回到了手机的主屏幕，如图 7.18 所示。

可以看到在切换关闭时，因为处在 BarAbilitySlice 的页面，所以 BarAbilitySlice 的状态先变为非激活状态，然后变为后台状态，不再展示在使用者面前，因为是直接关闭的，所以还会执行 onStop()，分别把两个页面都关闭，HiLog 信息如图 7.19 所示。

图 7.18 关闭程序后回到手机主屏幕

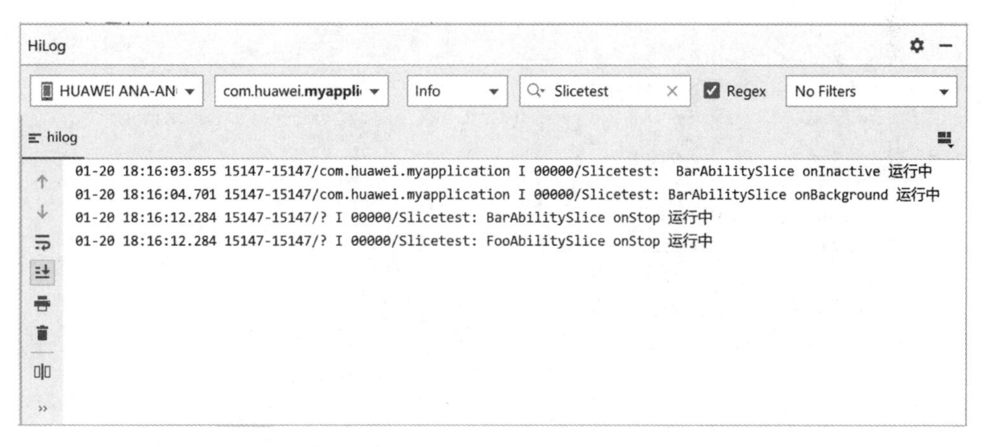

图 7.19 关闭程序后回到手机主屏幕的生命周期

7.2.4 Page 间导航

7.2.3 节中已经提到，系统管理或用户操作等行为均会引起 Page 实例在其生命周期的不同状态之间进行转换，而在这个过程中，一般需要在不同的 AbilitySlice 之间进行切换，因此，开发者需要掌握在不同的 AbilitySlice 之间进行导航的方法。此处分为两种情况进行讨论：同一 Page 内进行导航和不同 Page 间进行导航。

先从同一 Page 内说起。当发起导航的 AbilitySlice 和导航目标的 AbilitySlice 处于同一个 Page 时，可以通过 present()方法实现导航。7.2.3 节中的实例就是一个通过单击按钮从 FooAbilitySlice 导航到 BarAbilitySlice，回顾代码如下：

```
@Override
public void onStart(Intent intent) {
        super.onStart(intent);
        DirectionalLayout directionalLayout = (DirectionalLayout) LayoutScatter.getInstance
(this).parse(ResourceTable.Layout_ability_main,null,false);
         Button button = (Button) directionalLayout.findComponentById(ResourceTable.Id_
change);
        button.setClickedListener(new Component.ClickedListener() {
            @Override
            public void onClick(Component component) {
present(new BarAbilitySlice(),new Intent());
            }
        });
        super.setUIContent(directionalLayout);
}
```

效果在上个示例已有体现，实现了从 FooAbilitySlice 到 BarAbilitySlice 的跳转，因此不再赘述，接下来给出 BarAbilitySlice 的示例代码如下：

```
//BarAbilitySlice 的代码示例
public class BarAbilitySlice extends AbilitySlice {
    private HiLogLabel hiLogLabel = new HiLogLabel(HiLog.LOG_APP, 0, "Slicetest");

    @Override
    public void onStart(Intent intent) {
        super.onStart(intent);
        super.setUIContent(ResourceTable.Layout_se);
        HiLog.info(hiLogLabel, " BarAbilitySlice onStart 运行中");
    }

    @Override
    public void onActive() {
        super.onActive();
        HiLog.info(hiLogLabel, "BarAbilitySlice onActive 运行中");
    }

    @Override
```

```
    public void onInactive() {
        super.onInactive();
        HiLog.info(hiLogLabel, " BarAbilitySlice onInactive 运行中");
    }

    @Override
    public void onBackground() {
        super.onBackground();
        HiLog.info(hiLogLabel, "BarAbilitySlice onBackground 运行中");
    }

    @Override
    public void onForeground(Intent intent) {
        super.onForeground(intent);
        HiLog.info(hiLogLabel, "BarAbilitySlice onForeground 运行中");
    }

    @Override
    public void onStop() {
        super.onStop();
        HiLog.info(hiLogLabel, "BarAbilitySlice onStop 运行中");
    }
}
```

其中,BarAbilitySlice 根布局的 xml 文件代码如下:

```
//BarAbilitySlice 根布局的 xml 文件

<?xml version = "1.0" encoding = "utf - 8"?>
< DirectionalLayout
    xmlns:ohos = "http://schemas.huawei.com/res/ohos"
    ohos:height = "match_parent"
    ohos:width = "match_parent"
    ohos:orientation = "vertical">

< Text
        ohos:id = " $ + id:two"
        ohos:height = "match_content"
        ohos:width = "match_content"
        ohos:background_element = " $ graphic:background_ability_main"
        ohos:layout_alignment = "center"
        ohos:text = "BarAbilitySlice"
        ohos:text_size = "150"
    />
```

接下来对 FooAbilitySlice 和 BarAbilitySlice 进行改写,使用户从导航目标 AbilitySlice

返回时能够获得其返回结果。如果要得到结果，则应当使用 presentForResult()实现导航。用户从导航目标 AbilitySlice 返回时，系统将回调 onResult()来接收和处理返回结果，开发者需要重写该方法。返回结果由导航目标 AbilitySlice 在其生命周期内通过 setResult()进行设置。将 presentForResult()的第一个参数设置为 BarAbilitySlice，识别码设置为 0，代码如下：

```
//改写 FooAbilitySlice
@Override
public void onStart(Intent intent) {
        super.onStart(intent);
        DirectionalLayout directionalLayout = (DirectionalLayout) LayoutScatter.getInstance
(this).parse(ResourceTable.Layout_ability_main,null,false);
         Button button = (Button) directionalLayout.findComponentById(ResourceTable.Id_
change);
        button.setClickedListener(new Component.ClickedListener() {
            @Override
            public void onClick(Component component) {
presentForResult(new BarAbilitySlice(), new Intent(), 0);
            }
        });
        super.setUIContent(directionalLayout);
```

然后在 FooAbilitySlice 中新增 onResult()函数，处理返回的结果，代码如下：

```
//在 FooAbilitySlice 中新增 onResult 函数
@Override
protected void onResult(int requestCode, Intent resultIntent) {
        if (requestCode == 0) {
            HiLog.info(hiLogLabel, "0 号返回的 intent 结果为");
            HiLog.info(hiLogLabel, String.valueOf(resultIntent.getFlags()));
        }
        if (requestCode == 1) {
            HiLog.info(hiLogLabel, "1 号返回的 intent 结果为");
            HiLog.info(hiLogLabel, String.valueOf(resultIntent.getFlags()));
        }
    }
}
```

同时在 BarAbilitySlice 中的 onStart()函数中也要加上如下代码：

```
Intent intentresult = new Intent();
intentresult.setFlags(12345);
setResult(intentresult);
terminate();
```

在 BarAbilitySlice 中初始化了一个 Intent，Intent 可以包含信息，以便传递信息，在 setResult 中将 Intent 设置后，Intent 中的信息就会被返回给前一个 Slice，最后 terminate 表示关闭这个 Slice，返回上一个 Slice，运行后单击“跳转”按钮，可以看到 BarAbilitySlice 中的信息被成功传递回来了，如图 7.20 所示，因为 presentForResult()中传入的询问码为 0，所以返回后会执行 0 号操作。

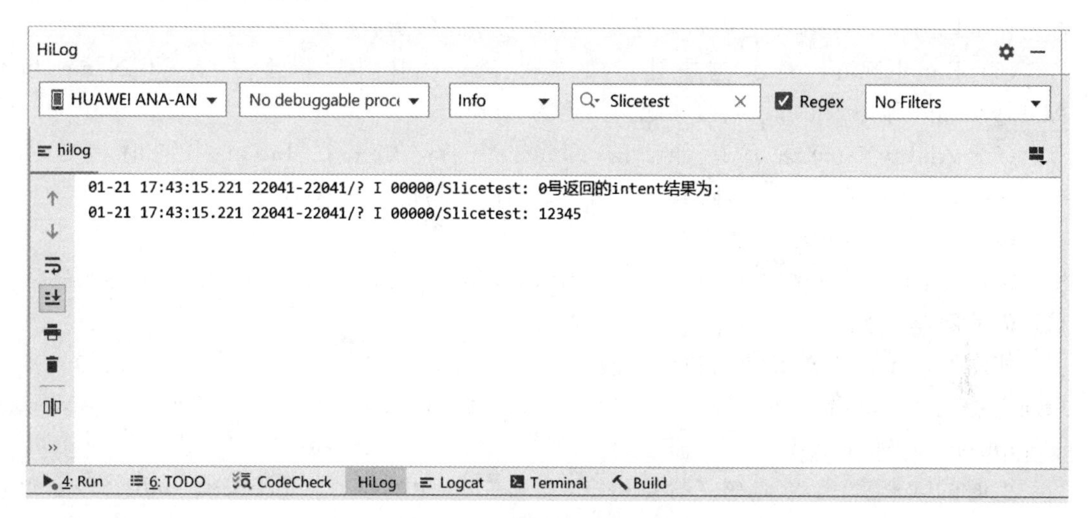

图 7.20　页面跳转后信息传递

系统为每个 Page 维护了一个 AbilitySlice 实例的栈，每个进入前台的 AbilitySlice 实例均会入栈。如果当开发者在调用 present()或 presentForResult()时指定的 AbilitySlice 实例已经在栈中存在，则栈中位于此实例之上的 AbilitySlice 均会出栈并终止其生命周期。

接下来讲述如何实现不同 Page 间的导航，不同 Page 中的 AbilitySlice 互不可见，因此无法通过 present()或 presentForResult()方法直接导航到其他 Page 的 AbilitySlice。AbilitySlice 作为 Page 的内部单元，以 Action 的形式对外暴露，因此可以通过配置 Intent 的 Action 导航到目标 AbilitySlice。Page 间的导航可以使用 startAbility()或 startAbilityForResult()方法，获得返回结果的回调为 onAbilityResult()。在 Ability 中调用 setResult()可以设置返回结果。

如果要实现不同 Page 间的导航，则必须介绍 Intent 的概念，因为其作为 Ability 之间传递信息的载体。当一个 Ability 需要启动另一个 Ability 时，可以通过 Intent 指定启动的目标，同时携带相关数据，或者当一个 AbilitySlice 需要导航到另一个 AbilitySlice 时，通过 Intent 指定导航目标，同时携带相关数据。Intent 的构成元素包括 Operation 与 Parameters。

其中，Operation 有 7 个子属性：

(1) Action 表示动作，通常使用系统预置 Action，应用也可以自定义 Action。例如 IntentConstants.ACTION_HOME 表示返回桌面动作。

(2) Entity 表示类别，通常使用系统预置 Entity，应用也可以自定义 Entity。例如 Intent.ENTITY_HOME 表示在桌面显示图标。

(3) Uri 表示 Uri 描述。如果在 Intent 中指定了 Uri，则 Intent 将匹配指定的 Uri 信息，包括 scheme、schemeSpecificPart、authority 和 path 信息。

(4) Flags 表示处理 Intent 的方式。例如 Intent.FLAG_ABILITY_CONTINUATION 标记在本地的一个 Ability 是否可以迁移到远端设备继续运行。

(5) BundleName 表示包描述。如果在 Intent 中同时指定了 BundleName 和 AbilityName，则 Intent 可以直接匹配到指定的 Ability。

(6) AbilityName 表示待启动的 Ability 名称。如果在 Intent 中同时指定了 BundleName 和 AbilityName，则 Intent 可以直接匹配到指定的 Ability。

(7) DeviceId 表示运行指定 Ability 的设备 ID。

Parameters 则是一种支持自定义的数据结构，开发者可以通过 Parameters 传递某些请求所需的额外信息。

使用 Intent 启动 Ability 可以分为两种方式，一种是显式启动，另一种是隐式启动。如果同时指定了 BundleName 与 AbilityName，则根据 Ability 的全称(例如 com.huaiwei.myapplication.MainAbility)可以直接启动应用，下面来看一个实例。

上面的实例里已经得到了 Mainability 和它包含的两个 Slice(BarAbilitySlice 和 FooAbilitySlice)，接下来将新建一个 SecondAbility，并在 BarAbilitySlice 中直接跳转到 SecondAbilitySlice，创建过程如图 7.21 所示。

SecondAbilitSlice 新建完毕后，修改它的显示内容为 Second Ability，代码如下：

```
<?xml version = "1.0" encoding = "utf - 8"?>
< DirectionalLayout
    xmlns:ohos = "http://schemas.huawei.com/res/ohos"
    ohos:height = "match_parent"
    ohos:width = "match_parent"
    ohos:orientation = "vertical">
< Text
        ohos:id = " $ + id:text_helloworld"
        ohos:height = "match_content"
        ohos:width = "match_content"
        ohos:background_element = " $ graphic:background_ability_second"
        ohos:layout_alignment = "horizontal_center"
ohos:text = "Second Ability"
        ohos:text_size = "150"
        />
</DirectionalLayout >
```

在 7.2.3 节创建的 FooAbilitySlice 中的 onStart()中加入代码如下：

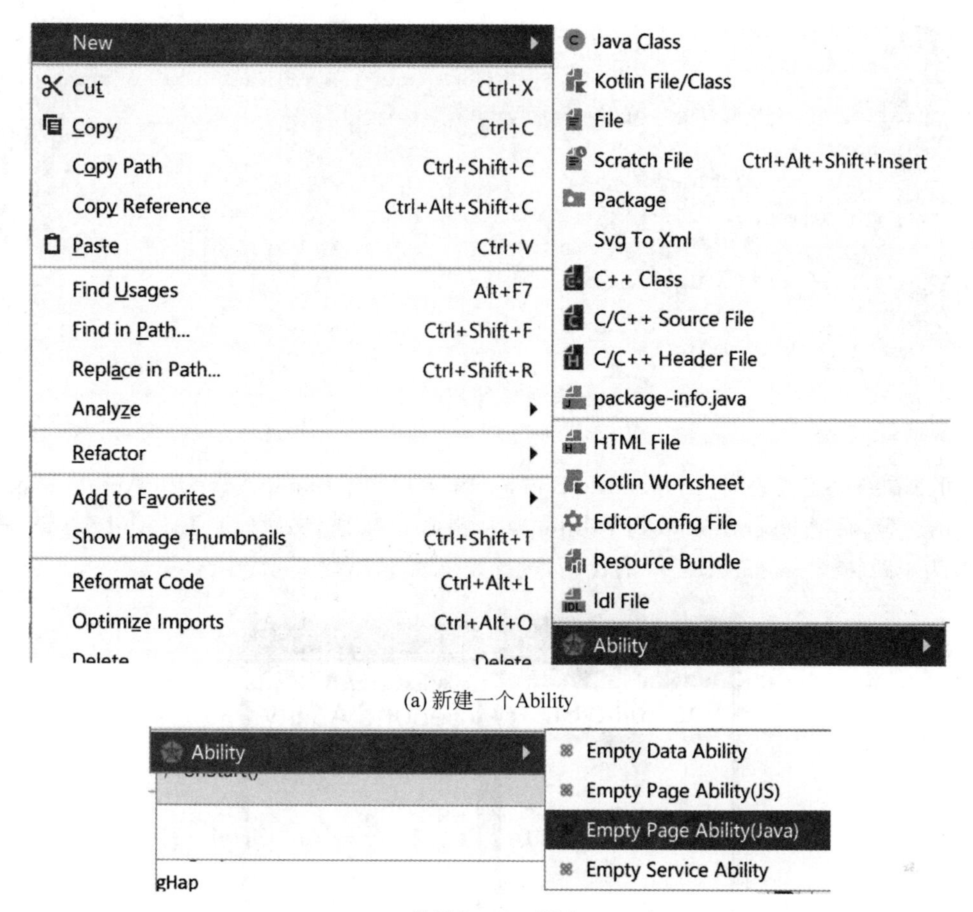

(a) 新建一个Ability

(b) 选择空Ability模板

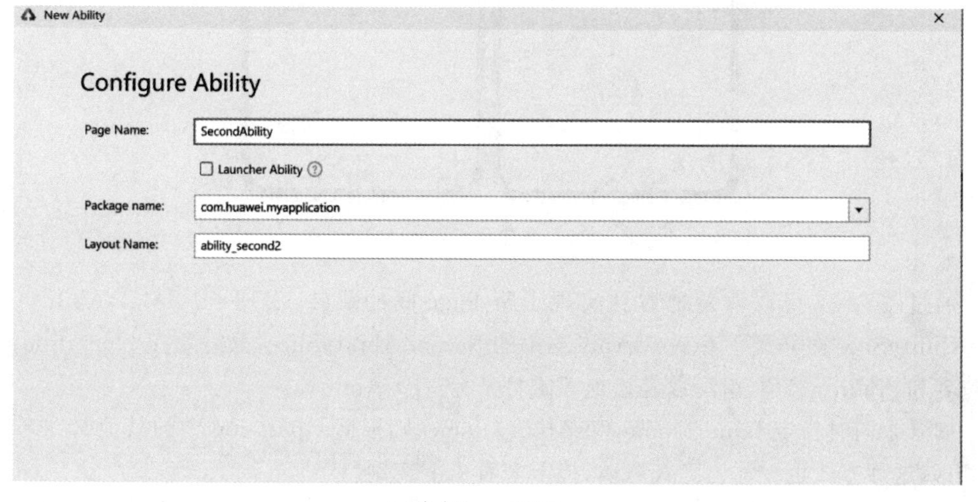

(c) 命名SecondAbility

图 7.21　新建一个 SecondAbility

```
//FooAbilitySlice.java
Operation operation = new Intent.OperationBuilder().withBundleName("com.huawei.
myapplication").withAbilityName("SecondAbility").build();
Intent intent1 = new Intent();
intent1.setOperation(operation);
button.setClickedListener(new Component.ClickedListener() {
    @Override
    public void onClick(Component component) {

    startAbility(intent1);
    }
});
```

可以看到上述代码利用 OperationBuilder 构筑了含有 BundleName 和 AbilityName 的 Operation，然后把 Operation 放到了 Intent 中，删掉了 7.2.3 节跳转到 BarAbilitySlice 的代码，改为了跳转到 SecondAbility 的代码，运行效果如图 7.22 所示。

图 7.22　跳转到 SecondAbility

通过图 7.22 可以清楚地看到实现了跨 Page 间的跳转，实现了从 MainAbility 中的 FooAbilitySlice 跳转到了 SecondAbility 中的 SecondAbilitySlice，只需要在 Operation 中指定包名和 Ability 名称，就可以轻松地显式打开另一个 Ability。

如果未同时指定 BundleName 和 AbilityName，则根据 Operation 中的其他属性来启动 Ability，这种方式也被称为隐式启动，接下来看一下实例。

首先，在 SecondAbility 中配置路由以便支持以此 action 导航到对应的 AbilitySlice，代码如下：

```
//SecondAbility.java
public class SecondAbility extends Ability {
    @Override
    public void onStart(Intent intent) {
        super.onStart(intent);
        super.setMainRoute(SecondAbilitySlice.class.getName());
addActionRoute("action.change",
                SecondAbilitySlice.class.getName());
    }
}
```

当然 addActionRoute()方法中使用的动作命名，同样需要在 config.json 中注册，打开 config.json 文件找到 SecondAbility 对应的配置，在 Skill 中添加对应的 action，将 action 命名为 change，代码如下：

```
{
        "orientation": "unspecified",
        "name": "com.huawei.myapplication.SecondAbility",
        "icon": "$media:icon",
"skills": [
            {
"entities": [
                "entity.system.home"
               ],
               "actions": [
                "action.change"
               ]
      }
        ],
        "description": "$string:secondability_description",
        "label": "entry",
        "type": "page",
        "launchType": "standard"
      }
```

这样，SecondAbility 的路由就配置完毕了，其他的 Ability 可以通过指定它的 action 来打开，在 FooAbilitySlice 中完成接下来的操作，修改代码如下：

```
//FooAbilitySlice.java
Operation operation = new Intent.OperationBuilder().withAction("action.change").build();
Intent intent1 = new Intent();
intent1.setOperation(operation);
button.setClickedListener(new Component.ClickedListener() {
       @Override
       public void onClick(Component component) {
```

```
startAbility(intent1);
        }
});
```

在 Operation 中配置 action，只不过这次使用的是 SecondAbility 中注册过的 action.change，可以看到使用 action 隐式打开 Ability 的方式也成功了，效果如图 7.23 所示。

图 7.23　使用 action 方式进行跳转

有些场景下，开发者需要在应用中使用其他应用提供的某种能力，而不感知提供该能力具体是哪一个应用。例如开发者需要通过浏览器打开一个链接，而不关心用户最终选择哪一个浏览器应用，则可以通过 Operation 的其他属性（除 BundleName 与 AbilityName 之外的属性）描述需要的能力。如果设备上存在多个应用提供同种能力，系统则弹出候选列表，由用户选择由哪个应用处理请求。至此，Page Ability 已经全部介绍完毕，开发者可以将前面介绍的 Component 应用在自己的 Page Ability 中，从而自定义想要的 UI 效果。

16min

7.3 线程

7.3.1 概述

在讲解 HarmonyOS 中的线程之前，先简单介绍关于线程的相关概念。

在介绍线程之前，首先需要了解进程。进程是程序的实体，是操作系统进行资源分配和调度的基本单位。通俗易懂地理解，进程是指系统中正在运行的一个应用程序，例如打开手机中的一个 App 就开启了一个进程，而线程是操作系统能够进行运算调度的最小单位。通常，线程是进程的实际运作单位，进程是线程的容器。一个线程是指进程中单一顺序的控制流，一个进程中可以并发多个线程，每个线程可以并行执行不同的任务。例如，打开微信程序后进行的查看朋友圈、收付款等多种操作都属于线程。实际上，在启动应用时，系统会为该应用创建一个称为“主线程”的执行线程，该线程随着应用创建或消失，是应用的核心线程。UI界面的显示和更新等操作都是在主线程上进行的，因此，主线程又称为UI线程，默认情况下，所有的操作都是在主线程上执行的，而其他比较耗时的任务，如文件下载、查询数据库等都可以通过创建其他线程来处理。

线程的主要特性有如下4点：

(1) 并发性：同一进程的多个线程可在一个或多个处理器上并发或并行地执行，从而进程之间的并发执行演变为不同进程的线程之间的并发执行。

(2) 共享性：同一个进程中的所有线程共享，但不拥有进程的状态和资源，且驻留在进程的同一个主存储地址空间中，可以访问相同的数据，所以需要有线程之间的通信和同步机制。稍后将介绍在HarmonyOS中如何进行线程的管理与通信。

(3) 动态性：线程是程序在相应数据集上的一次执行过程，由创建而产生，至撤销而消亡，有其生命周期，经历各种状态的变化。每个进程被创建时，至少同时为其创建一个线程，该线程就是主线程，并根据需要再创建其他线程。

(4) 结构性：线程是操作系统中的基本调度和分派单位，因此，它具有唯一的标识符和线程控制块，其中应包含调度所需的一切信息。

多线程编程具有高响应性，如果一个交互型应用程序采用多线程编程，则即使存在部分阻塞或需要执行冗长操作，应用程序仍可以继续执行，保持对用户的及时响应。例如浏览器可以将文件下载任务放置于后台，用户仍可以与浏览器进行交互。此外，由于线程能够共享所属进程的资源，所以创建和切换线程更加经济且高效。

7.3.2 线程管理

一个进程中可以并发多个线程，每个线程可以并行执行不同的任务。当应用的业务逻辑比较复杂并且需要创建多个线程来执行多个任务时，HarmonyOS提供了专门的任务分发器——TaskDispatcher接口来分发不同的任务，它是Ability进行任务分发的基本接口，隐藏任务所在线程的实现细节。

在HarmonyOS中，TaskDispatcher具有多种实现，每种实现对应着不同的任务分发器。为了保证应用的高响应性，在分发任务时可以指定任务的优先级，任务优先级按照获得执行的概率可分为HIGH(最高任务优先级)、DEFAULT(默认任务优先级)和LOW(低任务优先级)。由同一个任务分发器分发出的任务具有相同的优先级。需要说明的是，在UI线程上运行的任务默认以高优先级运行，如果某个任务无须等待结果，则可以用低优先级。

HarmonyOS提供的任务分发器主要包括以下4种：

(1) GlobalTaskDispatcher任务分发器，即全局并发任务分发器，由Ability执行getGlobalTaskDispatcher()获取。该任务分发器适用于任务之间没有联系的情况。一个应用只有一个GlobalTaskDispatcher，它在程序结束时才被销毁。

(2) ParallelTaskDispatcher 任务分发器，即并发任务分发器，由 Ability 执行 createParallelTaskDispatcher()创建并返回。ParallelTaskDispatcher 不具有全局唯一性，可以创建多个。

(3) SerialTaskDispatcher 任务分发器，即串行任务分发器，由 Ability 执行 createSerialTaskDispatcher()创建并返回。由该分发器分发的所有任务都按顺序执行，但执行这些任务的线程并不是固定的。其不适用于分发并行任务和相互之间没有依赖的任务，若要执行并行任务，则应使用 ParallelTaskDispatcher 或者 GlobalTaskDispatcher 进行任务分发，而不是创建多个 SerialTaskDispatcher。如果要分发的任务之间没有依赖，则使用 GlobalTaskDispatcher 进行任务分发，它的创建和销毁由开发者自己管理，开发者在使用期间需要持有该对象引用。

(4) SpecTaskDispatcher：专有任务分发器，绑定到专有线程上的任务分发器。前面所提到的主线程就是专有线程，相应的 UITaskDispatcher 和 MainTaskDispatcher 均属于 SpecTaskDispatcher。一般情况下建议使用 UITaskDispatcher，由它分发的任务均在主线程上按序执行。具体地，UITaskDispatcher 由 Ability 执行 getUITaskDispatcher()创建并返回，在应用程序结束时被销毁。MainTaskDispatcher 则由 Ability 执行 getMainTaskDispatcher()创建并返回。

接下来将介绍如何在 HarmonyOS 中利用上述任务分发器进行不同的任务分发。新建一个以 Phone 为设备、Java 为模板、名为 TaskDispatcher 的项目，进入 entry→main→java→com.huawei.taskdispatcher→slice→MainAbilitySlice 文件，覆写 onStart()方法，利用上述任务分发器实现任务分发功能。

1. 同步派发任务(syncDispatch)

派发任务并在当前线程等待任务执行完成，当前线程在给定任务完成之前被阻塞。具体实现代码如下：

```
//实现同步派发任务
//以 HiLog 形式显示任务的执行顺序
HiLogLabel label = new HiLogLabel(HiLog.LOG_APP ,200, "OSLog");
@Override
public void onStart(Intent intent) {
  super.onStart(intent);
  super.setUIContent(ResourceTable.Layout_ability_main);
  HiLog.fatal(label, "运行结果");

  //同步派发任务
  a0();

//异步派发任务
  a();
```

```
//异步延迟派发任务
    b();

//任务组：分组执行任务
    c();

  //同步设置屏障任务
d();

//异步设置屏障任务
    e();

    //执行多次任务
    f();

//取消任务：在任务执行前取消任务
     g();
}

//同步派发任务的实现
public void a0(){
TaskDispatcher globalTaskDispatcher = getGlobalTaskDispatcher(TaskPriority.DEFAULT);
     globalTaskDispatcher.syncDispatch(new Runnable() {
            @Override
            public void run() {
                HiLog.fatal(label, "sync task1 run");
            }
      });
      HiLog.fatal(label, "after sync task1");

      globalTaskDispatcher.syncDispatch(new Runnable() {
            @Override
            public void run() {
                HiLog.fatal(label, "sync task2 run");
            }
        });
      HiLog.fatal(label, "after sync task2");

      globalTaskDispatcher.syncDispatch(new Runnable() {
            @Override
            public void run() {
                HiLog.fatal(label, "sync task3 run");
            }
        });
      HiLog.fatal(label, "after sync task3");
}
//其他任务实现代码
...
```

在本例中，利用 GlobalTaskDispatcher 任务分发器同步派发了 3 个任务。首先 Ability 使用 getGlobalTaskDispatcher()获取优先级别为 DEFAULT 的任务分发器，接着利用 syncDispatch()方法同步分发了 3 个任务：task1、task2 和 task3，3 个任务实现了 Runnable 接口中的 run()方法。3 个任务的执行顺序以 HiLog 形式进行显示，结果如图 7.24 所示。

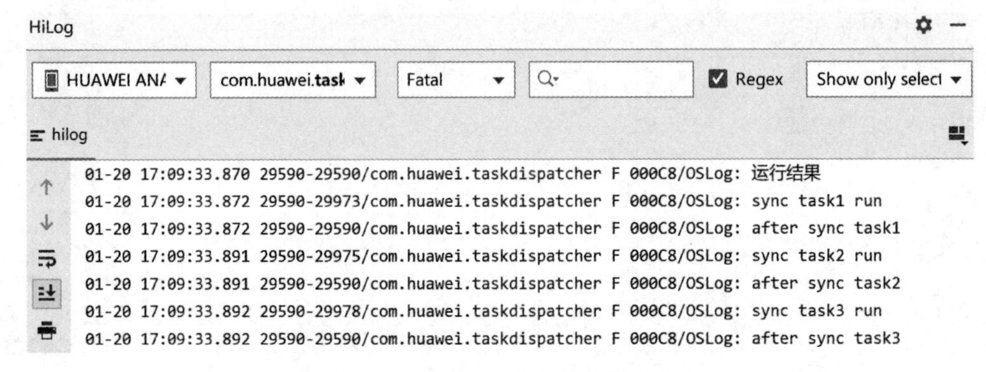

图 7.24 同步派发任务的执行结果

由图 7.24 可见，3 个任务同步执行。需要注意的是，如果对 syncDispatch 使用不当，将会导致死锁。可能导致死锁发生的情形包括以下几种：

(1) 在专有线程上，利用该专有任务分发器进行 syncDispatch。

(2) 在被某个串行任务分发器(dispatcher_a)派发的任务中，再次利用同一个串行任务分发器(dispatcher_a)对象派发任务。

(3) 在被某个串行任务分发器(dispatcher_a)派发的任务中，经过数次派发任务，最终又利用该(dispatcher_a)串行任务分发器派发任务。例如：dispatcher_a 派发的任务使用 dispatcher_b 进行任务派发，在 dispatcher_b 派发的任务中又利用 dispatcher_a 进行派发任务。

(4) 串行任务分发器(dispatcher_a)派发的任务中利用串行任务分发器(dispatcher_b)进行同步派发任务，同时 dispatcher_b 派发的任务中利用串行任务分发器(dispatcher_a)进行同步派发任务。在特定的线程执行顺序下将导致死锁。

2. 异步派发任务(asyncDispatch)

异步与同步相对，当一个异步过程调用发出后，调用者在没有得到结果之前，就可以继续执行后续操作。使用 GlobalTaskDispatcher 派发异步任务的代码如下：

```
//异步派发任务
public void onStart(Intent intent) {
...

//异步派发任务
        a();
```

```
    ...
}

//异步派发任务的实现
public void a(){
 TaskDispatcher globalTaskDispatcher = getGlobalTaskDispatcher(TaskPriority.DEFAULT);
    globalTaskDispatcher.asyncDispatch(new Runnable() {
            @Override
            public void run() {
                HiLog.fatal(label, "sync task1 run");
            }
        });
     HiLog.fatal(label, "after sync task1");

     globalTaskDispatcher.asyncDispatch(new Runnable() {
            @Override
            public void run() {
                HiLog.fatal(label, "sync task2 run");
            }
     });
     HiLog.fatal(label, "after sync task2");

     globalTaskDispatcher.asyncDispatch(new Runnable() {
            @Override
            public void run() {
                HiLog.fatal(label, "sync task3 run");
            }
        });
     HiLog.fatal(label, "after sync task3");
}
```

3 个异步任务的执行结果如图 7.25 所示。对比图 7.25(a)和图 7.25(b)可以看出 3 个异步任务的执行顺序是随机的。

3. 异步延迟派发任务(delayDispatch)

异步执行，函数立即返回，内部会在延时指定时间后将任务派发到相应队列中。延时时间参数仅代表在这段时间以后任务分发器会将任务加入队列中，任务的实际执行时间可能晚于这个时间。具体比这个数值晚多久，取决于队列及内部线程池的繁忙情况。在下面的例子中，延时时间设置为 3.15s，代码如下：

```
//异步派发任务
public void onStart(Intent intent) {
...
```

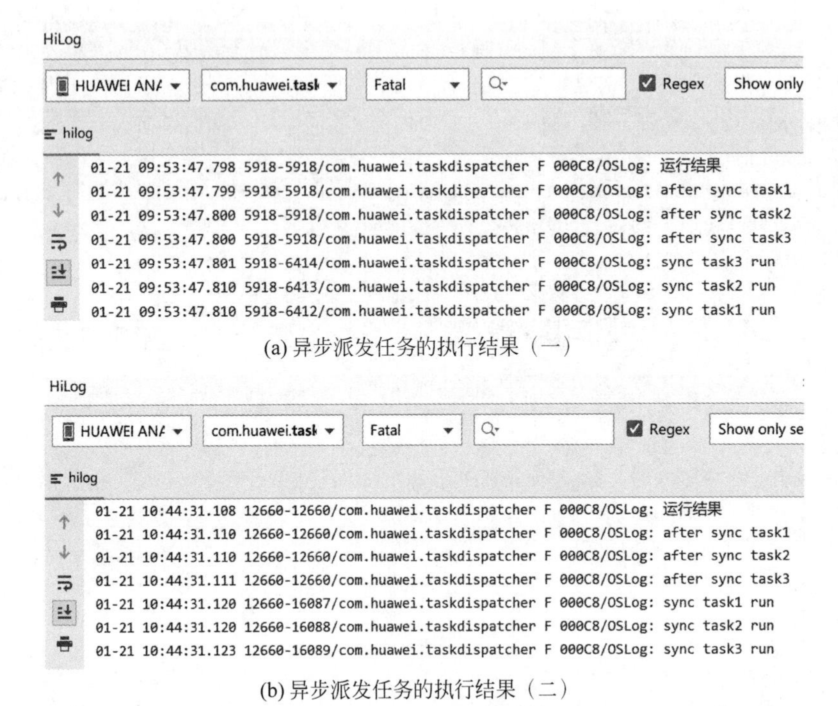

(a) 异步派发任务的执行结果（一）

(b) 异步派发任务的执行结果（二）

图 7.25 异步派发任务的执行结果

```
//异步延迟派发任务
        b();
}
//异步延迟派发任务的实现
public void b(){
     final long callTime = System.currentTimeMillis();
     final long delayTime = 3150;
TaskDispatcher globalTaskDispatcher = getGlobalTaskDispatcher(TaskPriority.DEFAULT);
     Revocable revocable = globalTaskDispatcher.delayDispatch(new Runnable() {
            @Override
            public void run() {
                HiLog.fatal(label, "delayDispatch task1 run");
                final long actualDelayMs = System.currentTimeMillis() - callTime;
                HiLog.fatal(label, "actualDelayTime >= delayTime: %{public}b", (actualDelayMs
>= delayTime));
            }
        }, delayTime);
        HiLog.fatal(label, "after delayDispatch task1");
}
```

执行结果如图 7.26 所示。

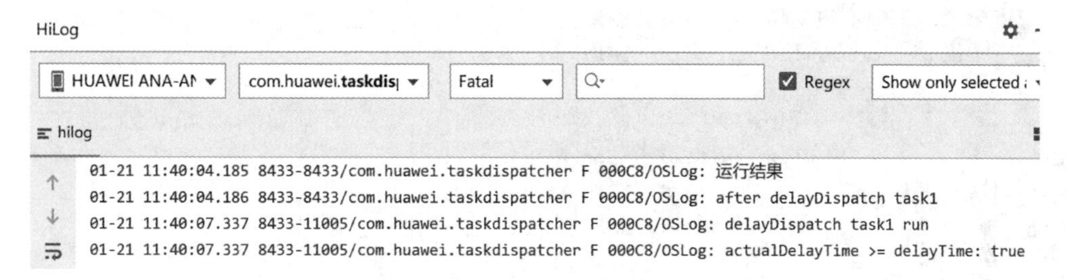

图 7.26 异步延迟派发任务执行结果

4. 任务组(Group)

表示一组任务,且该组任务之间有一定的联系,由 TaskDispatcher 执行 createDispatchGroup 创建并返回。以下是使用 ParallelTaskDispatcher 任务分发器进行任务分发的例子:首先创建并发任务分发器,接着创建一个任务组用于表示一组任务,然后利用 asyncGroupDispatch()方法异步添加两个相关任务 task1 和 task2 到任务组中,最后当所有任务完成后,调用 groupDispatchNotify()方法执行指定任务,代码如下:

```
//任务组
public void onStart(Intent intent) {
...

//任务组
        c();

  ...
}
//任务组的实现
public void c(){
//创建并发任务分发器
    String dispatcherName = "parallelTaskDispatcher";
    TaskDispatcher dispatcher = createParallelTaskDispatcher(dispatcherName, TaskPriority.
DEFAULT);

//创建任务组
    Group group = dispatcher.createDispatchGroup();

    //将任务 1 加入任务组,返回一个用于取消任务的接口
    dispatcher.asyncGroupDispatch(group, new Runnable(){
          @Override
          public void run() {
                HiLog.fatal(label, "download task1 is running");
          }
     });
```

```
    //将与任务1相关联的任务2加入任务组
    dispatcher.asyncGroupDispatch(group, new Runnable(){
        @Override
        public void run() {
            HiLog.fatal(label, "download task2 is running");
        }
    });

    //在任务组中的所有任务执行完成后执行指定任务
    dispatcher.groupDispatchNotify(group, new Runnable(){
        @Override
        public void run() {
            HiLog.fatal(label, "the close task is running after all tasks in the group are
completed");
        }
    });
}
```

以上示例的执行结果有两种情况，如图7.27所示。图7.27(a)中结果显示先执行task1，图7.27(b)中结果显示先执行task2，但无论先执行哪个任务，当任务组中的所有任务被执行完后便会执行指定的任务。

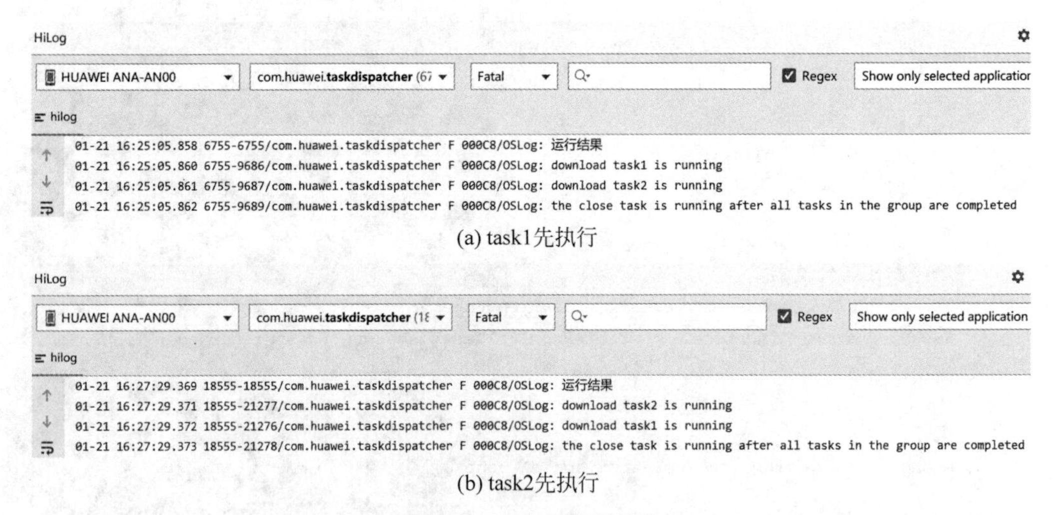

(a) task1先执行

(b) task2先执行

图7.27　任务组执行结果

5. 同步设置屏障任务(syncDispatchBarrier)

在任务组上设立任务执行屏障，同步等待屏障前的所有任务执行完成，再执行屏障后的指定任务。如下面的代码所示，task1和task2利用asyncGroupDispatch()方法异步添加到任务组中，接着利用syncDispatchBarrier()方法在任务组中为给定任务设置执行障碍，最后在任务组中的所有任务完成后执行指定任务，代码如下：

```
//同步设置屏障任务
public void onStart(Intent intent) {
...

//同步设置屏障任务
  d();

  ...
}
//同步设置屏障任务的实现
public void d(){
     String dispatcherName = "parallelTaskDispatcher";
 TaskDispatcher dispatcher = createParallelTaskDispatcher ( dispatcherName, TaskPriority.
DEFAULT);

     //创建任务组
     Group group = dispatcher.createDispatchGroup();
     //将任务加入任务组,返回一个用于取消任务的接口
     dispatcher.asyncGroupDispatch(group, new Runnable(){
            @Override
            public void run() {
                HiLog.fatal(label, "task1 is running");
            }
     });

     dispatcher.asyncGroupDispatch(group, new Runnable(){
            @Override
            public void run() {
                HiLog.fatal(label, "task2 is running");
            }
      });

      dispatcher.syncDispatchBarrier(new Runnable() {
            @Override
            public void run() {
                HiLog.fatal(label, "barrier");
            }
     });
     HiLog.fatal(label, "after syncDispatchBarrier");
}
```

执行结果如图 7.28 所示。由于 task1 和 task2 被异步添加到任务组中,所以二者是异步执行的,执行顺序并不固定,见图 7.28(a)和图 7.28(b)。

需要说明的是,在 GlobalTaskDispatcher 上同步设置任务屏障,将不会起到屏障作用。

(a) task1先执行

(b) task2先执行

图 7.28 同步设置屏障任务的执行结果

6. 异步设置屏障任务(asyncDispatchBarrier)

异步设置屏障任务与同步设置屏障任务过程相同，具体实现代码如下：

```
//异步设置屏障任务
public void onStart(Intent intent) {
...

//异步设置屏障任务
  e();

  ...
}
//异步设置屏障任务的实现
public void e(){
 TaskDispatcher dispatcher = createParallelTaskDispatcher("dispatcherName", TaskPriority.
DEFAULT);

   //创建任务组
   Group group = dispatcher.createDispatchGroup();

   dispatcher.asyncGroupDispatch(group, new Runnable(){
           @Override
           public void run() {
```

```
                HiLog.fatal(label, "task1 is running"); //1
            }
     });

    dispatcher.asyncGroupDispatch(group, new Runnable(){
            @Override
            public void run() {
                HiLog.fatal(label, "task2 is running"); //2
            }
    });

    dispatcher.asyncDispatchBarrier(new Runnable() {
            @Override
            public void run() {
                HiLog.fatal(label, "barrier"); //3
            }
    });

    HiLog.fatal(label, "after asyncDispatchBarrier"); //4
}
```

执行结果如图 7.29 所示，对比图 7.29(a)和图 7.29(b)可以发现，task1 和 task2 是异步执行的，执行顺序是随机的。由于屏障是异步设置的，所以主线程的执行也是随机的。

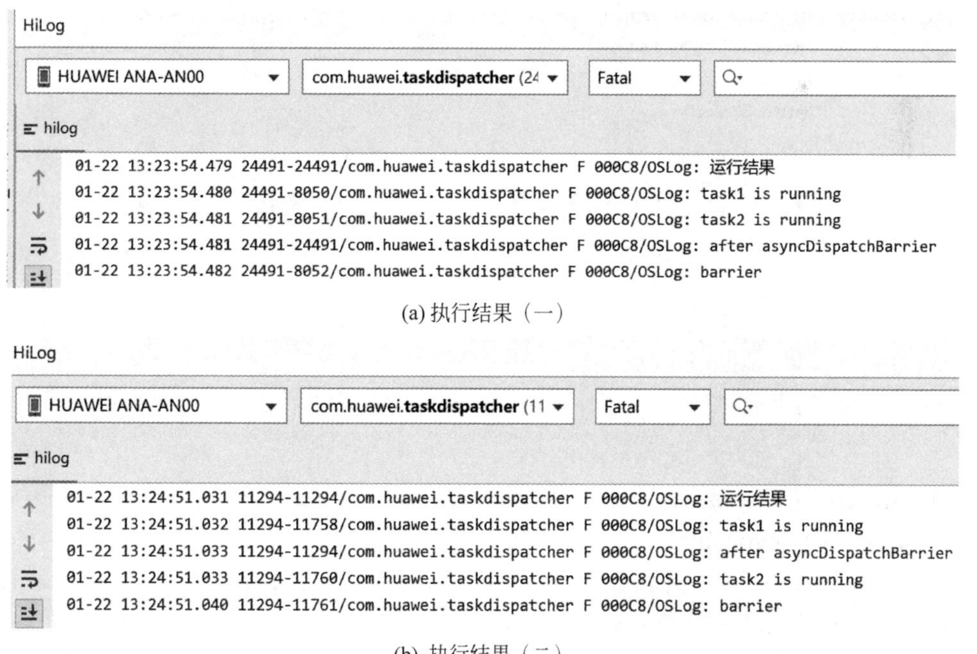

(a) 执行结果（一）

(b) 执行结果（二）

图 7.29 异步设置屏障任务的执行结果

需要说明的是，在GlobalTaskDispatcher上异步设置任务屏障不会起到屏障作用。此外，可以使用ParallelTaskDispatcher分离不同的任务组，达到微观并行、宏观串行的行为。

7. 执行多次任务(applyDispatch)

可以多次执行指定任务，该功能的实现主要依赖applyDispatch()方法。在下例中，执行10次指定的任务，代码如下：

```
//执行多次任务
public void onStart(Intent intent) {
...

//异步设置屏障任务
  f();

  ...
}
//异步设置屏障任务的实现
public void f(){
    final int total = 10;
final CountDownLatch latch = new CountDownLatch(total);
    final ArrayList<Long> indexList = new ArrayList<>(total);
TaskDispatcher dispatcher = getGlobalTaskDispatcher(TaskPriority.DEFAULT);

    //执行任务 total 次
dispatcher.applyDispatch((index) -> {
            indexList.add(index);
            HiLog.fatal(label, "执行: " + index);
            latch.countDown();
    }, total);

   //设置任务超时
    try {
            latch.await();
        } catch (InterruptedException exception) {
            HiLog.fatal(label, "latch exception");
        }
   HiLog.fatal(label, "list size matches, %{public}b", (total == indexList.size()));
}
```

执行结果如图7.30所示。

8. 取消任务(Revocable)

Revocable是取消一个异步任务的接口。异步任务包括通过asyncDispatch、delayDispatch、asyncGroupDispatch派发的任务。在下面的例子中，首先利用getUITaskDispatcher()获取绑定到UI线程的任务分发器，接着调用Revocable接口，取消由delayDispatch派发的异步任务。revoke()方法用于取消与Revocable对象相关的任务，需要注意的是若任务正在执行

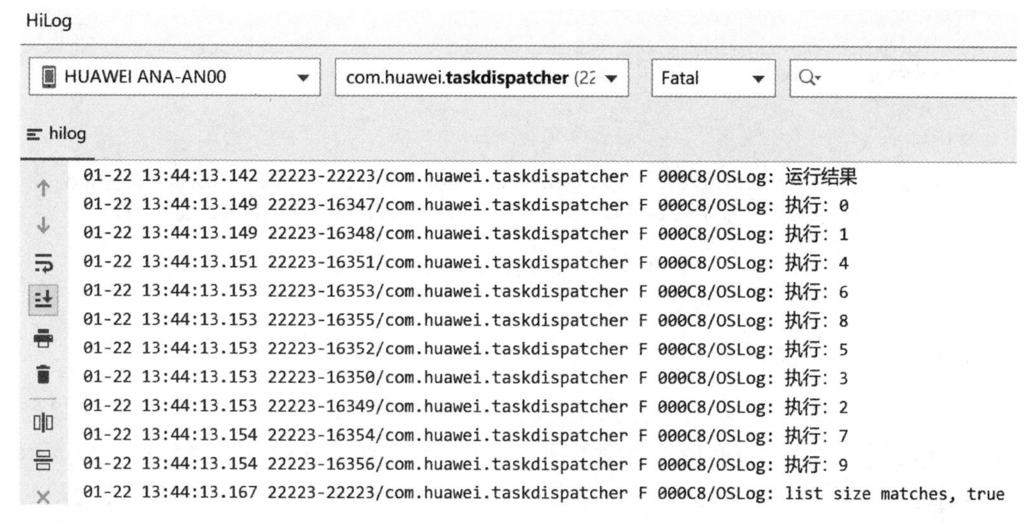

图 7.30　执行多次任务的结果

或已经执行结束，则无法取消该任务，返回值为 false，若任务成功取消，则返回值为 true，代码如下：

```
//取消任务
public void onStart(Intent intent) {
...

//取消任务
g();

    ...
}
//取消任务的实现
 public void g(){
  TaskDispatcher dispatcher = getUITaskDispatcher();
        Revocable revocable = dispatcher.delayDispatch(new Runnable() {
            @Override
            public void run() {
                HiLog.fatal(label, "delay dispatch");
            }
        }, 10);
        boolean revoked = revocable.revoke();
        HiLog.fatal(label, "%{public}b", revoked);
}
```

执行结果如图 7.31 所示，返回值为 true，故任务成功取消。

图 7.31 取消任务执行结果

7.3.3 线程通信

通过 7.3.2 节的学习,了解到在开发过程中,经常需要执行比较耗时的任务,可以通过创建其他线程来处理,从而避免当前线程受到阻塞,然而,线程之间如何进行通信成为亟待解决的问题。实际上,在 HarmonyOS 中提供了 EventHanlder 机制用于处理线程之间的通信,通过 EventRunner 创建新线程来执行耗时操作,从而既可以避免阻塞原来的线程又可以合理处理任务。例如主线程使用 EventHandler 创建子线程来执行文件下载这个耗时的操作,文件下载成功后,所建子线程再利用 EventHandler 机制通知主线程,主线程再进行 UI 更新。一般,EventHandler 有两个主要作用,一个是在不同线程间分发和处理 InnerEvent 事件或 Runnable 任务,另一个是延迟处理 InnerEvent 事件或 Runnable 任务。

EventHandler 的运作机制如图 7.32 所示。

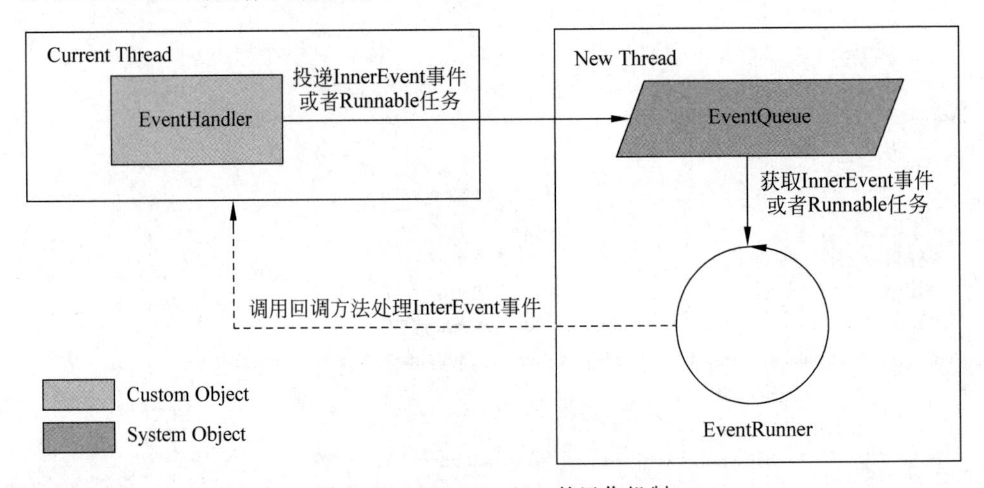

图 7.32 EventHanlder 的运作机制

由图 7.32 可见,使用 EventHandler 实现线程间通信的主要流程如下:

(1) 每个 EventHandler 和指定的 EventRunner 所创建的新线程绑定,并且该新线程内部有一个事件队列 EventQueue。EventHandler 投递具体的 InnerEvent 事件或者 Runnable 任务到 EventQueue。

(2) EventRunner 是一种事件循环器，可以循环从事件队列 EventQueue 中获取 InnerEvent 事件或者 Runnable 任务。

(3) EventRunner 处理事件或任务：

- 如果 EventRunner 取出的事件为 InnerEvent 事件，则触发 EventHandler 的处理方法，即在 EventRunner 所在线程执行 processEvent 回调，在新线程上处理该事件。
- 如果 EventRunner 取出的事件为 Runnable 任务，则 EventRunner 直接在新线程上处理 Runnable 任务，即在 EventRunner 所在线程执行 Runnable 的 run 回调。

需要说明的是，进行线程间的通信时，EventHandler 只能和 EventRunner 所创建的线程进行绑定，但绑定之前需要判断 EventRunner 是否成功创建线程，只有确保获取的 EventRunner 实例非空时，才可以使用 EventHandler 绑定 EventRunner。此外，EventHandler 和 EventRunner 之间是多对一的关系，即一个 EventHandler 只能同时与一个 EventRunner 绑定，而一个 EventRunner 可以创建多个 EventHandler。

在开发过程中，EventHandler 进行任务投递时，EventHandler 的优先级可在 IMMEDIATE(表示事件被立即投递)、HIGH(表示事件先于 LOW 优先级投递)、LOW(表示事件优于 IDLE 优先级投递，事件的默认优先级是 LOW)、IDLE(表示在没有其他事件的情况下才投递该事件)中选择，并设置合适的 delayTime。EventRunner 的工作模式可以分为托管模式和手动模式。两种模式是在调用 EventRunner 的 create()方法时，通过选择不同的参数实现的。具体地，参数为 true 时选定托管模式，参数为 false 时选定手动模式。托管模式和手动模式的区别是：托管模式不需要开发者调用 run()和 stop()方法去启动和停止 EventRunner，而手动模式则需要开发者自行调用 EventRunner 的 run()方法和 stop()方法来确保线程的启动和停止。当 EventRunner 实例化时，系统调用 run()来启动 EventRunner。当 EventRunner 不被引用时，系统调用 stop()来停止 EventRunner。

接下来，以托管模式投递 InnerEvent 为例展示线程间的通信过程。新建一个以 Phone 为设备、Java 为模板、名为 TaskDispatcher 的项目，进入 entry→main→java→包名→slice→MainAbilitySlice 文件，覆写 onStart()方法，此外，在 onStart()方法之外新建一个类继承 EventHandler，代码如下：

```
//托管模式实现线程通信
@Override
public void onStart(Intent intent) {
  super.onStart(intent);
  super.setUIContent(ResourceTable.Layout_ability_main);

  //1.创建 EventRunner A
  EventRunner runner = EventRunner.create(true);      //内部会新建一个线程

  //2.创建类继承 EventHandler,在 onStart()之外进行创建
```

```
    //3.实例化 EventHandler 并绑定 EventRunner
    MyEventHandler handler = new MyEventHandler(runner);

    //4.新建两个 InnerEvent 并发送至事件队列
    int eventId1 = 0;
    int eventId2 = 1;
    long param = 0;
    Object object = null;
    InnerEvent event1 = InnerEvent.get(eventId1, param, object);
    InnerEvent event2 = InnerEvent.get(eventId2, param, object);
    handler.sendEvent(event1, 0, EventHandler.Priority.IMMEDIATE);
    handler.sendEvent(event2, 2000, EventHandler.Priority.IMMEDIATE);

    //5.当 runnerA 没有任何对象引入时,线程会自动回收
    runner = null;
}

//6.创建类继承 EventHandler
public class MyEventHandler extends EventHandler {
        HiLogLabel label = new HiLogLabel(HiLog.LOG_APP ,200, "HarmonyOS Log");

        private MyEventHandler(EventRunner runner) {
            super(runner);
        }

        @Override
        public void processEvent(InnerEvent event) {
            super.processEvent(event);

            if (event == null) {
                return;
            }

            int eventId = event.eventId;

            switch (eventId) {
                case 0: {
                    HiLog.fatal(label, "enventId0");
                    break;
                }
                case 1: {
                    HiLog.fatal(label, "enventId1");
                    break;
                }

                default:
                    break;
            }
        }
    }
```

实现过程如下：

（1）在 onStart()方法中，调用 create(true)方法创建一个拥有新线程的 EventRunner，命名为 runner，且 EventRunner 的工作模式为托管模式。

（2）在 onStart()方法外，创建一个新类继承 EventHandler，命名为 MyEventHandler，在类中覆写 processEvent()方法来处理 EventHanlder 投递的 InnerEvent 事件，根据不同的 eventId 进行不同的操作。

（3）实例化 MyEventHandler，并将其与新建的 EventRunner 绑定。

（4）创建 InnerEvent 事件并投递：①调用 InnerEvent. get(int eventId, long param, Object object)方法新建两个具有指定的事件 ID 和必须数据的 InnerEvent 实例。其中，eventId 是事件的 Id，在本例中两个事件的 Id 分别为 1 和 2。param 指示此 InnerEvent 中包含的长整数数据，可以将其设置为任何长值。object 指示此 InnerEvent 中包含的对象。②调用 handler. sendEvent(InnerEvent event, long delayTime, EventHandler. Priority priority)发送一个指定优先级的延时事件到事件队列。在本例中，event1 和 event2 的优先级均为 IMMEDIATE，但 event1 的延时为 0ms，而 event2 的延时为 2000ms。

（5）由于 EventRunner 取出的事件为 InnerEvent 事件，故触发 EventHandler 的 processEvent()方法，在新线程上处理该事件。

（6）当 runner 没有任何对象引入时，线程会自动回收。

运行上述代码，执行结果如图 7.33 所示。从图中可以看出，event2 比 event1 延时 2s 执行。

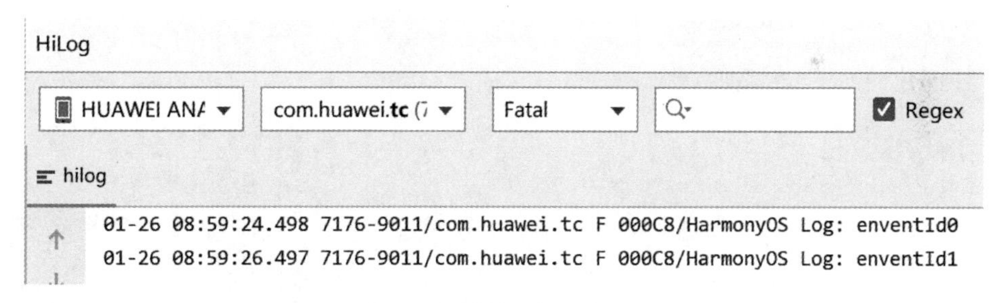

图 7.33　以托管模式投递 InnerEvent

7.4　ServiceAbility

ServiceAbility 是基于 Service 模板的 Ability（以下简称 Service），主要用于后台运行一些需要长时间运行的程序，不提供任何 UI 功能，例如在音乐播放应用中，需要一个能在后台运行的 Ability，在后台播放音乐的同时，UI 页面仍然能够显示页面并提供相应的功能。

Service 可由其他应用或 Ability 启动，即使用户切换到其他应用，Service 仍将在后台继续运行。同时 Service 是单实例的，在一个设备上，相同的 Service 只会存在一个实例。如果多个 Ability 共用这个实例，只有当与 Service 绑定的所有 Ability 都退出后，Service 才能够退出。由于 Service 是在主线程里执行的，因此，如果在 Service 里面的操作时间过长，开发者必须在 Service 里创建新的线程来处理，防止造成主线程阻塞，以及应用程序无响应。

7.4.1 创建并启动 Service

首先学习创建并启动 Service。打开 DevEco Studio，在想要新建 Service 的模块下右击，选择 New→Ability→Service Ability，如图 7.34 所示。

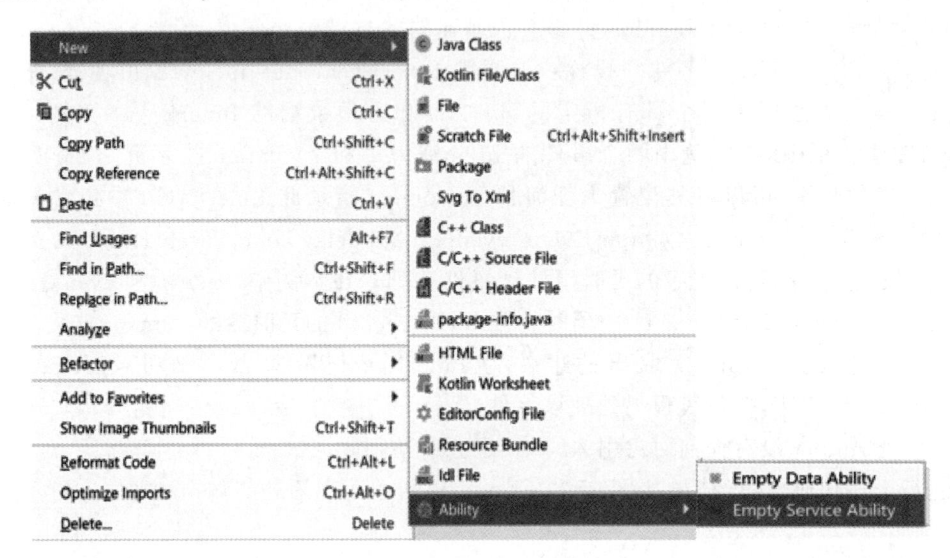

图 7.34 新建 ServiceAbility

打开创建的 Service 文件，代码如下：

```
//新建的 Service 中的代码
public class ServiceAbility extends Ability {
    private static final HiLogLabel LABEL_LOG = new HiLogLabel(3, 0xD001100, "Demo");

    @Override
    public void onStart(Intent intent) {
        HiLog.error(LABEL_LOG, "ServiceAbility::onStart");
        super.onStart(intent);
    }

    @Override
    public void onBackground() {
        super.onBackground();
        HiLog.info(LABEL_LOG, "ServiceAbility::onBackground");
    }

    @Override
```

```
    public void onStop() {
        super.onStop();
        HiLog.info(LABEL_LOG, "ServiceAbility::onStop");
    }

    @Override
    public void onCommand(Intent intent, boolean restart, int startId) {
    }

    @Override
    public IRemoteObject onConnect(Intent intent) {
        return null;
    }

    @Override
    public void onDisconnect(Intent intent) {
    }
}
```

接下来详细介绍 Service 中的内容,Service 也是一种 Ability,所以 Service 和 Ability 一样具有生命周期。Ability 为 Service 提供了以下生命周期方法,用户可以重写这些方法来添加自己的处理。

1) onStart()

该方法在创建 Service 的时候调用,用于 Service 的初始化,在 Service 的整个生命周期只会调用一次。

2) onCommand()

在 Service 创建完成之后调用,该方法在客户端每次启动该 Service 时都会调用,用户可以在该方法中做一些调用统计、初始化类的操作。

3) onConnect()

在 Ability 和 Service 连接时调用,该方法返回 IRemoteObject 对象,用户可以在该回调函数中生成对应 Service 的进程间通信(Inter-Process Communication 或 Interprocess Communication,以下简写 IPC)通道,以便 Ability 与 Service 交互。Ability 可以多次连接同一个 Service,系统会缓存该 Service 的 IPC 通信对象,只有当第一个客户端连接 Service 时,系统才会调用 Service 的 onConnect 方法来生成 IRemoteObject 对象,而后系统会将同一个 RemoteObject 对象传递至其他连接同一个 Service 的所有客户端,而无须再次调用 onConnect 方法。

4) onDisconnect()

在组件与绑定的 Service 断开连接时调用。

5) onStop()

在 Service 销毁时调用。Service 应通过实现此方法来清理任何资源,如关闭线程、注册侦听器等。

需要注意的是创建之后一定要注册 Service,但是 DevEco Studio 在创建 Service 时已经自动在应用配置文件中进行注册了,注册类型 type 已经被自动设置为 service,注册方式和 PageAbility 的方式类似,下面给出一个具体的注册示例,代码如下:

```
"module": {
…
  "abilities": [
    {
      "name": "com.harmony.myapplication.MainAbility",
      "description": "$string:mainability_description",
      "icon": "$media:icon",
      "label": "MyApplication",
      "launchType": "standard",
      "orientation": "landscape",
      "visible": false,
      "skills": [
        {
          "actions": [
            "action.system.home"
          ],
          "entities": [
            "entity.system.home"
          ]
        }
      ],
"type": "service",
      "visible": true,
      "formEnabled": false
    }
  ]
}
```

可以看到,Service 的注册方法需要把 type 标签改为 service,这是 ServiceAbility 独有的标签,其他的标签属性在第 1 章应用配置文件中有详细介绍。

创建完 Service 后,开始学习如何启动 Service。因为 Service 也是 Ability 的一种,开发者同样可以通过将 Intent 传递给 startAbility()方法来启动另外一个 Ability,以此来启动 Service。不仅支持启动本地 Service,还支持启动远程 Service。

下面看一个启动本地设备 Service 的示例,代码如下:

```
//启动本地设备 Service
Intent intent = new Intent();
Operation operation = new Intent.OperationBuilder()
        .withDeviceId("")
```

```
            .withBundleName("com.harmony.myapp")
            .withAbilityName("com.harmony.myapp.entry.ServiceAbility")
            .build();
    intent.setOperation(operation);
    startAbility(intent);
```

可以清楚地看到开发者可以通过构造包含DeviceId、BundleName与AbilityName的Operation对象设置目标Service信息。这3个参数的含义如下：

(1) DeviceId：表示设备ID。如果是本地设备，则可以直接留空。如果是远程设备，则可以通过harmonyos.distributedschedule.interwork.DeviceManager提供的getDeviceList获取设备列表，详见《API参考》。

(2) BundleName：表示包名称。

(3) AbilityName：表示待启动的Ability名称。

然后启动远程设施Service，示例代码如下：

```
    //启动远程设施Service
    Operation operation = new Intent.OperationBuilder()
            .withDeviceId("deviceId")
            .withBundleName("com.harmony.myapp")
            .withAbilityName("com.harmony.myapp.entry.ServiceAbility")
    .withFlags(Intent.FLAG_ABILITYSLICE_MULTI_DEVICE)
    //设置支持分布式调度系统多设备启动的标识
            .build();
    Intent intent = new Intent();
    intent.setOperation(operation);
    startAbility(intent);
```

可以看到远程启动和本地启动的区别在于Intent中多设置了一个FLAG_ABILITYSLICE_MULTI_DEVICE参数。

在执行上述代码后，Ability将通过startAbility()方法来启动Service。如果Service尚未运行，则系统会先调用onStart()来初始化Service，再回调Service的onCommand()方法来启动Service。如果Service正在运行，则系统会直接回调Service的onCommand()方法来启动Service。

Service一旦创建就会一直保持在后台运行，除非必须回收内存资源，否则系统不会停止或销毁Service。开发者可以在Service中通过terminateAbility()停止本Service或在其他Ability调用stopAbility()来停止Service。

停止Service同样支持停止本地设备Service和停止远程设备Service，使用方法与启动Service一样。一旦调用停止Service的方法，系统便会尽快销毁Service。

下面通过一个实例深入了解。这里采用7.1节的PageAbility开启一个Service服务。首先按照前文提到的方法新建一个Service，如图7.35所示，名字为ServiceAbility。

Configure Ability

Service Name: ServiceAbility

Visible

Package name: com.huawei.mytestapp

Enable background mode:

图 7.35　Service 新建页面

打开创建的 Service，覆写其中的函数，通过 log 输出以追踪 Service 的运行情况，其中 public IRemoteObject onConnect(Intent intent)函数将在 7.4.2 节进行介绍，可以暂不处理。Service 中的代码如下：

```
//ServiceAbility.java
public class ServiceAbility extends Ability {
    private static final HiLogLabel LABEL_LOG = new HiLogLabel(3, 0xD001100, "Demo");

    @Override
    public void onStart(Intent intent) {
HiLog.fatal(LABEL_LOG, "ServiceAbility::onStart");
        super.onStart(intent);
    }

    @Override
    public void onBackground() {
        super.onBackground();
HiLog.fatal(LABEL_LOG, "ServiceAbility::onBackground");
    }

    @Override
    public void onStop() {
        super.onStop();
HiLog.fatal(LABEL_LOG, "ServiceAbility::onStop");
    }

    @Override
    public void onCommand(Intent intent, boolean restart, int startId) {
HiLog.fatal(LABEL_LOG, "ServiceAbility::onCommand");
    }

    @Override
    public IRemoteObject onConnect(Intent intent) {
        HiLog.fatal(LABEL_LOG, "ServiceAbility::onConnect");
        return null;
```

```
    }

    @Override
    public void onDisconnect(Intent intent) {
HiLog.fatal(LABEL_LOG, "ServiceAbility::onDisconnect");
    }
}
```

接下来按照 7.1 节的方法创造一个 PageAbility，Page 中创建一个用来启动 Service 的 testSlice，其中代码如下：

```
//用来启动 Service 的 testSlice 代码
public class testSlice extends AbilitySlice {
    @Override
    public void onStart(Intent intent) {
        super.onStart(intent);
        Intent intent1 = new Intent();
        Operation operation = new Intent.OperationBuilder()
                .withDeviceId("")
                .withBundleName("com.harmony.myapplication")
                .withAbilityName("com.harmony.myapplication.ServiceAbility")
                .build();
        intent1.setOperation(operation);
        startAbility(intent1);
    }
}
```

需要注意的是创建的 Page 和 Service 一定要在配置文件里注册，否则会报错，右击新建 Service 时，IDE 会完成自动注册，打开 config.json 发现已经配置成功，如图 7.36 所示。

```
{
  "name": "com.harmony.myapplication.ServiceAbility",
  "icon": "$media:icon",
  "description": "$string:serviceability_description",
  "type": "service"
}
```

图 7.36 config.json 中的配置

设置完两个 Ability 后，需要对路由进行配置规划，具体实现代码如下：

```
public class MainAbility extends Ability {
    @Override
    public void onStart(Intent intent) {
        super.onStart(intent);
        super.setMainRoute(testSlice.class.getName());
    }
}
```

启动程序，查看 HiLog 输出面板，输出效果如图 7.37 所示。

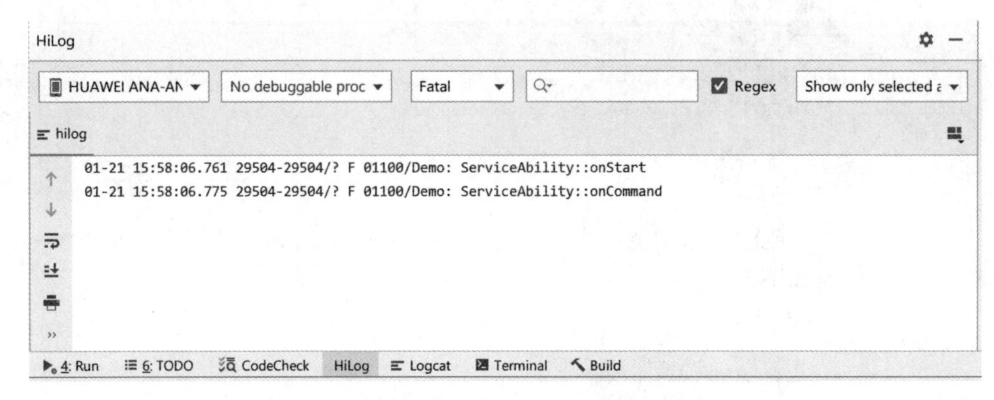

图 7.37　HiLog 输出信息

由图 7.37 可见，ServiceAbility 已经成功启动并且完成了 onStart 和 onCommand 这两个生命周期。

7.4.2　连接 Service

如果 Service 需要与 Page Ability 或其他应用的 ServiceAbility 进行交互，则应创建用于连接的 Connection。Service 支持其他 Ability 通过 connectAbility()方法与其进行连接。

在使用 connectAbility()处理回调时，需要传入目标 Service 的 Intent 与 IAbilityConnection 的实例。IAbilityConnection 提供了两种方法供开发者实现：onAbilityConnectDone()用来处理连接的回调，onAbilityDisconnectDone()用来处理断开连接的回调。

处理链接的示例代码如下：

```
//创建连接回调实例
private IAbilityConnection connection = new IAbilityConnection() {
    //连接到 Service 的回调
    @Override
    public void onAbilityConnectDone(ElementName elementName, IRemoteObject iRemoteObject,
int resultCode) {
  //在这里开发者可以获得服务器端传过来 IRemoteObject 对象，从中解析出服务器端传过来的信息
    }

    //断开与连接的回调
    @Override
    public void onAbilityDisconnectDone(ElementName elementName, int resultCode) {
    }
};
```

同时，Service 侧也需要在 onConnect()时返回 IRemoteObject，从而定义与 Service 进

行通信的接口。onConnect()需要返回一个IRemoteObject对象，HarmonyOS提供了IRemoteObject的默认实现，用户可以通过继承RemoteObject来创建自定义的实现类。

Service侧把自身的实例返回调用侧，代码如下：

```
//创建自定义 IRemoteObject 实现类
private class MyRemoteObject extends RemoteObject {
    public MyRemoteObject() {
        super("MyRemoteObject");
    }
}

//把 IRemoteObject 返回客户端
@Override
protected IRemoteObject onConnect(Intent intent) {
    return new MyRemoteObject();
}
```

回到7.4.1节的示例，可以发现之前创建的Service的public IRemoteObject onConnect(Intent intent)并没有进行覆写，这种方法用于进行通信，返回数据给其他Ability的接口，在其中进行HiLog输出，代码如下：

```
@Override
public IRemoteObject onConnect(Intent intent) {
        HiLog.info(LABEL_LOG, "ServiceAbility::onConnect");
        return new MyRemoteObject();
}
```

这样当程序成功执行到这一步时可以查看HiLog信息，然后在PageAbility中重新定义连接回调实例，代码如下：

```
IAbilityConnection connection = new IAbilityConnection() {
            //连接到 Service 的回调
            @Override
            public void onAbilityConnectDone(ElementName elementName, IRemoteObject iRemoteObject,
int resultCode) {
                HiLog.info(LABEL_LOG,"ServiceAbility 连接成功");
            }
            //断开与连接的回调
            @Override
            public void onAbilityDisconnectDone(ElementName elementName, int resultCode) {
                HiLog.info(LABEL_LOG,"ServiceAbility 断开连接成功");
            }
};
```

这里新建两个 Button 用来启动连接和断开连接,代码如下:

```
//启动与断开连接
ShapeElement shapeElement = new ShapeElement();
        shapeElement.setRgbColor(new RgbColor(255,255,0));

        Button button2 = new Button(this);
        button2.setText("connectAbility");
        button2.setTextSize(100);
        button2.setBackground( shapeElement);
        button2.setTouchEventListener(new Component.TouchEventListener() {
            @Override
            public boolean onTouchEvent(Component component, TouchEvent touchEvent) {
                connectAbility(intent1, connection);
                return true;
            }
        });
        Button button3 = new Button(this);
        button3.setText("disconnectAbility");
        button3.setTextSize(100);
        button3.setBackground( shapeElement);
        button3.setTouchEventListener(new Component.TouchEventListener() {
            @Override
            public boolean onTouchEvent(Component component, TouchEvent touchEvent) {
                disconnectAbility( connection);
                return true;
            }
        });
```

启动程序,查看 HiLog 输出信息如图 7.38 所示,发现成功显示了 onConnent,触发了 Service 中的 onConnent 生命周期,Service 也连接成功。

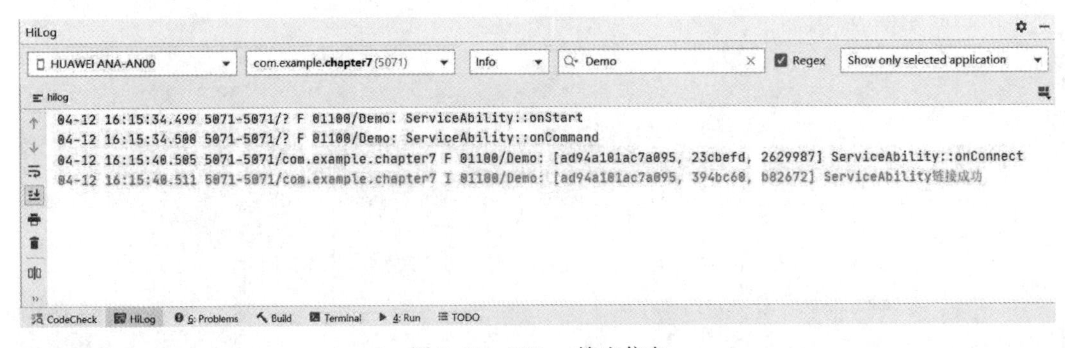

图 7.38　HiLog 输出信息

此外,还可以再加入 disconnectAbility(connection)来解除连接,其中只用一个 IAbilityconnection 参数就可以实现,因为目标 Service 已经确定不需要 Intent。可以看到效果如图 7.39 所示。

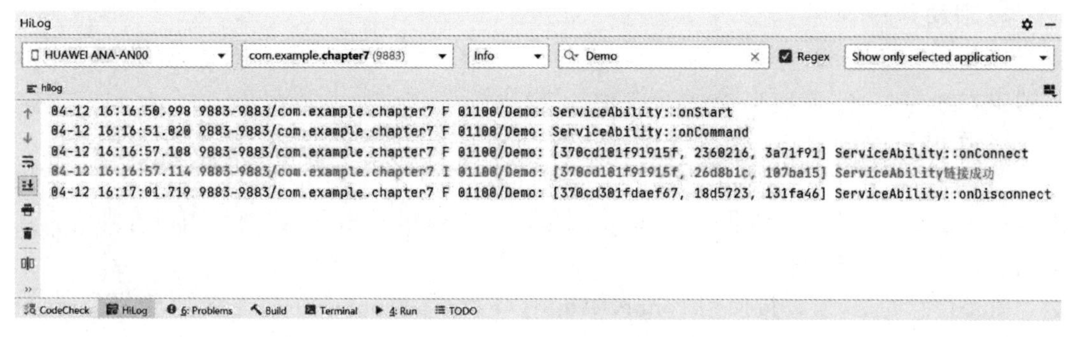

图 7.39　HiLog 输出信息

可以看到 Service 在断开连接时进行了 onDisconnect 的生命周期。

7.4.3　Service 的生命周期

与 Page 类似，Service 也拥有生命周期，接下来详细说明一下 Service 生命周期的运行流程，如图 7.40 所示。根据调用方法的不同，其启动和销毁生命周期有以下两种路径：

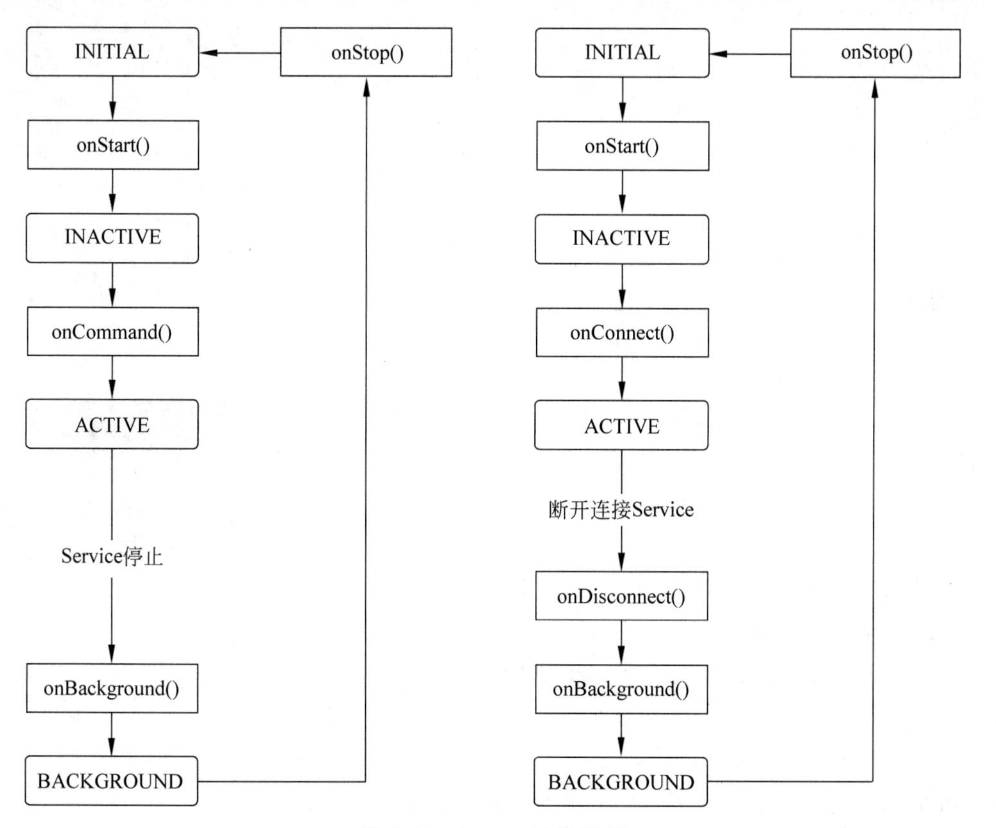

图 7.40　Service 生命周期

1）启动 Service 时

该 Service 在其他 Ability 调用 startAbility()时创建，然后保持运行。其他 Ability 通过调用 stopAbility()来停止 Service，Service 停止后，系统会将其销毁。

2）连接 Service 时

该 Service 在其他 Ability 调用 connectAbility()时创建，客户端可通过调用 disconnectAbility()断开连接。多个客户端可以绑定到相同 Service，而且当所有绑定全部取消后，系统即会销毁该 Service。

值得注意的是，当直接使用 connectAbility(intent1, connection)连接时，如果 Service 没有被创建，那么系统会自动创建并连接 Service，通过这种方式创建的 Service 在断开连接时会自动销毁。在上面的实例中，连接断开时，Service 仅执行到了 onDisconnent。

下面直接使用 connectAbility 来创建并连接 Service，首先定义一个 Button 用来确保 Service 被停止，代码如下：

```
//创建 Button 用于停止 Service
Button button4 = new Button(this);
button4.setText("stopService");
button4.setTextSize(100);
button4.setBackground( shapeElement);
button4.setTouchEventListener(new Component.TouchEventListener() {
  @Override
  public boolean onTouchEvent(Component component, TouchEvent touchEvent) {
        stopAbility(intent1);
        return true;
  }
});
```

运行后得到 UI 页面如图 7.41 所示。

此时 HiLog 输出信息如图 7.42 所示，可以看到，生命周期的顺序是 Service 先运行 onStart，接着运行 onCommand。

这时单击 StopService，停止 Service 运行，HiLog 输出如图 7.43 所示。

从图 7.43 可以看到，Service 在停止时其生命周期由 onBackground 变为 onStop。

可以看出 Service 处在完全停止的状态，如果这时单击 connentAbility，输出信息如图 7.44 所示。

由图 7.44 可以清楚地看到 Service 的生命周期为 onStart→OnConnect，此时已经成功连接，如果在 Service 中写入了相应的功能，Service 就会成功运行，从而实现对应的能力。

继续单击 disconnentAbility，HiLog 输出如图 7.45 所示。

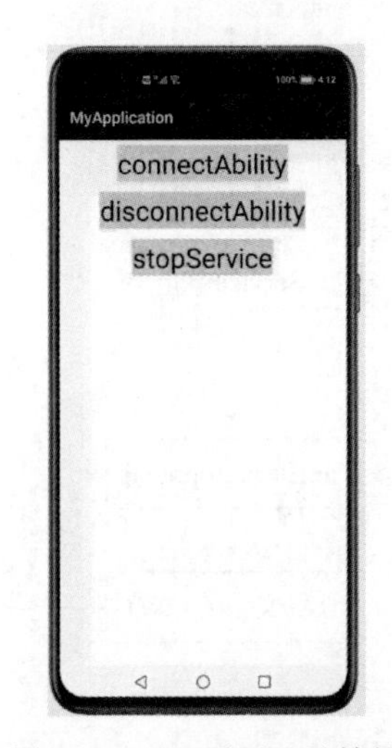

图 7.41 UI 页面显示效果

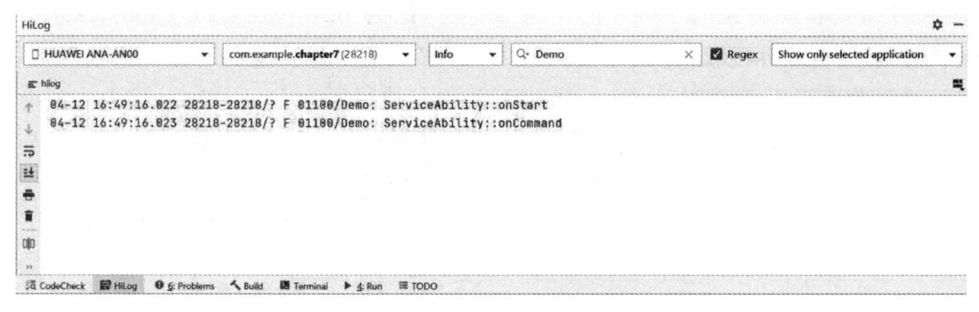

图 7.42 HiLog 输出信息

```
04-12 16:49:16.022 28218-28218/? F 01100/Demo: ServiceAbility::onStart
04-12 16:49:16.023 28218-28218/? F 01100/Demo: ServiceAbility::onCommand
04-12 16:49:23.860 28218-28218/com.example.chapter7 F 01100/Demo: ServiceAbility::onBackground
04-12 16:49:23.861 28218-28218/com.example.chapter7 F 01100/Demo: ServiceAbility::onStop
```

图 7.43 HiLog 输出信息

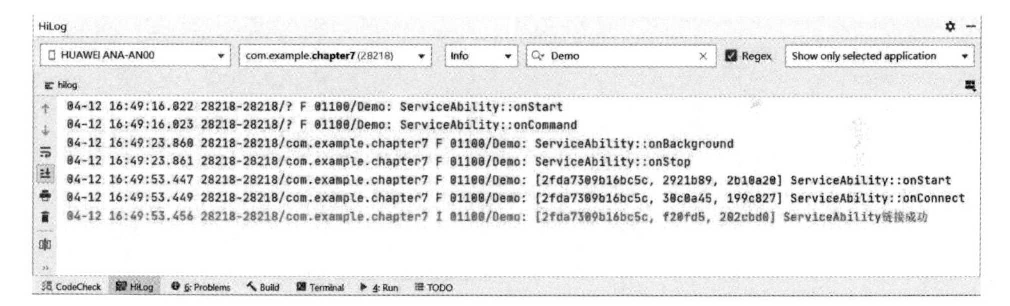

图 7.44 HiLog 输出信息

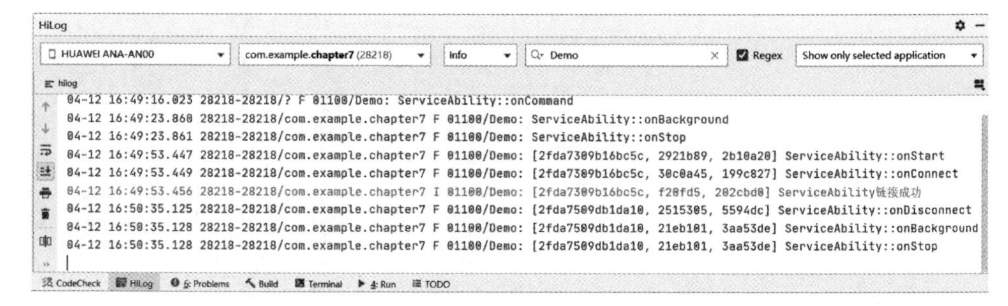

图 7.45 HiLog 输出信息

由图 7.45 可以看到，Service 的生命周期变化为 onDisConnect→onBackground→onStop，Service 在断开连接的同时进行了关闭，并多出了 onBackground→onStop 的生命周期，上一个实例在断开连接时 Service 仍在运行，仅仅执行了 onDisConnect 这个生命周期。

至此 ServiceAbility 也介绍完毕，开发者可以用 Service 构建一些后台操作，例如后台播放音乐，以及后台下载数据，同时还不会影响主 UI 线程的操作。

DataAbility 将在第 8 章进行讲解，在学习 DataAbility 之前读者需要先了解一下数据库的基础知识。

第8章

数 据 管 理

HarmonyOS应用数据管理不仅支持单设备的各种结构化数据的持久化，还支持跨设备之间数据的同步、共享及搜索功能，因此，开发者基于HarmonyOS应用数据管理功能，能够实现应用程序数据在不同终端设备之间的无缝衔接，从而保证用户在跨设备使用数据时所用数据的一致性。

本章将介绍HarmonyOS如何提供本地应用数据管理、分布式数据服务、分布式文件服务等知识。

8.1 本地应用数据管理

15min

8.1.1 SQLite数据库

HarmonyOS提供单设备上结构化数据的存储和访问能力。其中，本地数据库中的关系型数据库（Relational Database，RDB）和对象关系映射数据库（Object Relational Mapping Database，ORMDB）均使用SQLite作为持久化存储引擎。即二者均基于SQLite数据库实现了对数据的管理和操作，因此，在正式介绍HarmonyOS中的数据库之前，先占用些许篇幅简单介绍一下SQLite数据库的相关概念和特点。了解SQLite数据库的读者可以跳过此节。

SQLite数据库是由D. Richard Hipp于2000年发布的开源数据库，是遵守ACID的关系型数据库管理系统。其中，ACID是指原子性（Atomicity）、一致性（Consistency）、隔离性（Isolation）和持久性（Durability）的缩写，这是数据库事务正确执行的4个基本要素。一个支持事务的数据库，在事务过程中想要保证数据的正确性就必须满足这4个基本要素，下面简要介绍4个基本要素：

（1）原子性（Atomicity）：一个事务要么全部提交成功，要么全部失败回滚，不能只执行其中的一部分操作。事务在执行过程中发生错误，会被回滚到事务开始前的状态，就像这个事务从来没有执行过一样。

（2）一致性（Consistency）：事务的执行不能破坏数据库数据的完整性和一致性，一个事务在执行之前和执行之后，数据库都必须处于一致性状态。

(3) 隔离性(Isolation)：事务的隔离性确保在并发环境中,并发的事务相互隔离,即事务的执行不被其他事务干扰。

(4) 持久性(Durability)：事务完成后,对数据库的更改会被持久保存在数据库中,且不会被回滚。

除此之外,SQLite之所以风靡,还在于其有以下一些特性：

(1) SQLite是轻量级的嵌入式数据库,可移植性高,可以被嵌入在多种系统中,如Linux、Mac OS X、Windows、Android、iOS及HarmonyOS。

(2) SQLite是一个零配置的数据库,即无须在系统中配置或管理。

(3) SQLite不需要外部依赖,即不需要为其提供单独的服务器进程或操作系统。

(4) SQLite可以为独立应用和设备提供本地化存储。

(5) SQLite不需要管理,在无须数据库管理员支持的场景中表现极佳,使用方便,应用设备广泛,如：广泛应用于移动电话、摄像机、智能手表、汽车、医疗器械、机器人等。

(6) 完整的SQLite数据库是单个磁盘文件,可以兼容多个操作系统平台,被用于实现系统间的数据传输。具体地,发送者将数据导入一个SQLite文件并发送该文件,接收者就可以用SQL提取需要的数据。

(7) SQLite提供了简单和易于使用的API,强调易用性和可靠性。

8.1.2 关系型数据库

在介绍HarmonyOS中的关系型数据库之前,先来回顾一下什么是关系模型。简单来讲,关系模型指的就是二维表格模型,即关系模型把数据看作一个二维表格,通过行号+列号来确定唯一数据,因此,关系型数据库就是由二维表及其之间的联系所组成的一个数据组织,即关系型数据库是一种基于关系模型来管理数据的数据库,以行和列的形式存储数据。

HarmonyOS中,关系型数据库的运作机制如图8.1所示。

由图8.1可知,HarmonyOS中,关系型数据库的底层使用了SQLite作为持久化存储引擎,因此关系型数据库支持SQLite所具有的数据库特性。对上层应用提供了通用的操作接口,支持应用进行一系列增、删、改、查操作。由此可见,HarmonyOS提供的关系型数据库功能更加完善,查询效率更高,不仅支持直接运行用户输入的SQL语句,还支持接口调用进行一系列对数据库的操作来满足复杂的场景需要。

下面通过一个实例实现关系型数据库,创建Phone设备下的Java模板工程,首先打开entry→src→main→java→com. huawei. dbtest→slice→MainAbilitySclice文件,并在onStart()方法中实现关系型数据库的创建及对其关系表进行操作,操作流程如下：

```
public void onStart(Intent intent) {
    super.onStart(intent);

    //新建数据库与表
    ...
```

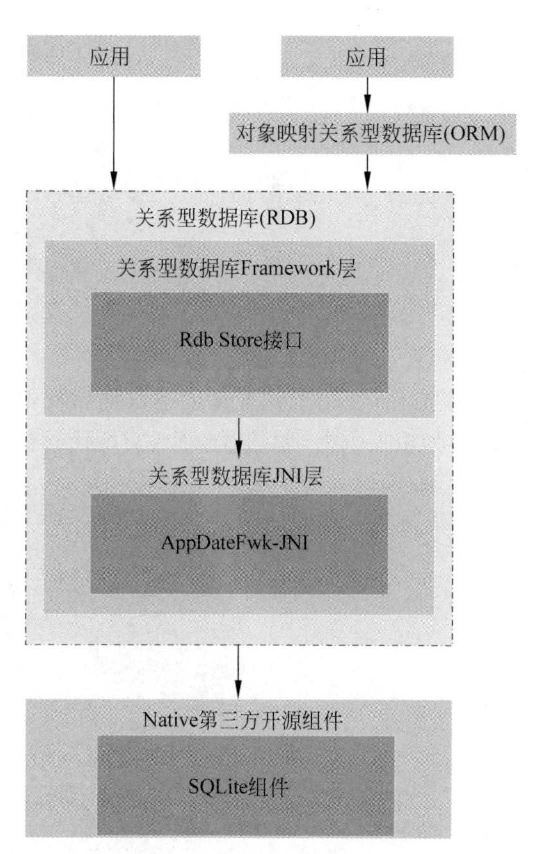

图 8.1 HarmonyOS 中关系型数据库的运作机制

```
    //向表中添加数据,以 HiLog 形式显示插入数据后的表结构
    ...

    //按条件查询表中数据并显示表
    ...

    //按条件更新数据,显示更新数据后的表
    ...

    //按条件删除表中数据,显示删除数据后的表
    ...

    super.setUIContent(ResourceTable.Layout_ability_main);
}
```

下面具体展开讲解关系型数据库的创建和操作方法。

1. 数据库创建

(1) 在创建关系型数据库之前,需要先配置与数据库相关的信息,包括设置数据库名、存储模式、日志模式、同步模式、是否为只读，以及对数据库加密等，代码如下：

```
//新建数据库与表

//1. 配置数据库相关信息
StoreConfig config = StoreConfig.newDefaultConfig("RdbStoreTest.db");

//2. 初始化数据库表结构和相关属性
...
```

其中，StoreConfig 类用于管理关系数据库的配置，StoreConfig. newDefaultConfig(String name)通过指定数据库名称，使用默认配置创建数据库配置。

(2) 初始化数据库表结构和相关数据，创建数据库，实现代码如下：

```
//初始化数据库表结构和相关属性
RdbOpenCallback callback = new RdbOpenCallback() {
      @Override
      public void onCreate(RdbStore store) {
//建表
 store.executeSql("CREATE TABLE IF NOT EXISTS test (id INTEGER PRIMARY KEY AUTOINCREMENT, name
TEXT NOT NULL, age INTEGER, salary REAL)");
      }

      @Override
      public void onUpgrade(RdbStore store, int oldVersion, int newVersion) {
      }
};

//建库
DatabaseHelper helper = new DatabaseHelper(this.getContext());
RdbStore store = helper.getRdbStore(config, 1, callback, null);
```

其中，RdbOpenCallback 类用于管理数据库的创建和升级等，并提供了 onCreate(RdbStore store)、onUpgrade()等方法。

这里重点讲解 onCreate(RdbStore store)方法，该方法在数据库创建时被回调，开发者可以在该方法中初始化表结构，并将一些应用可能使用的初始化数据添入表中。RdbStore 接口类提供了用于管理关系数据库的方法，可以实现创建、查询、更新和删除关系型数据库。executeSql(String sql)方法执行 SQL 语句。在本例中，store.executeSql 语句判断数据库中是否存在表名为 test 的表，若不存在，则新建 test 表，并添加相关属性：id(主键)、name(非空)、age、salary。DatabaseHelper 是数据库场景中的主入口类，提供了构造和删除关系型数据库、对象关系映射数据库的方法。getRdbStore()方法实现根据配置创建或打开数据库。

至此，名为 RdbStoreTest.db 的关系型数据库新建完成，且数据库中包含了新建的关系表 test 表。

2. 数据库插入

(1) 在操作之前需要说明的是，数据以 ValuesBucket 形式存储，因此在执行插入操作之前，首先需要构造要插入的数据，实现代码如下：

```
//向表中添加数据,以 HiLog 形式显示插入数据后的表结构

//1. 构造要添加的数据

        //新增数据 1
ValuesBucket values1 = new ValuesBucket();
        values1.putInteger("id", 1);
        values1.putString("name", "aaa");
        values1.putInteger("age", 26);
        values1.putDouble("salary", 1000.5);

        //新增数据 2
ValuesBucket values2 = new ValuesBucket();
        values2.putInteger("id", 2);
        values2.putString("name", "bbb");
        values2.putInteger("age", 25);
        values2.putDouble("salary", 1233.5);

        //新增数据 3
ValuesBucket values3 = new ValuesBucket();
        values3.putInteger("id", 3);
        values3.putString("name", "ccc");
        values3.putInteger("age", 25);
        values3.putDouble("salary", 1500.3);

        //2. 插入数据并以 HiLog 形式显示表
        ...
```

其中，putInteger(String columnName, Integer value)方法实现的功能是将整数值 value 放入当前 ValuesBucket 并将其与给定的列名 columnName 相关联。putString()、putDouble()和 putByteArray()方法实现的功能同理。以 values.putInteger("id", 1)为例，设置插入数据的 id 属性值为 1。

(2) 对数据执行插入操作，实现代码如下：

```
//2. 插入数据并以 HiLog 形式显示表

store.insert("test", values1);
store.insert("test", values2);
store.insert("test", values3);
```

```
//显示数据
HiLogLabel logLabel = new HiLogLabel(HiLog.LOG_APP,0,"RDB");
//显示表头
String s = new String("id");
s = s.concat("\tname");
s = s.concat("\tage");
s = s.concat("\t\tsalary");
HiLog.fatal(logLabel,"插入数据后查询表:");
HiLog.fatal(logLabel,s);

//显示表中所有数据并按 id 升序排列
String[] columns1 = new String[]{"id", "name", "age", "salary"};
RdbPredicates rdbPredicates = new RdbPredicates("test").orderByAsc("id");
ResultSet resultSet = store.query(rdbPredicates, columns1);
boolean next = resultSet.goToNextRow();
while(next){
String s1 = new String(resultSet.getString(resultSet.getColumnIndexForName("id")));
s1 = s1.concat("\t\t");
s1 = s1.concat(resultSet.getString(resultSet.getColumnIndexForName("name")));
s1 = s1.concat("\t\t");
s1 = s1.concat(resultSet.getString(resultSet.getColumnIndexForName("age")));
s1 = s1.concat("\t\t");
s1 = s1.concat(resultSet.getString(resultSet.getColumnIndexForName("salary")));
HiLog.fatal(logLabel,s1);
next = resultSet.goToNextRow();
}
```

关系型数据库提供了插入数据的接口，通过 insert(String table, ValuesBucket initialValues)方法将数据插入关系表中，并通过返回值判断是否插入成功，插入成功时返回最新插入数据所在的行号，失败则返回－1。

读者可以插入多条不同的数据，但注意每次插入之前都要将数据构造成 ValuesBucket 形式。以 HiLog 形式显示插入 3 条数据后的 test 表，此处涉及了数据库的查询操作，稍后会进行详细讲解。插入结果如图 8.2 所示。从图中可以看出 3 条数据均成功插入表中。

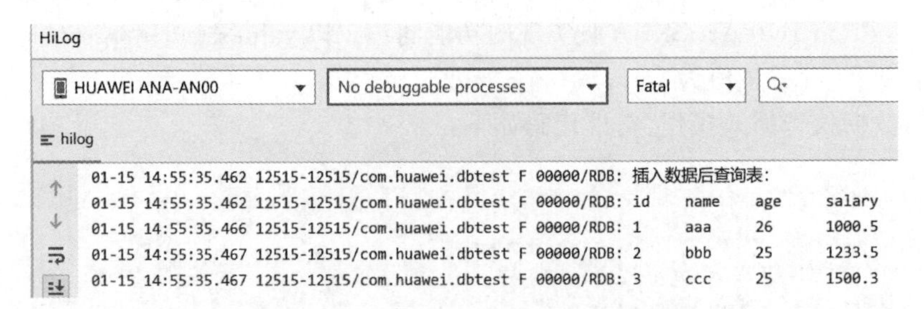

图 8.2 在新建的 test 表中插入 3 条数据

3. 查询数据并显示

(1) 指定查询的返回列,实现代码如下:

```
//按条件查询表中数据并显示
...
//1. 指定查询的返回列
String[] columns = new String[] {"id", "name", "age", "salary"};
//2. 设置查询条件
...
//3. 查询数据并显示查询结果
...
```

在本例中返回表中的全部列,也可以只返回部分列。

(2) 构造用于查询的谓词对象,设置查询条件。谓词是数据库中用来代表数据实体的性质、特征或者数据实体之间关系的词项,主要用来定义数据库的操作条件。

AbsRdbPredicates 类用于为关系数据库设置数据库操作条件的抽象父类,其实现类包括 RdbPredicates 和 RawRdbPredidates。其中,RdbPredicates 类提供了 equalTo、notEqualTo、groupBy、orderByAsc、beginsWith 等方法,使开发者无须编写复杂的 SQL 语句,仅通过调用方法即可自动完成 SQL 语句拼接,方便用户聚焦业务操作。RawRdbPredidates 可满足复杂 SQL 语句的场景,支持开发者自己设置 where 条件子句和 whereArgs 参数,但不支持 equalTo 等条件接口的使用。下面是使用 RdbPredicates 设置查询条件的例子,实现代码如下:

```
//2. 设置查询条件,查询 age = 25 的数据
RdbPredicates rdbPredicates1 = new RdbPredicates("test").equalTo("age",25);
```

本例中查询条件为 age,其属性值等于 25。此外,RdbPredicates 还提供了 notEqualTo、between、in、greaterThan 及 lessthan 等众多接口,便于设置查询条件。可以在官方 API 文档中查看相应的使用方法。

(3) 查询数据。关系型数据库提供了两种查询数据的方式。一种是通过直接调用查询接口。使用该接口,会将包含查询条件的谓词自动拼接成完整的 SQL 语句进行查询操作,无须用户传入原生的 SQL。例如,利用 query 接口根据指定的条件查询数据库中的数据,实现代码如下:

```
store.query(rdbPredicates, columns);
```

另一种方式是执行用于查询的原生 SQL 语句,感兴趣的读者可以自行进行测试,此处就不介绍 SQL 语句的使用了。

(4) 调用结果集接口,遍历并返回结果。关系型数据库提供了查询返回的结果集 ResultSet,指向查询结果中的一行数据,供用户对查询结果进行遍历和访问。需要说明的

是，ResultSet 利用 goToNextRow()方法进行数据遍历。具体代码如下：

```
//3. 查询数据并显示查询结果
HiLog.fatal(logLabel,"查询年龄为 25 的数据：");
HiLog.fatal(logLabel,s);
ResultSet resultSet1 = store.query(rdbPredicates1, columns);
boolean next1 = resultSet1.goToNextRow();
while(next1){
    String s1 = newString(resultSet1.getString(resultSet1.getColumnIndexForName("id")));
    s1 = s1.concat("\t\t");
    s1 = s1.concat(resultSet1.getString(resultSet1.getColumnIndexForName("name")));
    s1 = s1.concat("\t\t");
    s1 = s1.concat(resultSet1.getString(resultSet1.getColumnIndexForName("age")));
    s1 = s1.concat("\t\t");
    s1 = s1.concat(resultSet1.getString(resultSet1.getColumnIndexForName("salary")));
    HiLog.fatal(logLabel,s1);
    next1 = resultSet1.goToNextRow();
}
```

显示数据时，利用 getString()方法获取当前行指定索列的值，以 String 类型返回。getColumnIndexForName(String columnName)基于指定的列名获取列索引。例如，获取当前查询结果的 name 属性值的代码如下：

```
String name = resultSet.getString(resultSet.getColumnIndexForName("name"));
```

查询表中 age 属性值为 25 的全部数据，结果如图 8.3 所示。

```
01-15 14:55:35.467 12515-12515/com.huawei.dbtest F 00000/RDB: 查询年龄为25的数据:
01-15 14:55:35.467 12515-12515/com.huawei.dbtest F 00000/RDB: id    name    age    salary
01-15 14:55:35.469 12515-12515/com.huawei.dbtest F 00000/RDB: 2     bbb     25     1233.5
01-15 14:55:35.469 12515-12515/com.huawei.dbtest F 00000/RDB: 3     ccc     25     1500.3
```

图 8.3　查询表中 age 属性值为 25 的数据

4. 更新数据

数据更新的方法为 int update(ValuesBucket values, AbsRdbPredicates predicates)。其中，values 是以 ValuesBucket 形式存储并需要更新的数据，predicates 为指定的更新条件，因此该方法将满足更新条件 predicates 的数据更新为传入的新数据 values。该方法的返回值表示更新操作影响的行数。如果更新失败，则返回 0。为了更好地让读者理解，本例中将 id=3 的数据进行更新，具体实现代码如下：

```
//按条件更新数据，显示更新数据后的表

//1. 将新数据以 ValuesBucket 形式存储
ValuesBucket values4 = new ValuesBucket();
values4.putInteger("id", 3);
```

```
values4.putString("name", "ddd");
values4.putInteger("age", 41);
values4.putDouble("salary", 1300.2);

//2. 设置更新条件:更新 id = 3 的数据
RdbPredicates rdbPredicatesupdate = newRdbPredicates("test").equalTo("id",3);

//3. 进行更新操作
store.update(values4,rdbPredicatesupdate);

//4. 显示进行数据更新后的表
HiLog.fatal(logLabel,"查询更新后的表,查询结果按 id 升序排列:");
HiLog.fatal(logLabel,s);
ResultSet resultSet3 = store.query(rdbPredicates, columns);
boolean next3 = resultSet3.goToNextRow();
while(next3){
    String s1 = new String(resultSet3.getString(resultSet3.getColumnIndexForName("id")));
    s1 = s1.concat("\t\t");
    s1 = s1.concat(resultSet3.getString(resultSet3.getColumnIndexForName("name")));
    s1 = s1.concat("\t\t");
    s1 = s1.concat(resultSet3.getString(resultSet3.getColumnIndexForName("age")));
    s1 = s1.concat("\t\t");
    s1 = s1.concat(resultSet3.getString(resultSet3.getColumnIndexForName("salary")));
    HiLog.fatal(logLabel,s1);
    next3 = resultSet3.goToNextRow();
}
```

使用 HiLog 方式显示更新后的 test 表,如图 8.4 所示,与图 8.3 所示数据相比,id=3 的数据被成功修改。

```
01-15 14:55:35.474 12515-12515/com.huawei.dbtest F 00000/RDB: 查询更新后的表, 查询结果按id升序排列:
01-15 14:55:35.474 12515-12515/com.huawei.dbtest F 00000/RDB: id    name    age    salary
01-15 14:55:35.475 12515-12515/com.huawei.dbtest F 00000/RDB: 1     aaa     26     1000.5
01-15 14:55:35.475 12515-12515/com.huawei.dbtest F 00000/RDB: 2     bbb     25     1233.5
01-15 14:55:35.475 12515-12515/com.huawei.dbtest F 00000/RDB: 3     ddd     41     1300.2
```

图 8.4　更新 id=3 的数据

5. 删除数据

当需要删除表中的数据时,可以调用删除接口 int delete(AbsRdbPredicates predicates),通过 AbsRdbPredicates 指定删除条件。本例中将删除 id=2 的数据,实现代码如下:

```
//删除数据并显示删除数据后的表

//设置删除条件: id = 2
RdbPredicates rdbPredicatesdel = new RdbPredicates("test").equalTo("id",2);
```

```
//删除满足条件的数据
store.delete(rdbPredicatesdel);

HiLog.fatal(logLabel,"查询进行数据删除后的表,查询结果按 id 升序排列:");
HiLog.fatal(logLabel,s);
ResultSet resultSet4 = store.query(rdbPredicates, columns);
boolean next4 = resultSet4.goToNextRow();
while(next4){
      String s1 = new String(resultSet4.getString(resultSet4.getColumnIndexForName("id")));
      s1 = s1.concat("\t\t");
      s1 = s1.concat(resultSet4.getString(resultSet4.getColumnIndexForName("name")));
      s1 = s1.concat("\t\t");
      s1 = s1.concat(resultSet4.getString(resultSet4.getColumnIndexForName("age")));
      s1 = s1.concat("\t\t");
      s1 = s1.concat(resultSet4.getString(resultSet4.getColumnIndexForName("salary")));
      HiLog.fatal(logLabel,s1);
      next4 = resultSet4.goToNextRow();
}
```

删除数据后的表如图 8.5 所示。对比图 8.5 与图 8.4 可知数据删除成功。

```
01-15 14:55:35.477 12515-12515/com.huawei.dbtest F 00000/RDB: 查询进行数据删除后的表，查询结果按id升序排列:
01-15 14:55:35.477 12515-12515/com.huawei.dbtest F 00000/RDB: id      name     age      salary
01-15 14:55:35.478 12515-12515/com.huawei.dbtest F 00000/RDB: 1       aaa      26       1000.5
01-15 14:55:35.478 12515-12515/com.huawei.dbtest F 00000/RDB: 3       ddd      41       1300.2
```

图 8.5　删除 id=2 的数据

8.1.3 对象关系映射数据库

HarmonyOS 对象关系映射数据库是 HarmonyOS 中的本地数据库之一，是在 SQLite 数据库的基础上提供的一个抽象层。它屏蔽了底层 SQLite 数据库的 SQL 操作，将实例对象映射到关系上，并提供增、删、改、查等一系列面向对象的接口，使应用程序使用操作实例对象的语法来操作关系型数据库，因此，开发者可以不必编写那些复杂的 SQL 语句，从而既可以提升效率也能聚焦于业务开发。HarmonyOS 中对象关系映射数据库的运作机制如图 8.6 所示。

从图 8.6 可以看出，对象关系映射数据库的 3 个主要组件包括：

(1) 数据库：通过@Database 注解，且继承自 OrmDatabase 的类，对应关系型数据库。

(2) 实体对象：通过@Entity 注解，且继承自 OrmObject 的类，对应关系型数据库中的表。

(3) 对象数据操作接口：包括数据库操作的入口 OrmContext 类和谓词接口(OrmPredicate)等。

创建 Phone 设备下的 Java 模板新项目。在进行开发之前，首先要进行 build.gradle 文件的配置，否则无法识别相关数据库的类的包。

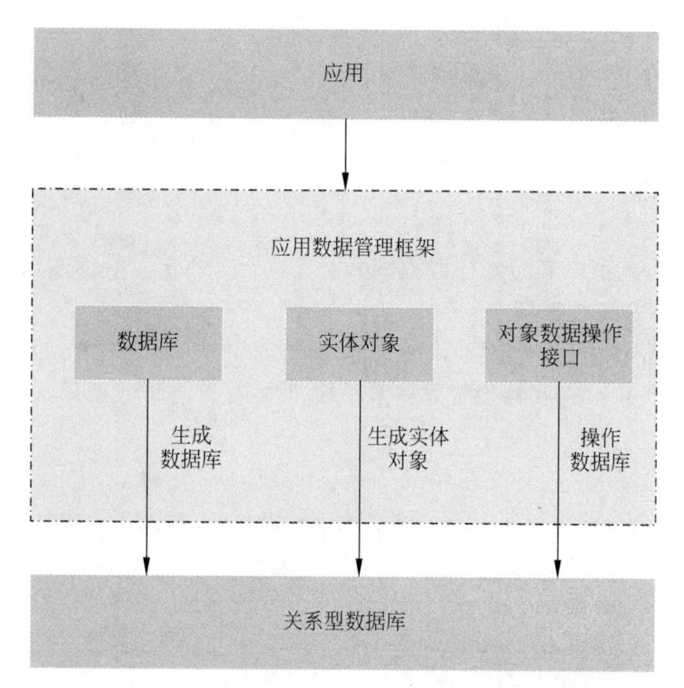

图 8.6　对象关系映射数据库的运作机制

一般情况下，使用的注解处理器的模块为 com. huawei. ohos. hap 模块，需要在模块的 build. gradle 文件的 ohos 节点中添加以下配置，代码如下：

```
compileOptions{
    annotationEnabled true
}
```

如果使用注解处理器的模块为 com. huawei. ohos. library，则需要在模块的 build. gradle 文件 dependencies 节点中配置注解处理器，此外，还需要在本地的 HUAWEI SDK 中找到 orm_annotations_java. jar、orm_annotations_processor_java. jar 和 javapoet_java. jar 这 3 个 jar 包的对应目录，并将这 3 个 jar 包的路径导入，代码如下：

```
dependencies {
    compile files("orm_annotations_java.jar 的路径","orm_annotations_processor_java.jar 的路径","javapoet_java.jar 的路径")
    annotationProcessor files("orm_annotations_java.jar 的路径","orm_annotations_processor_java.jar 的路径","javapoet_java.jar 的路径")
}
```

如果使用注解处理器的模块为 java-library，则还需要导入 ohos. jar 的路径，具体代码如下：

```
dependencies {
    compile files("ohos.jar 的路径","orm_annotations_java.jar 的路径","orm_annotations_processor_java.jar 的路径","javapoet_java.jar 的路径")
    annotationProcessor files("orm_annotations_java.jar 的路径","orm_annotations_processor_java.jar 的路径","javapoet_java.jar 的路径")
}
```

配置完成之后，就可以开始新建数据库并对其进行操作了。具体步骤如下：

1. 新建数据库及属性配置

创建数据库时，首先需要定义一个表示数据库的类，继承 OrmDatabase，再通过@Database 注解内的 entities 属性指定哪些数据模型类属于这个数据库，version 属性指明数据库版本号。

在本例中，首先新建一个数据库类 BookStore.java，选择 entry→src→main→com.huawei.ormdb，右击选择并新建 Class，命名为 BookStore，如图 8.7 所示。

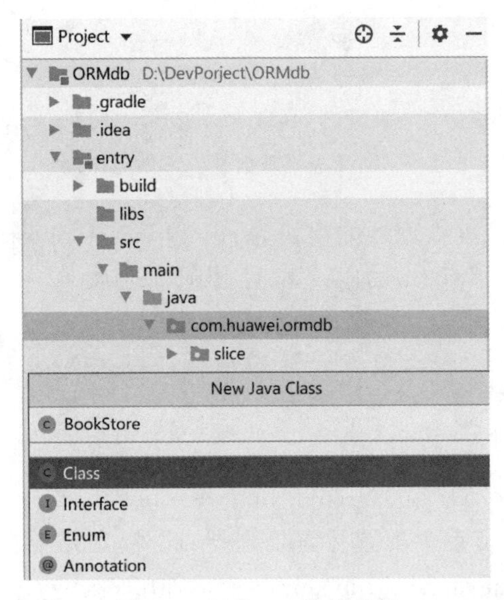

图 8.7 新建一个数据库类

数据库类 BookStore.java 包含了 User 和 Book 两个表，版本号为 1，将数据库类设置为虚类。具体代码如下：

```
import ohos.data.orm.OrmDatabase;
import ohos.data.orm.annotation.Database;

@Database(entities = {User.class, Book.class}, version = 1)
    public abstract class BookStore extends OrmDatabase {
}
```

2. 构造数据表

构造数据表，即创建数据库实体类并配置对应的属性（如对应表的主键、外键等）。可通过创建一个继承了OrmObject并用@Entity注解的类，获取数据库实体对象，也就是表的对象。需要注意的是，数据表必须与其所在的数据库在同一个模块中。以新建的User表为例，新建的User类与BookStore类位于同一模块下，相应的目录如图8.8所示，新建过程同BookStore类，此处不再赘述。

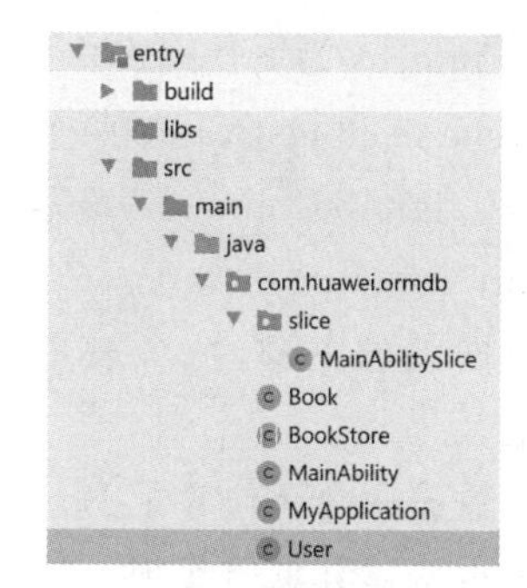

图8.8　BookStore数据库与User表位于同一模块

具体的构建过程代码如下：

```
//新建 User 表
import ohos.data.orm.OrmObject;
import ohos.data.orm.annotation.Entity;
import ohos.data.orm.annotation.Index;
import ohos.data.orm.annotation.PrimaryKey;
@Entity(tableName = "user", ignoredColumns = {"ignoreColumn1", "ignoreColumn2"},
indices = {@Index(value = {"firstName", "lastName"}, name = "name_index", unique = true)})
public class User extends OrmObject {
    //此处将 userId 设为自增的主键.注意只有在数据类型为包装类型时,自增主键才能生效
    @PrimaryKey(autoGenerate = true)
    private Integer userId;
    private String firstName;
    private String lastName;
    private int age;
    private double balance;
    private int ignoreColumn1;
    private int ignoreColumn2;

    //设置字段的 getter 和 setter 方法,此处仅给出示例,读者可以根据所设置的属性自行
    //补全方法
    public void setBalance(double balance) {
        this.balance = balance;
    }

   ...

    public Integer getUserId() {
        return userId;
    }

   ...
}
```

在新建的User类中进行属性配置，tableName = "user"即在对应数据库内的表名为

user，indices 为属性列表，@Index 注解的内容对应数据表索引的属性，本例中 indices 为 firstName 和 lastName 两个字段建立了复合索引，索引名为 name_index，并且索引值是唯一的。ignoreColumns 表示该字段不需要添加到 user 表的属性中，即类中定义的 ignoreColumn1 和 ignoreColumn2 不属于 user 表的属性。被@PrimaryKey 注解的变量对应数据表的主键，一个表里只能有一个主键，在本例的 user 表中，将 userId 设为自增的主键。

数据库内还包含了 Book 表，Book 表的构建和对其操作的实现过程同 User 表相同，此处不再赘述其具体实现。感兴趣的读者可以参考以下对 User 表的说明自行完成对 Book 表的构建和操作。

3. 创建数据库

在 MainAbilitySlice 的 onStart()方法中完成数据库的创建，使用对象数据操作接口 OrmContext 创建数据库，实现代码如下：

```
public void onStart(Intent intent) {
    super.onStart(intent);
    super.setUIContent(ResourceTable.Layout_ability_main);

//使用对象数据操作接口 OrmContext 创建数据库
    DatabaseHelper helper = new DatabaseHelper(this);
    OrmContext context = helper.getOrmContext("BookStore", "BookStore.db", BookStore.class);

//对数据库进行操作
    //增加数据
      …

    //查询数据
      …

    //修改数据
      …

    //删除数据
      …
}
```

在 8.1.2 节创建了关系型数据库，相信读者对此并不陌生，但与构建关系型数据库不同的是，此处 new DatabaseHelper(context)方法中 context 的入参类型为 ohos.app.Context，必须直接传入 slice 而不能使用 slice.getContext()获取 context，否则会出现找不到类的报错。到这里，就成功创建了一个别名为 BookStore 且数据库文件名为 BookStore.db 的数据库。如果数据库已经存在，则执行上述代码并不会重复建库。通过 context.getDatabaseDir()可以获取创建的数据库文件所在的目录。

4. 数据操作

对象数据操作接口 OrmContext 提供了对数据库进行增、删、改、查的一系列方法，接下来进行详细介绍。

(1) 增加数据。例如，在名为 user 的表中，新建一个 User 对象并设置其属性。OrmContext 提供 insert()方法将对象插入数据库。执行完 insert()方法后，数据被保存在内存中，只有在 flush()被调用后才会将数据持久化到数据库中。以 HiLog 形式显示插入数据后的 user 表，如图 8.9 所示。此处涉及对数据库的查询操作，将在稍后进行详细讲解。具体代码如下：

```
//增加数据

//新建 User 对象
User user = new User();
//设置对象属性
user.setUserId(0);
user.setFirstName("aa");
user.setLastName("AA");
user.setAge(22);
user.setBalance(120.51);

//再新建两个 User 对象,设置属性后将其持久化到数据库中,此处读者可以自行添加代码
...
//将新建对象插入并持久化到数据库中
boolean isSuccessed = context.insert(user);
isSuccessed = context.flush();

//以 HiLog 形式显示新增数据后的 user 表,并以 userID 升序显示
HiLogLabel logLabel = new HiLogLabel(HiLog.LOG_APP,0,"OrmDB");
String s = new String("userID");
s = s.concat(" firstName");
s = s.concat(" lastName");
s = s.concat(" age");
s = s.concat(" balance");
HiLog.fatal(logLabel,"显示插入数据后的 user 表: ");
HiLog.fatal(logLabel,s);

//查询数据
OrmPredicates query = context.where(User.class).orderByAsc("userID");
List<User> users = context.query(query);
int i = 0;
while (i != users.size()) {
            int id = users.get(i).getUserId();
            String s1 = "\t" + id + "\t\t";
            s1 = s1.concat(users.get(i).getFirstName() + "\t");
```

```
            s1 = s1.concat("\t\t" + users.get(i).getLastName());
            s1 = s1.concat("\t " + users.get(i).getAge() + "\t");
            s1 = s1.concat(" " + users.get(i).getBalance());
            HiLog.fatal(logLabel, s1);
            i = i + 1;
    }
```

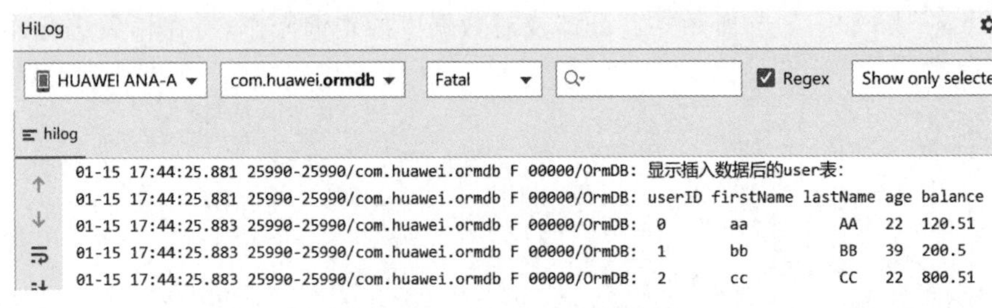

图 8.9 向 user 表中添加数据

(2) 查询数据。OrmContext 提供 query()方法查询满足指定条件的对象实例。例如，在 user 表中查询 age=22 的对象，在查询之前，依旧需要先设置谓词 query1，利用 query()方法查找 user 表中满足 query1 的数据，得到一个 User 的列表 user1，遍历 user1，获取 user1 中各个对象的属性值，实现代码如下：

```
//查询 age = 22 的数据并显示

//设置谓词
OrmPredicates query1 = context.where(User.class).equalTo("age", "22");
//查询
List<User> users1 = context.query(query1);

//将查询结果以 HiLog 形式显示
HiLogLabel logLabel1 = new HiLogLabel(HiLog.LOG_APP,0,"OrmDB");
HiLog.fatal(logLabel1,"显示 user 表中 age 值为 22 的数据:");
HiLog.fatal(logLabel1,s);
i = 0;
while (i != users1.size()){
      int id = users1.get(i).getUserId();
      String s1 = "\t" + id +"\t\t";
      s1 = s1.concat(users1.get(i).getFirstName() + "\t");
      s1 = s1.concat("\t\t" + users1.get(i).getLastName());
      s1 = s1.concat("\t " + users1.get(i).getAge() + "\t");
      s1 = s1.concat(" " + users1.get(i).getBalance());
      HiLog.fatal(logLabel1,s1);
      i = i + 1;
}
```

查询结果如图 8.10 所示。

```
01-15 17:44:25.883 25990-25990/com.huawei.ormdb F 00000/OrmDB: 显示user表中age=22的数据:
01-15 17:44:25.883 25990-25990/com.huawei.ormdb F 00000/OrmDB: userID firstName lastName age balance
01-15 17:44:25.884 25990-25990/com.huawei.ormdb F 00000/OrmDB:  0       aa        AA    22  120.51
01-15 17:44:25.884 25990-25990/com.huawei.ormdb F 00000/OrmDB:  2       cc        CC    22  800.51
```

图 8.10 查询 user 表中 age=22 的数据

(3) 修改数据。修改数据包括两种方式，一种是通过直接传入 OrmObject 对象的接口来更新数据，需要先从表中查到需要更新的结果对象列表，然后修改选定对象的值，再调用更新接口将数据持久化到数据库中。例如，将 user 表中 age=39 的数据的 firstName 更新为 sd，需要先查找 user 表中对应的数据，得到一个结果列表，然后选择列表中需要更新的 User 对象，设置更新值，并调用 update 接口传入被更新的 User 对象。最后调用 flush 接口将数据持久化到数据库中。具体代码如下：

```
//更新数据
OrmPredicates predicates = context.where(User.class);
predicates.equalTo("age", 39);
//获取满足条件的数据集
List<User> users2 = context.query(predicates);
//选定要更新的数据
User userUD = users2.get(0);
//设置更新
userUD.setFirstName("sd");
context.update(userUD);
context.flush();

HiLog.fatal(logLabel, "显示更新数据后的 user 表: ");
HiLog.fatal(logLabel, s);
query = context.where(User.class).orderByAsc("userID");
users = context.query(query);
i = 0;
while (i != users.size()) {
     int id = users.get(i).getUserId();
     String s1 = "\t" + id + "\t\t";
     s1 = s1.concat(users.get(i).getFirstName() + "\t");
     s1 = s1.concat("\t\t" + users.get(i).getLastName());
     s1 = s1.concat("\t " + users.get(i).getAge() + "\t");
     s1 = s1.concat(" " + users.get(i).getBalance());
     HiLog.fatal(logLabel, s1);
i = i + 1;
}
```

另一种方式，可以通过传入谓词的接口来更新和删除数据，方法与 OrmObject 对象的接口类似，只是无须 flush 就可以将数据持久化到数据库中。通过这种方式将 user 表中 age=39 的数据的 firstName 更新为 sd，具体代码如下：

```
//使用 valuesBucket 来改变数据
ValuesBucket valuesBucket = new ValuesBucket();
valuesBucket.putString("firstName", "sd");
OrmPredicates update = context.where(User.class).equalTo("age", 39);
context.update(update, valuesBucket);
```

分别利用两种方式进行数据更新,更新后的结果如图 8.11 所示。

```
01-15 17:44:25.887 25990-25990/com.huawei.ormdb F 00000/OrmDB: 显示更新数据后的user表:
01-15 17:44:25.887 25990-25990/com.huawei.ormdb F 00000/OrmDB: userID firstName lastName age balance
01-15 17:44:25.887 25990-25990/com.huawei.ormdb F 00000/OrmDB:  0       aa          AA   22  120.51
01-15 17:44:25.887 25990-25990/com.huawei.ormdb F 00000/OrmDB:  1       sd          BB   39  200.5
01-15 17:44:25.887 25990-25990/com.huawei.ormdb F 00000/OrmDB:  2       cc          CC   22  800.51
```

图 8.11 数据更新结果

(4) 删除数据。同修改数据一样,也包括两种方式。区别只是删除数据不需要更新对象的值。这里通过直接传入 OrmObject 对象的接口删除数据的方式删除满足 firstName="aa"的第一个 User 对象,并用 flush 接口将数据持久化到数据库中,代码如下:

```
//删除满足 firstName = "aa"的第一个对象
OrmPredicates del = context.where(User.class).equalTo("firstName", "aa");
List<User> usersdel = context.query(del);
User userd = usersdel.get(0);
context.delete(userd);
context.flush();
HiLog.fatal(logLabel, "显示删除数据后的 user 表: ");
HiLog.fatal(logLabel, s);
query = context.where(User.class).orderByAsc("userID");
users = context.query(query);
i = 0;
while (i != users.size()) {
    int id = users.get(i).getUserId();
    String s1 = "\t" + id + "\t\t";
    s1 = s1.concat(users.get(i).getFirstName() + "\t");
    s1 = s1.concat("\t\t" + users.get(i).getLastName());
    s1 = s1.concat("\t " + users.get(i).getAge() + "\t");
    s1 = s1.concat(" " + users.get(i).getBalance());
    HiLog.fatal(logLabel, s1);
    i = i + 1;
}
```

显示删除数据后的表,如图 8.12 所示。

```
01-15 17:44:25.890 25990-25990/com.huawei.ormdb F 00000/OrmDB: 显示删除数据后的user表:
01-15 17:44:25.890 25990-25990/com.huawei.ormdb F 00000/OrmDB: userID firstName lastName age balance
01-15 17:44:25.891 25990-25990/com.huawei.ormdb F 00000/OrmDB:  1       sd          BB   39  200.5
01-15 17:44:25.891 25990-25990/com.huawei.ormdb F 00000/OrmDB:  2       cc          CC   22  800.51
```

图 8.12 删除数据后的结果显示

8.1.4 轻量级偏好数据库

HarmonyOS 中，还有轻量级偏好数据库的本地数据库。区别于关系数据库，轻量级偏好数据库不保证遵循 ACID 特性，也不采用关系模型来组织数据，而是以键值对存储数据，提供轻量级 Key-Value 操作，Key 是关键字，Value 是值。此外，轻量级偏好数据库支持本地应用存储少量数据，数据既存储在本地文件中，也加载在内存中，因此访问速度更快，效率更高。轻量级偏好数据库的运作机制如图 8.13 所示。

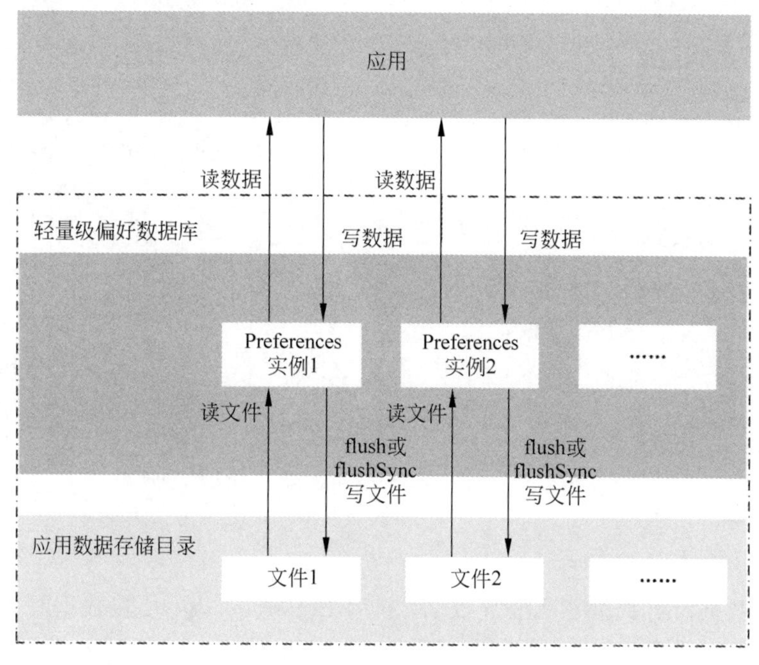

图 8.13 轻量级偏好数据库的运作机制

由图 8.13 可知，轻量级偏好数据库向本地应用提供了操作偏好型数据库的 API，支持本地应用读写少量数据及观察数据变化。具体地，应用可以借助 DatabaseHelper API 将指定文件内容加载到文件所对应的 Preferences 实例，每个文件最多有一个 Preferences 实例。接着系统会通过静态容器将该实例存储在内存中，直到应用主动从内存中移除该实例或者删除该文件。获取文件对应的 Preferences 实例后，应用可以借助 Preferences API，从 Preferences 实例中读取数据或者将数据写入 Preferences 实例，通过 flush 或者 flushSync 将 Preferences 实例持久化。

创建 Phone 设备下的 Java 模板项目，接下来进行轻量级偏好数据库的建立及操作。依旧通过覆写 MainAbilitySlice 中的 onStart()方法实现，具体过程如下。

(1) 新建 Preferences 实例。读取指定文件，将数据加载到 Preferences 实例，用于数据操作。其中，fileName 为文件名，其取值不能为空，也不能包含路径，在本例中文件名为 test。实现代码如下：

```
//获取 Preferences 实例

public void onStart(Intent intent) {
        super.onStart(intent);
        super.setUIContent(ResourceTable.Layout_ability_main);

DatabaseHelper DatabaseHelper = new DatabaseHelper(this);
        String fileName = "test";
        //获取 Preferences 实例
        Preferences preferences = DatabaseHelper.getPreferences(fileName);

        //插入数据
        ...

        //读取数据
        ...

        //注册观察者,进行数据更新
        ...

        //移动文件
        ...

        //删除文件
        ...

}
```

(2) 插入数据。将数据写入指定文件，利用 Preferences API 将数据写入文件对应的 Preferences 实例，通过 flush()或者 flushSync()将 Preferences 实例异步或同步进行持久化。插入的数据的值可以是 Int 类型、String 类型、Float 类型等，不同的数据类型对应不同的插入方法：putInt()、putString()、putFloat()等，代码如下：

```
//插入数据,进行异步持久化

preferences.putInt("Id", 1);                    //key = "Id", value = 1
preferences.putString("Name", "Alice");         //key = "Name", value = "Alice"
preferences.flush();
```

(3) 从指定文件读取数据。借助 Preferences API 读取基于指定文件所对应的 Preferences 实例。例如读取上述插入的数据并以 HiLog 形式显示读取结果，代码如下：

```
//读取插入的数据
int valueId = preferences.getInt("Id", 0);                        //得到 key = "Id"对应的数据值
String valueName = preferences.getString("Name","NoName");
```

```
//HiLog 形式显示读取结果
HiLogLabel hiLogLabel = new HiLogLabel(HiLog.LOG_APP,0,"PDB");
HiLog.fatal(hiLogLabel,"添加数据后的结果：");
String showId = "Id: " + valueId;
String showName = "Name: " + valueName;
HiLog.fatal(hiLogLabel, showId);
HiLog.fatal(hiLogLabel, showName);
```

需要说明的是，getInt("Id", 0)获取 key＝"Id"所对应的 Int 类型的值，若所对应的值类型不为 Int 类型或者对应的值为 null，则返回默认值 0。同理，getString("Name","No")返回的是 key＝"Name"所匹配的 String 类型的值，否则返回默认值 No。显示结果如图 8.14 所示。

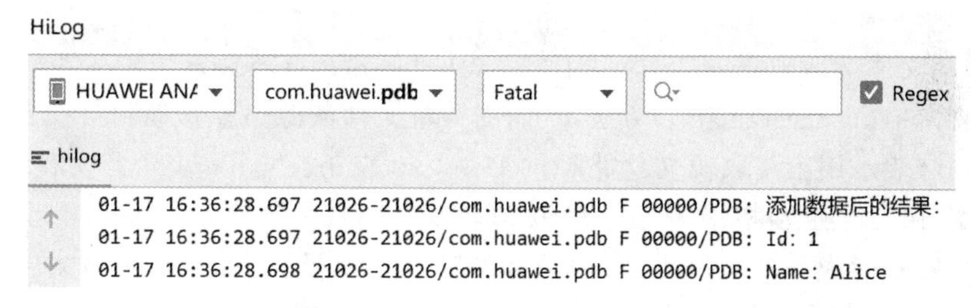

图 8.14　读取新添加的数据

(4) 注册观察者。Preferences 对象支持注册观察者，观察者对象需要实现 Preferences.PreferencesObserver 接口。Preferences 对象进行更新并通过 flushSync()方法或 flush()方法持久化更新操作后，该对象注册的所有观察者的 onChange()方法会被回调。即观察者一旦被注册，就能观察到注册该观察者的 Preferences 对象的所有持久化的更新操作。将 key＝"Id"所对应的值更新为 2，具体代码如下：

```
//注册观察者
preferences.registerObserver(new Preferences.PreferencesObserver(){
    @Override
    public void onChange(Preferences preferences, String s) {
        HiLog.fatal(hiLogLabel,"检测到数据改变");

        //显示更新后的 Preferences 对象
        int valueId = preferences.getInt("Id", 0);
        String showId = "Id: " + valueId;
        String valueName = preferences.getString("Name","NoName");
        String showName = "Name: " + valueName;
        HiLog.fatal(hiLogLabel, showId);
        HiLog.fatal(hiLogLabel, showName);
    }
```

```
});

//数据更新
preferences.putInt("Id", 2);

//持久化更新
preferences.flush();
```

注册观察者后进行更新的结果如图 8.15 所示。从显示结果可以看出,观察者观察到了其被注册之后所有的更新操作并成功显示更新后的结果。

```
01-17 16:36:28.698 21026-21026/com.huawei.pdb F 00000/PDB: 检测到数据改变
01-17 16:36:28.698 21026-21026/com.huawei.pdb F 00000/PDB: Id: 2
01-17 16:36:28.698 21026-21026/com.huawei.pdb F 00000/PDB: Name: Alice
```

图 8.15 注册观察者并检测到数据更新

(5) 移动文件。HarmonyOS 提供了 movePreferences(Context sourceContext, String sourceName, String targetName)方法,从而支持将文件从源路径移动到目标路径。其中,sourceContext 表示用于获取源文件路径的 Context,SourceName 表示源文件名或者源文件的绝对路径,不能为相对路径,其取值不能为空。当 SourceName 为文件名时,sourceContext 不能为空,targetName 为目标文件名。需要注意的是,移动文件时,应用不允许再操作该文件数据,否则会出现数据一致性问题。具体代码如下,图 8.16 为移动文件后的 HiLog 输出结果。

```
//移动文件
String srcFile = "test"; //
String targetFile = "targetFile";        //targetFile 表示目标文件名,其取值不能为空也不能
                                         //包含路径
//移动
boolean result = DatabaseHelper.movePreferences(this, srcFile, targetFile);

//显示是否移动成功
HiLog.fatal(hiLogLabel,"是否移动成功?(true/false): ".concat(String.valueOf (result)));
```

```
01-17 16:36:28.698 21026-21026/com.huawei.pdb F 00000/PDB: 是否移动成功? (true/false): true
```

图 8.16 移动文件

(6) 删除文件。HarmonyOS 同样支持从内存中移除指定文件对应的 Preferences 单实例,并删除指定文件及其备份文件、损坏文件。需要注意的是删除指定文件时,应用不允许再使用该实例进行数据操作,否则会出现数据一致性问题。具体代码如下,删除文件的结果如图 8.17 所示。

```
01-17 16:36:28.700 21026-21026/com.huawei.pdb F 00000/PDB: 是否删除成功? (true/false): true
```

图 8.17 删除文件

```
//删除文件
result = DatabaseHelper.deletePreferences(targetFile);
HiLog.fatal(hiLogLabel, "是否删除成功?(true/false): ".concat(String.valueOf(result)));
```

8.2　分布式服务

14min

8.2.1　多设备协同权限

在 HarmonyOS 中，不仅支持对本地数据的管理，还支持跨设备之间数据的同步、共享等操作。分布式数据服务支持用户跨设备同步数据，从而保证用户多种终端设备上的数据的一致性。分布式文件服务支持用户相同账号下同一应用文件的跨设备访问，从而使多设备之间具有文件共享能力。

需要注意的是，在使用 HarmonyOS 所提供的分布式服务之前，用户必须在多种设备上登录相同账号，且必须开启应用的多设备协同的相关权限和系统的多设备协同连接。

首先，需要打开手机系统中的多设备协同连接，进入设置→更多连接→多设备协同，开启多设备协同开关，如图 8.18 所示。

图 8.18　开启系统多设备协同连接

随后需要开启应用的相关权限。应用的权限管理需要在config.json文件中进行配置。可以在entry→src→main→config.json中找到项目对应的config.json文件,如图8.19所示。

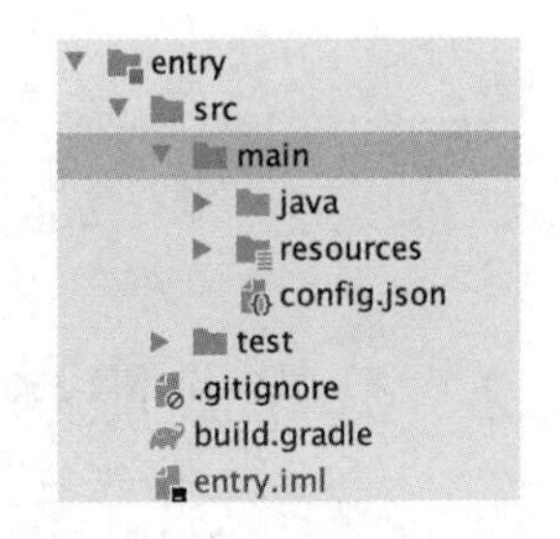

图8.19 config文件位置

通过阅览config.json文件可以发现该文件由若干个字段组成,例如app、deviceConfig、abilities等。如需要定义权限,则需要在这些字段的基础上添加reqPermissions字段,并且将对应的权限信息添加进去。下面以多设备协同能力为例,展示如何添加权限,代码如下:

```
"reqPermissions": [{
        "name": "ohos.permission.DISTRIBUTED_DATASYNC",
        "grantMode": "user_grant",
        "reason": "",
        "usedScene": {
            "ability": [
                "com.huawei.myapplication.MainAbility"
            ],
            "when": "always"
        }
    }
]
```

将上述代码同abilities、deviceConfig一样,并列添加至config.json文件中,即可完成多设备协同的权限信息声明。

将上面的reqPermissions信息添加进去之后,应用就具备了开启多设备协同的能力,但是默认情况下,该能力还是关闭状态,需要在设置中打开。进入设置→隐私→权限管理→权限→其他权限→多设备协同,开启应用进行多设备协同访问的权限,如图8.20所示。

除了在设置中直接开启权限,还可以让应用主动查询权限开启状态。若没有开启,则主动询问是否开启权限,并直接在应用内打开所需权限。这种主动的权限开启方式可以通过在程序内添加代码实现。同样以多设备协同权限为例,代码如下:

```
String[] permission = {"ohos.permission.DISTRIBUTED_DATASYNC"};
for (int i = 0;i < permission.length;i++){
    if(verifyCallingOrSelfPermission(permission[i]) != 0){
        if(canRequestPermission(permission[i])){
            requestPermissionsFromUser(permission, 0);
        }
    }
}
```

当该段代码成功运行时,若应用没有开启设备协同权限,则可以看到如图8.21所示的页面。单击"始终允许"按钮后即可将该权限成功打开。

← 多设备协同

允许本设备以下应用访问其他设备上的音视频与运动健康信息，以进行多个设备间的协同操作。

图 8.20　开启多设备协同访问权限

图 8.21　询问是否开启访问权限

8.2.2　分布式数据服务

在 HarmonyOS 中，分布式数据服务(Distributed Data Service，DDS)为应用程序提供了不同设备间数据库数据分布式服务的能力。分布式数据服务支持应用程序通过调用分布式数据接口将数据保存到分布式数据库中，也支持在通过了可信认证的设备之间进行应用数据同步，从而保持数据的一致性。分布式数据服务主要实现对用户设备中应用程序的数据内容的分布式同步。当设备 1 上的应用在分布式数据库中增、删、改数据后，设备 2 上的应用也可以获取该数据库的变化。分布式数据服务被广泛应用于分布式图库、信息、通讯录、文件管理器等场景中。

实际上，在终端开发中，HarmonyOS 分布式数据服务并不怎么需要结构化的关系型数据库，因此分布式数据库采用的是 KV 数据模型，对外提供 KV 类型的访问接口，value 值支持多种数据类型。HarmonyOS 中，分布式数据库包括单版本分布式数据库和设备协同分布式数据库。单版本是指数据在本地以单个 KV 条目为单位保存，对每个 Key 最多只保存一个条目项，当数据在本地被用户修改时，无论是否进行了同步，均直接在这个条目上进行修改。同步也以此为基础，按照它在本地被写入或更改的顺序将当前最新一次修改逐条同步至远端设备。设备协同分布式数据库建立在单版本分布式数据库之上，在应用程序存入的 KV 数据中的 Key 前面拼接了本设备的 DeviceID 标识符，从而保证各设备产生的数据严格隔离，底层按照设备的维度管理这些数据。设备协同分布式数据库支持以设备的维度查询分布式数据，但是不支持修改远端设备同步过来的数据。

此外，由于在分布式数据服务中通常涉及多个设备，因此保持数据的一致性是至关重要的。HarmonyOS 分布式数据库的一致性可以分为强一致性、弱一致性和最终一致性。其中，强一致性是指某一设备成功增、删、改数据后，组网内设备对该数据的读取操作都将得到更新后的值；弱一致性是指某一设备成功增、删、改数据后，组网内设备可能能够读取到本次更新的数据，也可能读取不到，不能保证在多长时间后每个设备的数据一定是一致的；最终一致性是指某一设备成功增、删、改数据后，组网内设备可能读取不到本次更新的数据，但在某个时间窗口之后组网内设备的数据能够达到一致状态。其中，强一致性对分布式数据的管理要求非常高，适用于服务器的分布式场景，而移动端设备不常在线且无中心，因此不支持强一致性，而只支持最终一致性。

HarmonyOS 分布式数据服务的运作机制如图 8.22 所示。应用程序通过调用分布式数据服务接口实现分布式数据库创建、访问、订阅功能，服务接口通过操作服务组件提供的能力，将数据存储至存储组件，存储组件调用同步组件实现数据同步，同步组件使用通信适配层将数据同步至远端设备，远端设备通过同步组件接收数据，并更新至本端存储组件，通过服务接口提供给应用程序使用。

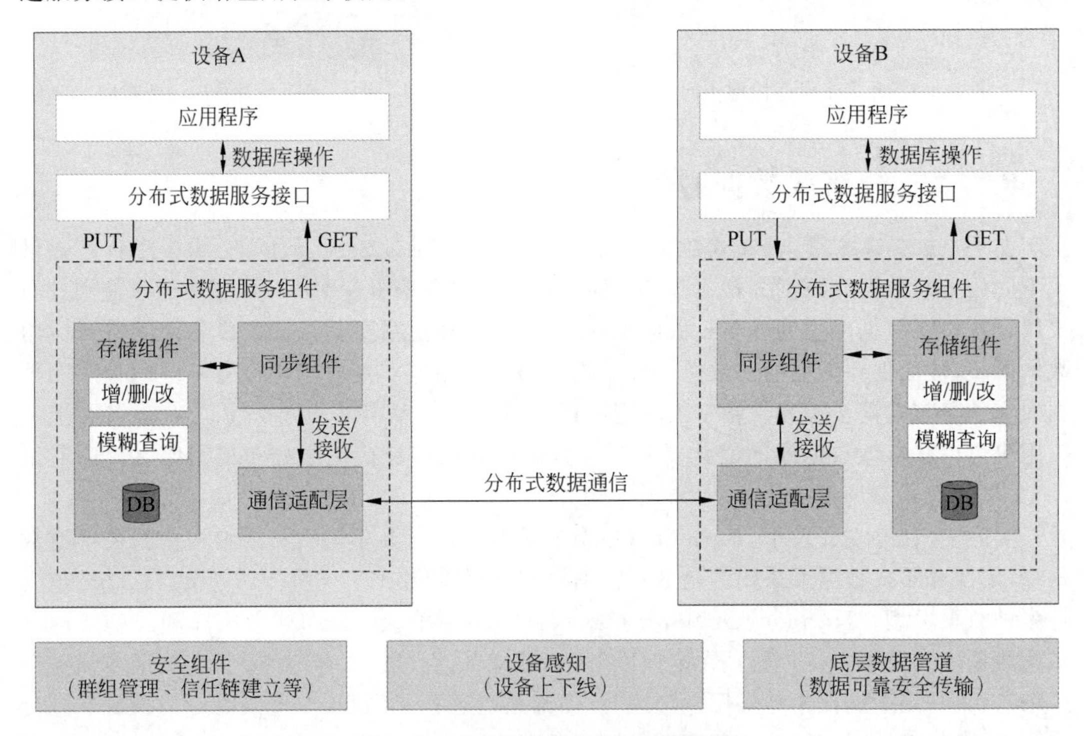

图 8.22　分布式数据服务的运作机制

接下来，将通过一个实例介绍如何创建分布式数据库及如何对其进行操作。分别在两个手机上安装应用程序，Phone A 可以实现注入数据、修改数据、删除数据功能，Phone B 可以实现数据查询及显示功能。

创建 Phone 设备下的 Java 模板新项目，在不同的 slice 中实现各自的功能。在工程目录下的 MainAbilitySlice 文件中实现 Phone A 的功能，在 slice 目录下新建 Java Class，并命名为 MainAbility2Slice，从而实现 Phone B 的功能。在 Phone A 上安装应用程序时需要在 MainAbility 中的 onStart( ) 方法中设置 super.setMainRoute(MainAbilitySlice.class.getName( ))，在 Phone B 上安装应用程序时需要重新设置 super.setMainRoute(MainAbility2Slice.class.getName())。

这里以 Phone A 上的应用程序为例，讲解 MainAbilitySlice 文件中的示例代码。首先声明应用程序的布局，每个功能通过单击布局中的 Button 组件触发完成，因此，需要添加对应的 Button 组件，代码如下：

```
//声明布局
public void onStart(Intent intent) {
      super.onStart(intent);

      //线性布局
      DirectionalLayout directionalLayout = new DirectionalLayout(this);
      DirectionalLayout.LayoutConfig layoutConfig = new
      DirectionalLayout.LayoutConfig ( ComponentContainer. LayoutConfig. MATCH _ CONTENT,
ComponentContainer.LayoutConfig.MATCH_CONTENT);

      //创建数据库
      ...

      ShapeElement shapeElement = new ShapeElement();
      shapeElement.setShape(ShapeElement.RECTANGLE);
      shapeElement.setRgbColor(new RgbColor(0,0,255));
      layoutConfig.setMargins(20,20,20,20);
      layoutConfig.alignment = LayoutAlignment.HORIZONTAL_CENTER;

      //功能按钮,设置相关属性,以注入数据按钮为例
      Button insert = new Button(this);
      insert.setBackground(shapeElement);
      insert.setText("注入数据");
      insert.setTextSize(100);
      insert.setLayoutConfig(layoutConfig);
      insert.setClickedListener(new Component.ClickedListener() {
             @Override
             public void onClick(Component component) {
             //向数据库中增加数据
                 ...
             }
        });
      directionalLayout.addComponent(insert);

      //添加其余功能按钮,实现修改数据和删除数据功能
      ...

super.setUIContent(directionalLayout);
}
```

Phone A 中的应用界面如图 8.23 所示。

Button 组件触发操作事件之前，应该先创建好数据库，接下来，介绍创建数据库的实现。

(1) 首先，根据配置构造分布式数据库管理类实例。具体地，先根据应用上下文创建 KvManagerConfig 对象，该对象提供 KvManager 实例的配置信息，包括调用者的包名称和分布式网络类型，随后创建分布式数据库管理器 KvManager 的实例。

(2) 获取/创建单版本分布式数据库。首先声明需要创建的单版本分布式数据库 ID 描述，接着创建单版本分布式数据库，跨设备自动同步数据库功能是默认开启的，具体代码如下：

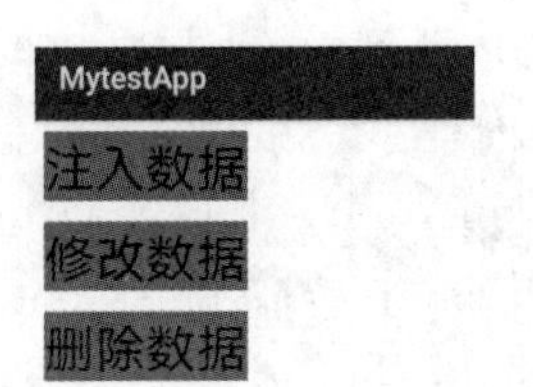

图 8.23 Phone A 的应用界面

```
//创建单版本分布式数据库

//创建 KvManagerConfig 对象
KvManagerConfig config = new KvManagerConfig(this);
//创建分布式数据库管理器实例
KvManager kvManager = KvManagerFactory.getInstance().createKvManager(config);

//创建单版本分布式数据库
//数据库的配置信息
Options CREATE = new Options();
CREATE.setCreateIfMissing(true).setEncrypt(false).setKvStoreType(KvStoreType.SINGLE_
VERSION).setAutoSync(true);
//声明新建数据库的 ID
String storeID = "test";
//创建单版本分布式数据库
SingleKvStore singleKvStore = kvManager.getKvStore(CREATE, storeID);
```

其中，Options 类提供了用于创建分布式数据库的配置选项，例如当前不存在分布式数据库的情况下是否新建数据库、是否加密该数据库及数据库类型，getKvStore(CREATE, storeID)根据 CREATE 配置信息创建和打开标识符为 storeID 的分布式数据库。需要说明的是，数据库创建时会生成数据库的句柄。在该句柄(如上例中的 singleKvStore)的生命周期内无须重复创建数据库，可直接使用句柄对数据库进行数据的插入等操作。如果不主动调用 kvManager.closeKvStore()方法而关闭句柄，则该句柄会一直存在，直到程序结束。

(3) 添加数据。在分布式数据库中，数据是以 KV 数据模型存储的，因此在添加数据之前，需要先构造数据。接着，将构造好的数据写入数据库中。该功能在实现注入数据的 Button 中触发实现，故将该功能实现覆写在其 onClick()方法中。本例中添加了三条数据，代码如下：

```
insert.setClickedListener(new Component.ClickedListener() {
    @Override
    public void onClick(Component component) {
        //向数据库中增加数据
        String key = "User";
        String value = "Zhang san";
        singleKvStore.putString(key, value);

        String key2 = "User2";
        String value2 = "lisi";
        singleKvStore.putString(key2, value2);

        String key3 = "User3";
        String value3 = "wangwu";
        singleKvStore.putString(key3, value3);
    }
});
```

注意,在本例中,value是String类型,因此在插入数据时使用的是putString()方法,在实际操作时可以根据value的实际类型选择相应的插入方法,如putBoolean(String key, boolean value)、putInt(String key, int value)、putString(String key, String value)等。在Phone A上可以单击注入数据的按钮即可添加以上三条数据。

(4) 查询数据。查询数据的功能需要在Phone B中实现,因此需要覆写MainAbility2Slice中的onStart()方法。在查询数据库中的数据时需要先构造查询的Key,这样才能从数据库快照中获取数据。在Phone B上实现查询上述新添加的数据,具体实现代码如下:

```
//在Phone B中数据查询

//声明布局
public void onStart(Intent intent) {
    super.onStart(intent);
    HiLogLabel hiLogLabel = new HiLogLabel(HiLog.LOG_APP,0,"ddsTest");

    //声明线性布局,同Phone A
    ...

    //创建数据库,同Phone A
    ...

    //实例化ShapeElement并设置属性,同Phone A
    ...

    //添加Button,实现数据查询显示功能
Button query = new Button(this);
```

```
query.setText("查询数据");
        query.setBackground(shapeElement);
query.setTextSize(100);
query.setLayoutConfig(layoutConfig);
query.setClickedListener(new Component.ClickedListener() {
  @Override
public void onClick(Component component) {
            //查询显示所有数据,以 HiLog 形式显示
                HiLog.fatal(hiLogLabel, "数据查询: ");
                HiLog.fatal(hiLogLabel, "==============================");
String key = "User";
String valueshow = singleKvStore.getString(key);
                //以 HiLog 形式显示查询结果
HiLog.fatal(hiLogLabel, valueshow);

String key2 = "User2";
String valueshow2 = singleKvStore.getString(key2);
HiLog.fatal(hiLogLabel, valueshow2);

String key3 = "User3";
String valueshow3 = singleKvStore.getString(key3);
HiLog.fatal(hiLogLabel, valueshow3);
HiLog.fatal(hiLogLabel, "==============================");
}
});

      directionalLayout.addComponent(query);

super.setUIContent(directionalLayout);
}
```

同样地,当 getString()方法从单版本分布式数据库中查询字符串类型数据时,可以根据实际 Value 类型选择所需要的方法,如 getInt(String key)、getFloat(String key)等。在 Phone B 中单击“查询数据”按钮,查询数据库中的所有数据,查询结果如图 8.24 所示。

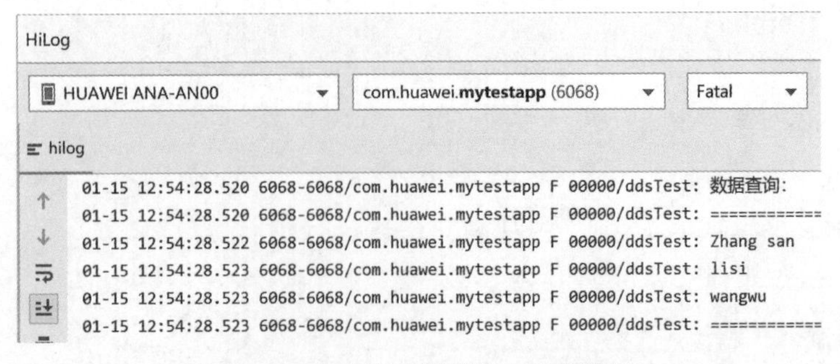

图 8.24 在分布式数据库中查询数据

(5) 修改数据。修改数据的方法与插入数据的方法相同,将要修改的数据的 key 所对应的 value 值修改后重新插入数据库即可。上述数据中将 key="User"对应数据的 value 改为 Alice,具体代码如下:

```
//在 Phone A 中修改数据
Button change = new Button(this);
change.setBackground(shapeElement);
change.setTextSize(100);
change.setText("修改数据");
change.setLayoutConfig(layoutConfig);
change.setClickedListener(new Component.ClickedListener() {
    @Override
    public void onClick(Component component) {
String key = "User";
        String value = "Alice";
singleKvStore.putString(key, value);
    }
});
directionalLayout.addComponent(change);
```

在 Phone A 中进行数据修改后,在 Phone B 中查询修改后的结果,结果如图 8.25 所示。对比图 8.25 可知数据修改成功。

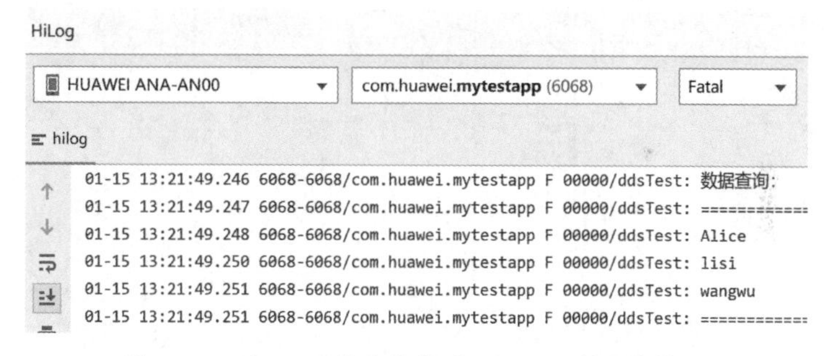

图 8.25 Phone A 修改数据后,Phone B 的查询结果

(6) 删除数据。SingleKvStore 数据库提供了 delete()方法用来删除数据。本例中 Phone A 将 key="User3"的数据删除,实现代码如下:

```
//在 Phone A 中实现删除数据功能
Button delete = new Button(this);
delete.setBackground(shapeElement);
delete.setTextSize(100);
delete.setText("删除数据");
delete.setLayoutConfig(layoutConfig);
delete.setClickedListener(new Component.ClickedListener() {
@Override
    public void onClick(Component component) {
```

```
String key3 = "User3";
            singleKvStore.delete(key3);
        }
 });
directionalLayout.addComponent(delete);
```

在 Phone B 中查询并显示修改后的所有数据，如图 8.26 所示。由于第三条数据已经被删除，所以显示完前两条数据之后 Phone B 上的应用程序会退出。

```
01-15 13:37:13.725 6068-6068/com.huawei.mytestapp F 00000/ddsTest: 数据查询:
01-15 13:37:13.725 6068-6068/com.huawei.mytestapp F 00000/ddsTest: ============
01-15 13:37:13.726 6068-6068/com.huawei.mytestapp F 00000/ddsTest: Alice
01-15 13:37:13.728 6068-6068/com.huawei.mytestapp F 00000/ddsTest: lisi
```

图 8.26　显示删除数据后的结果

(7) 此外，还可以进行数据库的关闭和删除操作。通过 deleteKvStore(String storeId) 方法将指定 ID 的数据库删除。在本例中，所建立的单版本分布式数据库 ID 为 test。将其删除的代码如下：

```
//在 Phone B 中删除数据库
Button delete = new Button(this);
delete.setBackground(shapeElement);
delete.setTextSize(100);
delete.setText("删除数据库");
delete.setLayoutConfig(layoutConfig);
delete.setClickedListener(new Component.ClickedListener() {
    @Override
    public void onClick(Component component) {
        kvManager.deleteKvStore("test");
    }
});
directionalLayout.addComponent(delete);
```

8.2.3　分布式文件服务

在正式讲解 HarmonyOS 中的分布式文件服务之前，先简单介绍一下相关概念。

1. 分布式文件系统(Distributed File System，DFS)

通过计算机网络将分布在不同地点的节点相连，利用网络进行节点间的通信和数据传输，从而将固定于某个地点的某个文件系统，扩展到任意多个地点/文件系统，众多的节点组成的文件系统网络即为分布式系统。简单来讲，DFS 为分布在网络上任意位置的资源提供一个逻辑上的树形文件系统结构，从而使用户访问分布在网络上的共享文件更加简便。分布式文件系统的突出优点就是可以屏蔽文件系统的物理位置，人们在使用分布式文件系统时，无须关心数据是存储在哪个节点上或者是从哪个节点获取的，而只需像使用本地文件系统一样管理和存储文件系统中的数据。

此外，分布式文件系统还包括以下特点：

(1) 冗余性：分布式文件系统可以提供冗余备份，当系统中某些节点出错时，整体文件服务不会停止，还能继续为用户提供服务，具有较高的容错性。

(2) 安全性：分布式文件系统的安全性离不开其冗余性，当出现故障的节点存储的数据损坏时可以由其他节点进行数据恢复。此外，在分布式文件系统中，大量数据被分散到不同的节点上进行存储，数据丢失的风险大大减小。

(3) 扩展性：分布式文件系统可以通过网络连接将大量的计算机连接到一起，任何计算机只需经过简单的配置就可以加入分布式文件系统中。

2. 分布式文件

分布式文件是指依赖于分布式文件系统并分散存储在多个用户设备上的文件，应用间的分布式文件目录互相隔离，不同应用的文件不能互相访问。

3. 文件元数据

文件元数据是用于描述文件特征的数据，包含文件名、文件大小、创建、访问、修改时间等信息。

在 HarmonyOS 中，分布式文件服务支持用户设备中的应用程序在同一账号下多设备之间进行文件共享。即应用程序可以屏蔽文件具体的存储位置，在多个设备之间无障碍访问文件。HarmonyOS 中分布式文件服务的运作机制如图 8.27 所示。

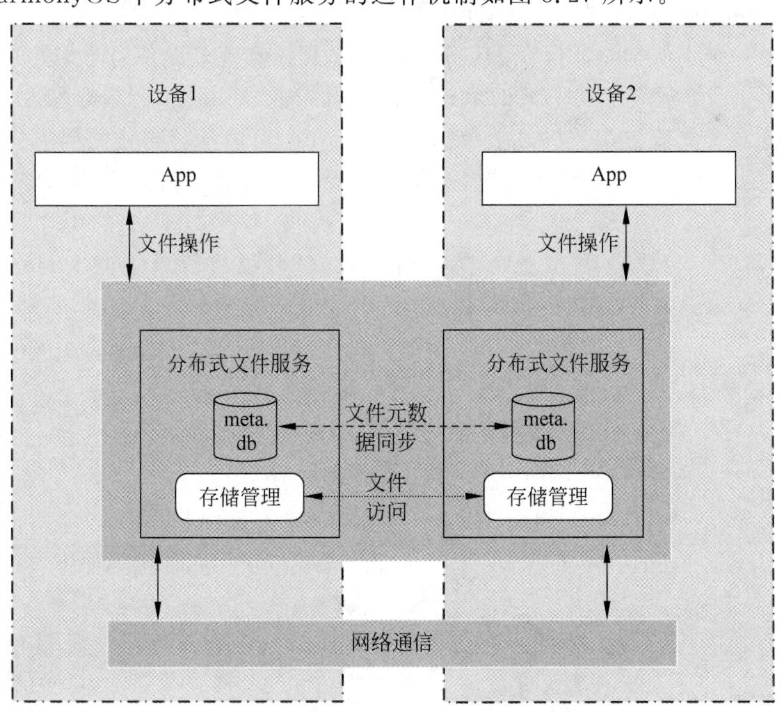

图 8.27　HarmonyOS 中分布式文件服务的运作机制

从图 8.27 可以看出，在 HarmonyOS 中，分布式文件服务采用无中心节点的设计，每个设备都存储一份全量的文件元数据和本设备上产生的分布式文件，元数据在多台设备间互相同步，当应用需要访问分布式文件时，分布式文件服务首先查询本设备上的文件元数据，获取文件所在的存储设备，然后对存储设备上的分布式文件服务发起文件访问请求，将文件内容读取到本地。

实际上，在 HarmonyOS 中实现分布式文件服务前，需要满足以下条件：

(1) 应用程序如需使用分布式文件服务完整功能，需要申请分布式数据管理权限，具体地，申请 ohos. permission. DISTRIBUTED_DATASYNC 权限，从而允许不同设备间的数据交换。

(2) 要实现分布式共享文件，则多个设备需登录同一个华为账号，打开蓝牙设备，连接同一个 WLAN 局域网。

(3) 存在多设备并写的场景，为了避免冲突，开发者需要对文件加锁保护，保证文件独享。非持锁情况下，并发写冲突时，后一次会覆盖前一次。

(4) 应用访问分布式文件时，文件所在设备不能离线，否则文件不能访问。

(5) 当网络情况较差时，访问存储在远端的分布式文件可能会长时间得不到响应甚至响应失败，因此需要应用考虑到对这种场景的处理。

(6) 当两台设备有同名文件时，如果元数据进行同步则会产生冲突，分布式文件服务会根据时间戳将文件按创建的先后顺序重命名，因此，为避免此类场景出现，应用在文件名上可以进行相应设备区分，例如，deviceID+时间戳。

下面通过一个实例学习如何在 HarmonyOS 中具体实现多设备间的文件共享。实例实现了在两个手机设备上进行分布式时间读写的功能，即在手机 A 上单击写入 Button，将当前时间写入分布式文档。在手机 B 上单击读取 Button，可以从分布式文档中获取手机 A 中写入的时间，反之亦然。需要注意的是，两个手机设备需要登录同一个华为账号，故需要开启多设备协同权限。

首先，创建 Phone 设备下的 Java 模板新项目，打开项目目录下的 MainAbilitySclice. java 文件，在 onStart()方法中声明布局，代码如下：

```
DirectionalLayout directionLayout = new DirectionalLayout(this);
directionLayout.setWidth(ComponentContainer.LayoutConfig.MATCH_PARENT);
directionLayout.setHeight(ComponentContainer.LayoutConfig.MATCH_PARENT);
directionLayout.setOrientation(Component.VERTICAL);
directionLayout.setPadding(32, 32, 32, 32);
```

在布局中添加 Text 组件用以显示提示信息及读取到的时间，代码如下：

```
Text text = new Text(this);
text.setText("初始文本");                          //设置初始显示文本
text.setTextSize(50);
DirectionalLayout.LayoutConfig layoutConfig = newDirectionalLayout.LayoutConfig
```

```
(ComponentContainer.LayoutConfig.MATCH_CONTENT,ComponentContainer.LayoutConfig.MATCH_CONTENT);
layoutConfig.alignment = LayoutAlignment.HORIZONTAL_CENTER;
text.setLayoutConfig(layoutConfig);
directionLayout.addComponent(text);
```

在本例中两个设备通过各自的 Button 组件实现时间的读写，因此，需要添加两个 Button 组件，首先添加写入时间的 Button 组件，代码如下：

```
//实现写入功能的 Button,用来读取当前时间,并写入分布式文档中
Button button1 = new Button(this);
layoutConfig.setMargins(0, 50,0,0);
button1.setLayoutConfig(layoutConfig);
button1.setText("写入现在时间");
button1.setTextSize(50);
ShapeElement background1 = new ShapeElement();
background1.setRgbColor(new RgbColor(0xFF51A8DD));
background1.setCornerRadius(25);
button1.setBackground(background1);
button1.setPadding(10, 10, 10, 10);
button1.setClickedListener(new Component.ClickedListener() {
        @Override
        public void onClick(Component Component) {
            goWrite(text);
//单击 Button,实现写入功能
        }
});
directionLayout.addComponent(button1);
```

添加读取时间的 Button 组件，代码如下：

```
//实现读取功能的 Button,从分布式文档中读取已经写入的时间
Button button2 = new Button(this);
layoutConfig.setMargins(0, 50,0,0);
button2.setLayoutConfig(layoutConfig);
button2.setText("读取上一个时间");
button2.setTextSize(50);
ShapeElement background2 = new ShapeElement();
background2.setRgbColor(new RgbColor(0xFF5100DD));
background2.setCornerRadius(25);
button2.setBackground(background1);
button2.setPadding(10, 10, 10, 10);
button2.setClickedListener(new Component.ClickedListener() {
        @Override
       //单击 Button,实现读取功能
        public void onClick(Component Component) {
            goRead(text);
```

```
        }
});
directionLayout.addComponent(button2);
```

写入功能是由 goWrite()方法实现的,分析其实现过程,代码如下:

```
//goWrite():写入 button1 的 onClick 事件执行的方法
private void goWrite(Text text) {
      String sharedFileName = sharedFileName(this);
      SimpleDateFormat simpleDateFormat = new SimpleDateFormat("yyyy-MM-dd HH:mm:ss");
      String str = simpleDateFormat.format(new Date().getTime());
                                                        //获取时间戳并转换成标准形式

//将时间写入分布式文件
      try{
            FileWriter fileWriter = new FileWriter(sharedFileName,false);
            fileWriter.write(str);
            fileWriter.close();
      } catch (IOException e) {
              e.printStackTrace();
      }
      text.setText("写入的时间: " + str);
                text.invalidate();
}
```

读取功能是由 goRead()方法实现的,其实现过程代码如下:

```
//goRead(): 读取 button2 的 onClick 事件执行的方法
private void goRead(Text text) {
      String sharedFileName = sharedFileName(this);

   //读取分布式文件中的数据,若读到则输出,若没读到则输出没读到
      try{
              FileReader fileReader = new FileReader(sharedFileName);
              BufferedReader br = new BufferedReader(fileReader);
              String b = br.readLine();
              text.setText("读取到的上一个写入的时间: " + b);
              text.invalidate();
              fileReader.close();
          } catch (IOException e) {
              e.printStackTrace();
              text.setText("没读到");
              text.invalidate();
          }
}
```

二者都是通过 shareFileName()方法获取分布式文件路径的，代码如下：

```
//获取分布式文件所在路径
public static String sharedFileName(Context context){
     File distDir = context.getDistributedDir();
//获取分布式文件目录
     //在分布式文件目录下新建一个名为 note.txt 的文件,读写都在这个文件中进行
     String filePath = distDir + File.separator + "note.txt";
     return filePath;
}
```

需要说明的是，利用 Context. getDistributedDir()接口可以获取属于自己的分布式目录，然后通过 libc 或 JDK 接口，可以在该目录下创建、删除、读写文件或目录。本例中在所获取的分布式目录下创建了一个读写文件 note. txt。

基于两个手机设备进行验证，在开始功能验证之前，两个手机需要登录同一个华为账号，并且需要开启多设备协同权限且开启多设备协同连接。具体的设置过程可参考前面对多设备协同权限设置的介绍。

设置完成后进行功能验证。首先，在未写入的情况下直接进行读取，读取结果如图 8.28 所示，未写入的情况下读取内容为空。

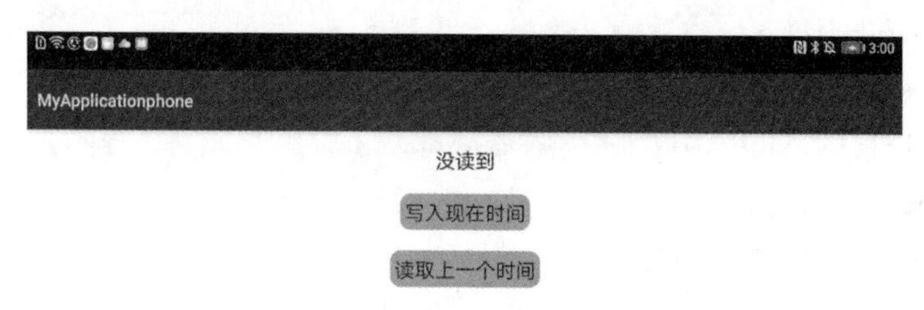

图 8.28　未写入情况下直接进行读取

由 Phone A 写入当前时间，Phone B 读取当前时间，读写效果如图 8.29 所示。

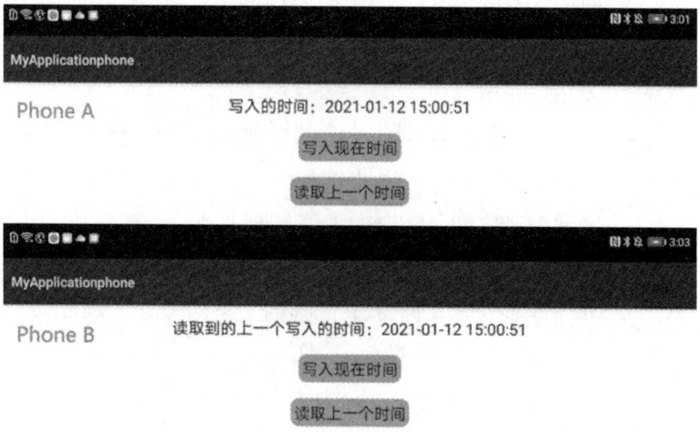

图 8.29　Phone A 写入当前时间，Phone B 读取 A 写入的时间

由 Phone B 写入当前时间，Phone A 进行读取时，读写效果如图 8.30 所示。

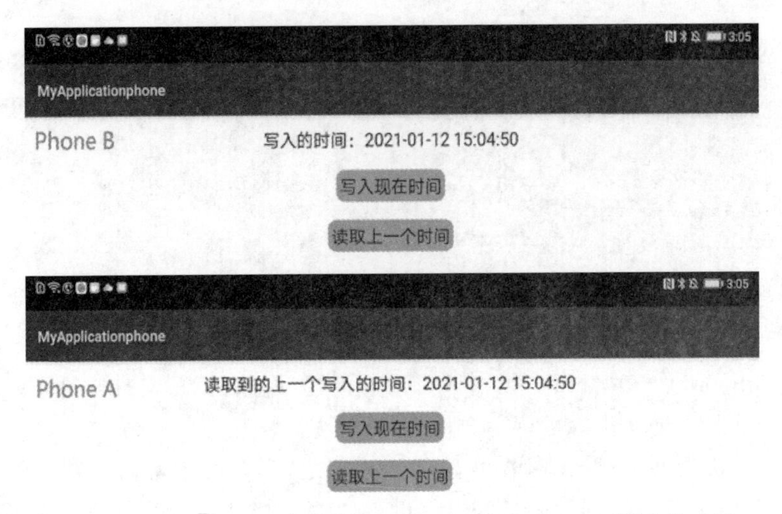

图 8.30　Phone B 写入当前时间，Phone A 读取 Phone B 写入的时间

至此，分布式文件服务功能成功实现。

8.3 DataAbility

15min

使用 Data 模板的 Ability（以下简称 Data）有助于应用管理其自身和其他应用存储数据的访问，为不同的软件之间共享数据提供统一的接口。Data 既可用于同设备不同应用的数据共享，也支持跨设备不同应用的数据共享。

数据的存放形式有两种，一种是数据库，另一种是磁盘上的文件。Data 对外提供对数据的增、删、改、查，以及打开文件等接口，这些接口的具体实现由开发者提供。也就是说如果想让其他的应用使用程序内的数据，就可以使用 Data 定义一个对外开放的接口，从而使其他的应用可以使用当前应用中的文件、数据库内存储的信息。

8.3.1 创建 Data

如果需要一个 Ability 进行同设备不同应用的数据共享，或者跨设备不同应用的数据共享，那么使用 Data 模板的 Ability 将是最佳的选择，Data 形式仍然是 Ability，因此需要为应用添加一个或多个 Ability 的子类，来提供程序与其他应用之间的接口。Data 为结构化数据和文件提供了不同 API 供用户使用，在创建 Data 时开发者需要首先确定好使用何种类型的数据。Data 支持两种数据形式，一种是文件数据，例如文本、图片、音乐等，另一种是结构化数据，如数据库等。

DataAbility 接收其他应用发送的请求，提供外部程序访问的入口，从而实现应用间的数据访问。接下来新建一个 DataAbility，如图 8.31 所示。

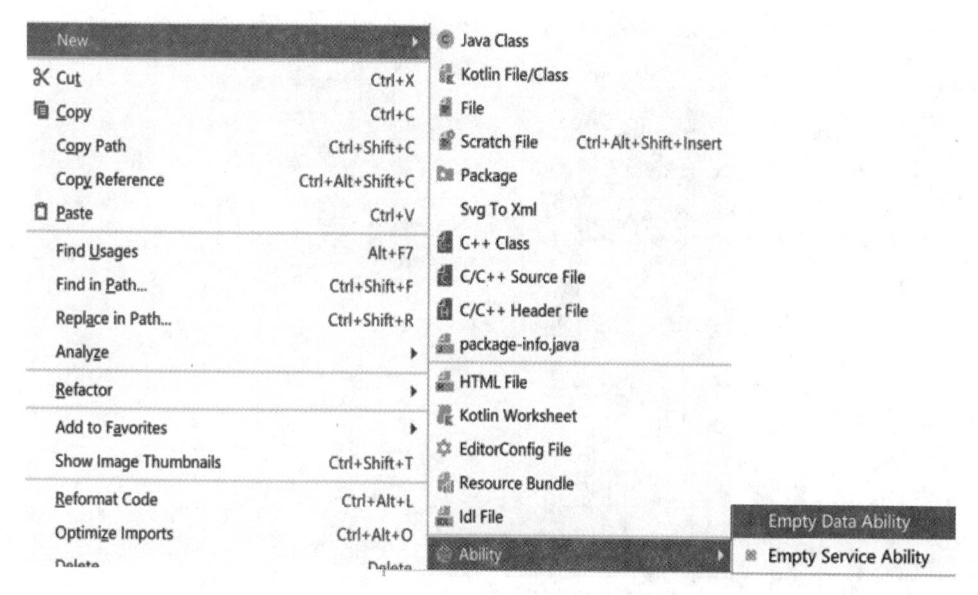

图 8.31 选择新建 DataAbility

右击想要新建 DataAbility 的模块，选择 New→Ability→Empty Data Ability，新建一个 DataAbility，代码如下：

```
//DataAbility.java
{
    private static final HiLogLabel LABEL_LOG = new HiLogLabel(3, 0xD001100, "Demo");

    @Override
    public void onStart(Intent intent) {
        super.onStart(intent);
        HiLog.info(LABEL_LOG, "ProviderAbility onStart");
    }

    @Override
    public ResultSet query(Uri uri, String[] columns, DataAbilityPredicates predicates) {
        return null;
    }

    @Override
    public int insert(Uri uri, ValuesBucket value) {
        HiLog.info(LABEL_LOG, "ProviderAbility insert");
        return 999;
    }

    @Override
    public int delete(Uri uri, DataAbilityPredicates predicates) {
```

```
        return 0;
    }

    @Override
    public int update(Uri uri, ValuesBucket value, DataAbilityPredicates predicates) {
        return 0;
    }

    @Override
    public FileDescriptor openFile(Uri uri, String mode) {
        return null;
    }

    @Override
    public String[] getFileTypes(Uri uri, String mimeTypeFilter) {
        return new String[0];
    }

    @Override
    public PacMap call(String method, String arg, PacMap extras) {
        return null;
    }

    @Override
    public String getType(Uri uri) {
        return null;
    }
}
```

Data初始化完毕了，接下来要选择数据的存取方式，可以选择通过文件的方式来存取数据或者通过数据库的方式进行数据的存取，选择不同的方式需要改写的接口也不一样，在接口改写完成后，其他的应用就可以通过这些接口获得想要的数据。

8.3.2 文件存取

下面将详细讲述如何操作文件中的数据。

如果开发者需要访问文件的数据，那么首先需要在Data中重写FileDescriptor openFile(Uri uri, String mode)方法来操作文件，代码如下：

```
//重写 FileDescriptor openFile()方法用于操作文件
@Override
    public FileDescriptor openFile(Uri uri, String mode) throws FileNotFoundException {
//创建 messageParcel
        MessageParcel messageParcel = MessageParcel.obtain();
```

```
    File file = new File(uri.getDecodedPathList().get(1));
    if (mode == null || !"rw".equals(mode)) {
        file.setReadOnly();
    }
    FileInputStream fileIs = new FileInputStream(file);
    FileDescriptor fd = null;
    try {
        fd = fileIs.getFD();
    } catch (IOException e) {
        e.printStackTrace();
    }
    //绑定文件描述符
    return messageParcel.dupFileDescriptor(fd);
    }
```

其中，uri 参数为客户端传入请求目标路径，Data 的提供方和使用方都通过 URI (Uniform Resource Identifier)来标识一个具体的数据，例如数据库中的某个表或磁盘上的某个文件。HarmonyOS 的 URI 仍基于 URI 通用标准，格式如图 8.32 所示。

Scheme://[authority]/[path][?query][#fragment]

协议方案名　设备ID　资源路径　查询参数　访问的子资源

图 8.32　URI 格式

其中，每个参数的具体含义为

- Scheme：协议方案名，固定为 dataability，代表 Data Ability 所使用的协议类型。
- authority：设备 ID。如果为跨设备场景，则为目标设备的 ID；如果为本地设备场景，则不需要填写。
- path：资源的路径信息，代表特定资源的位置信息。
- query：查询参数。
- fragment：可以用于指示要访问的子资源。

第二个参数 mode 为开发者对文件的操作选项，可选方式包含 r(读)、w(写)、rw(读写)等。

可以看到本例中使用了 MessageParcel 类提供的一个静态方法，用于获取 MessageParcel 实例，然后通过 dupFileDescriptor()函数复制待操作文件流的文件描述符，并将其返回，供远端应用使用，这样就可以达成使用根据传入的 uri 打开对应文件的目的了。

现在可以访问文件数据的 Data 创建完毕了，接下来需要访问和进行使用。

可以通过 DataAbilityHelper 类访问当前应用或其他应用提供的共享数据。DataAbilityHelper 作为客户端，与提供方的 Data 进行通信。Data 接收到请求后，执行相应的处理，并返回结果。DataAbilityHelper 提供了一系列与 Data Ability 对应的方法。

首先通过 DataAbilityHelper helper = DataAbilityHelper.creator(this) 创建一个

DataAbilityHelper，然后通过 FileDescriptor openFile(Uri uri，String mode)方法来操作文件，其中的 uri 和 mode 的含义上文已经介绍，代码如下：

```
FileInputStream fileIs = new FileInputStream(file);
FileDescriptor fd = null;
try {
    fd = helper.openFile(uri, "r");
} catch (DataAbilityRemoteException e) {
    e.printStackTrace();
}
```

获取文件流后，就可以对文件进行操作了。

8.3.3 数据库操作

介绍完操作文件的存储格式后，下面来介绍操作数据库的存储格式。

如果用数据库存储文件，在应用启动时调用 onStart()方法创建 Data 实例时，应该创建数据库连接，并获取连接对象，以便后续和数据库进行操作。为了避免影响应用启动速度，应当尽可能将非必要的耗时任务推迟到使用时执行，而不是在此方法中执行所有初始化。

在本例中重写 onStart()方法，新建一个对象关系映射(ORM)型数据库，在 ORM 数据库中插入 3 个 User 类的数据(User 类的定义方式详见 8.1.3 节)，代码如下：

```
//数据库中插入数据
private static final String DATABASE_NAME = "UserDataAbility.db";
private static final String DATABASE_NAME_ALIAS = "UserDataAbility";
private OrmContext ormContext = null;

@Override
public void onStart(Intent intent) {
     super.onStart(intent);
        DatabaseHelper manager = new DatabaseHelper(this);
        ormContext = manager.getOrmContext(DATABASE_NAME_ALIAS, DATABASE_NAME, BookStore.
class);
        User user = new User();              //插入数据 1
        user.setUserId(0);
        user.setFirstName("Zhang");
        user.setLastName("San");
        user.setAge(25);
        user.setBalance(100.51);
        ormContext.insert(user);
        User user2 = new User();             //插入数据 2
        user2.setUserId(1);
        user2.setFirstName("Li");
        user2.setLastName("Si");
```

```
        user2.setAge(26);
        user2.setBalance(120.51);
        ormContext.insert(user2);
        User user3 = new User();                        //插入数据 3
        user3.setUserId(2);
        user3.setFirstName("Wang");
        user3.setLastName("Wu");
        user3.setAge(27);
        user3.setBalance(140.51);
        ormContext.insert(user3);
        ormContext.flush();
        HiLog.info(LABEL_LOG, "ProviderAbility onStart");
}
```

在 DataAbility 中定义了 6 种方法供用户对数据库表数据的增、删、改、查进行处理。这 6 种方法在 Ability 中已默认实现，开发者可按需重写，示例代码如下：

```
//数据库表数据的增、删、改、查
ResultSet query(Uri uri, String[] columns, DataAbilityPredicates predicates)
int insert(Uri uri, ValuesBucket value)
int batchInsert(Uri uri, ValuesBucket[] values)
int delete(Uri uri, DataAbilityPredicates predicates)
int update(Uri uri, ValuesBucket value, DataAbilityPredicates predicates)
DataAbilityResult[] executeBatch(ArrayList<DataAbilityOperation> operations)
```

第 1 种方法 query()用来查询数据库数据，该方法接收 3 个参数，分别是查询的目标路径、查询的列名，以及查询条件，查询条件由类 DataAbilityPredicates 构建。本例中根据传入的列名和查询条件查询用户表，代码如下：

```
//query(): 查询数据库数据
public ResultSet query(Uri uri, String[] columns, DataAbilityPredicates predicates) {
        if (ormContext == null) {
            HiLog.error(LABEL_LOG, "failed to query, ormContext is null");
            return null;
        }
        //查询数据库
        OrmPredicates ormPredicates = DataAbilityUtils.createOrmPredicates(predicates,
User.class);
        ResultSet resultSet = ormContext.query(ormPredicates, columns);
        if (resultSet == null) {
            HiLog.fatal(LABEL_LOG, "resultSet is null");
        }
        //返回结果
        return resultSet;
    }
```

第 2 种方法 insert()用来向数据库插入数据，该方法接收两个参数，分别是插入的目标路径和插入的数据值。其中，插入的数据由 ValuesBucket 封装，服务器端可以从该参数中解析出对应的属性，然后插入数据库中。此方法返回一个 int 类型的值用于标识结果。例如本例中接收传过来的用户信息并把它保存到数据库中，代码如下：

```
//insert()：数据库插入数据
@Override
public int insert(Uri uri, ValuesBucket value) {
    //参数校验
    if (ormContext == null) {
        HiLog.error(LABEL_LOG, "failed to insert, ormContext is null");
        return -1;
    }

    //构造插入数据
    User user = new User();
    user.setUserId(value.getInteger("userId"));
    user.setFirstName(value.getString("firstName"));
    user.setLastName(value.getString("lastName"));
    user.setAge(value.getInteger("age"));
    user.setBalance(value.getDouble("balance"));

    //插入数据库
    boolean isSuccessed = true;
    isSuccessed = ormContext.insert(user);
    if (!isSuccessed) {
        HiLog.error(LABEL_LOG, "failed to insert");
        return -1;
    }
    isSuccessed = ormContext.flush();
    if (!isSuccessed) {
        HiLog.error(LABEL_LOG, "failed to insert flush");
        return -1;
    }
    DataAbilityHelper.creator(this, uri).notifyChange(uri);
    int id = Math.toIntExact(user.getRowId());
    return id;
}
```

第 3 种方法 batchInsert()为批量插入方法，接收一个 ValuesBucket 数组用于单次插入一组对象，它的作用是提高插入多条重复数据的效率，该方法系统已实现，我们就不需要改变它了。

delete()方法是用来删除数据库数据的。删除条件由类 DataAbilityPredicates 构建，服务器端在接收到该参数之后可以从中解析出要删除的数据，然后到数据库中执行。根据传入的条件删除用户表数据的代码如下：

```
//delete()：删除数据库数据
public int delete(Uri uri, DataAbilityPredicates predicates) {
if (ormContext == null) {
            HiLog.error(LABEL_LOG, "failed to delete, ormContext is null");
            return -1;
        }

        OrmPredicates ormPredicates  =  DataAbilityUtils. createOrmPredicates ( predicates,
User.class);
        int value = ormContext.delete(ormPredicates);
        DataAbilityHelper.creator(this, uri).notifyChange(uri);
        return value;
}
```

update()被用来执行更新操作。用户可以在 ValuesBucket 参数中指定要更新的数据，在 DataAbilityPredicates 中构建更新的条件等。本例中更新用户表数据的示例代码如下：

```
//update()：更新数据库数据
public int update(Uri uri, ValuesBucket value, DataAbilityPredicates predicates) {
    if (ormContext == null) {
            HiLog.error(LABEL_LOG, "failed to update, ormContext is null");
            return -1;
        }

        OrmPredicates ormPredicates  =  DataAbilityUtils. createOrmPredicates ( predicates,
User.class);
        int index = ormContext.update(ormPredicates, value);
        HiLog.info(LABEL_LOG, "UserDataAbility update value:" + index);
        DataAbilityHelper.creator(this, uri).notifyChange(uri);
        return index;
}
```

上述代码完成了 DataAbility 的构建，注意查看配置文件，可以看到 Data 的 uri 为 dataability://com.huawei.dataabillitytest.DataAbility，需要的权限是 com.huawei.dataabillitytest.DataAbility.DATA，如果有需要自定义的需求，则可以修改 uri 或者添加更多的需求，代码如下：

```
{
    "visible": true,
    "permissions": [
        "com.huawei.dataabillitytest.DataAbility.DATA"
    ],
    "name": "com.huawei.dataabillitytest.DataAbility",
```

```
    "icon": "$media:icon",
    "description": "$string:dataability_description",
    "type": "data",
    "uri": "dataability://com.huawei.dataabillitytest.DataAbility"
}
```

接下来新建一个项目 Datatest，用于测试能否访问上述创建的基于数据库存储的 DataAbility。

首先在 UI 的主 layout 中定义界面布局，初始化 HilogLabel，然后定义 columns 供数据库查询，这里只查询 User 中的 userId、firstName、lastName、age 4 个属性，初始化的 ShapeElement 被设定为蓝色并作为 Button 背景，具体代码如下：

```
//创建布局用于测试 DataAbility
private static final HiLogLabel LABEL_hiLogLabel = new HiLogLabel(HiLog.LOG_APP, 0, "Demo");
private DirectionalLayout myLayout = new DirectionalLayout(this);
    private DirectionalLayout.LayoutConfig layoutConfig = new DirectionalLayout.
LayoutConfig (ComponentContainer.LayoutConfig.MATCH_CONTENT, ComponentContainer.
LayoutConfig.MATCH_CONTENT);
    @Override
    public void onStart(Intent intent) {
    super.onStart(intent);
layoutConfig.setMargins(20,20,20,20);
    layoutConfig.alignment = LayoutAlignment.HORIZONTAL_CENTER;
    myLayout.setLayoutConfig(layoutConfig);
    layoutConfig.width = ComponentContainer.LayoutConfig.MATCH_CONTENT;
    layoutConfig.height = ComponentContainer.LayoutConfig.MATCH_CONTENT;

    String[] columns = new String[] {"userId", "firstName", "lastName", "age"};
ShapeElement shapeElement = new ShapeElement();
shapeElement.setShape(ShapeElement.RECTANGLE);
shapeElement.setRgbColor(new RgbColor(0,0,255));
```

接下来定义 4 个 Button，分别代表查询、插入、删除、更新操作，更多的方法，例如 executeBatch()批量操作方法和 batchInsert()批量插入方法此处不再赘述，操作方法与上述 4 个类似。接下来逐一修改代码，定义其中的操作，首先在 query 中定义查询操作，代码如下：

```
//定义 Button 进行查询操作
Button query = new Button(this);
query.setText("query");
query.setTextSize(150);
query.setLayoutConfig(layoutConfig);
query.setBackground(shapeElement);
```

```
query.setClickedListener(new Component.ClickedListener() {
    @Override
    public void onClick(Component component) {
        DataAbilityHelper helper = DataAbilityHelper.creator(MainAbilitySlice.super.
getContext());

        //构造查询条件
        DataAbilityPredicates predicates = new DataAbilityPredicates();
        predicates.between("userId", 0, 5);
        //进行查询
        ResultSet resultSet = null;
        try {
            resultSet = helper.query(Uri.parse("dataability:///com.huawei.dataabillitytest.
DataAbility"),columns,predicates);
        } catch (DataAbilityRemoteException e) {
            e.printStackTrace();
        }
        //处理结果
        resultSet.goToFirstRow();
        HiLog.fatal(hiLogLabel,"show name:");
        do{
            HiLog.fatal(hiLogLabel,resultSet.getString(1).concat(" ").concat(resultSet.
getString(2)));
        }while(resultSet.goToNextRow());
        HiLog.fatal(hiLogLabel,"=====================================");
    }
});
myLayout.addComponent(query);
```

在本例的query操作中设置谓词为查询前6个数据，从Data构建时可知数据库一共只有3个User数据，所以能够查询到全部数据。如果不满足这样的查询条件，则在使用查询功能时可以随意修改相应的查询条件。需要注意的是uri要和Data中配置的uri保持一致，但是在中间两个“//”的部分需要额外再加一个“/”。接下进行insert操作，代码如下：

```
//定义Button进行插入操作
Button insert = new Button(this);
insert.setText("insert");
insert.setTextSize(150);
insert.setLayoutConfig(layoutConfig);
insert.setBackground(shapeElement);
insert.setTouchEventListener(new Component.TouchEventListener() {
    @Override
    public boolean onTouchEvent(Component component, TouchEvent touchEvent) {
        switch (touchEvent.getAction()) {
            case TouchEvent.PRIMARY_POINT_DOWN:
```

```
            DataAbilityHelper helper = DataAbilityHelper.creator(MainAbilitySlice.super.
getContext());
            //构造插入数据
            ValuesBucket valuesBucket = new ValuesBucket();
            valuesBucket.putString("firstName", "Zhao");
            valuesBucket.putString("lastName", "Liu");
            valuesBucket.putInteger("age", 20);
            valuesBucket.putDouble("balance",300.3);
            try {
                helper. insert (Uri. parse ( " dataability:///com. huawei. dataabillitytest.
DataAbility"), valuesBucket);
            } catch (DataAbilityRemoteException e) {
                e.printStackTrace();
            }
            break;
            case TouchEvent.OTHER_POINT_UP:
                break;
            default:
                break;
            }
            return true;
    }
});
myLayout.addComponent(insert);
```

在本例的 insert 操作中，插入第 4 个 User 数据，定义 valuesBucket 用来表示插入数据，通过 valuesBucket 将 User 需要定义的属性(具体属性参照对象关系映射数据库)定义完成后，才能使用 Data 接口插入数据，而 update 和 insert 操作类似，代码如下：

```
//定义 Button 进行更新操作
Button update = new Button(this);
update.setText("update");
update.setTextSize(150);
update.setLayoutConfig(layoutConfig);
update.setBackground(shapeElement);
update.setClickedListener(new Component.ClickedListener() {
    @Override
    public void onClick(Component component) {
          DataAbilityHelper helper = DataAbilityHelper.creator(MainAbilitySlice.super.
getContext());

        //构造更新条件
        DataAbilityPredicates predicates = new DataAbilityPredicates();
        predicates.equalTo("userId",1);
```

```
        //构造更新数据
        ValuesBucket valuesBucket = new ValuesBucket();
        valuesBucket.putString("firstName", "John");
        valuesBucket.putString("lastName", "Tom");
        valuesBucket.putInteger("age", 20);
        valuesBucket.putDouble("balance",233.3);
        try {
            helper.update (Uri.parse ("dataability:///com.huawei.dataabillitytest.
DataAbility"), valuesBucket, predicates);
        } catch (DataAbilityRemoteException e) {
            e.printStackTrace();
        }
    }
});
myLayout.addComponent(update);
super.setUIContent(myLayout);
```

可以看到在 update 操作中将条件设置为 userId 等于 1，即通过 update 操作指定修改 userId 为 1 的数据，删除操作与此类似，代码如下：

```
//定义 Button 进行删除操作
Button delete = new Button(this);
delete.setText("delete");
delete.setTextSize(150);
delete.setLayoutConfig(layoutConfig);
delete.setBackground(shapeElement);
delete.setClickedListener(new Component.ClickedListener() {
    @Override
    public void onClick(Component component) {
        DataAbilityHelper helper = DataAbilityHelper.creator(MainAbilitySlice.super.
getContext());

        //构造删除条件
        DataAbilityPredicates predicates = new DataAbilityPredicates();
        predicates.between("userId", 1,2);
        try {
            helper.delete (Uri.parse ("dataability:///com.huawei.dataabillitytest.
DataAbility"),predicates);
        } catch (DataAbilityRemoteException e) {
            e.printStackTrace();
        }
    }
});
myLayout.addComponent(delete);
```

delete 操作中构筑了 userId 从 1 到 2 的条件，这样在调用了 Data 的接口后就会把相应的两条数据删除掉。记住，想要访问对应的 Data，还需要在配置文件中增加权限，权限的名字必须与之前创建 Data 时申明需要的权限一致，代码如下：

```
"abilities": [
    {
        "skills": [
          {
            "entities": [
              "entity.system.home"
            ],
            "actions": [
              "action.system.home"
            ]
          }
        ],
    "reqPermissions": [
      {
        "name": "com.huawei.dataabilitytest.DataAbility.DATA"
      },
      {
        "name": "ohos.permission.READ_USER_STORAGE"
      },
      {
        "name": "ohos.permission.WRITE_USER_STORAGE"
      }
    ],
    "orientation": "landscape",
    "formEnabled": false,
    "name": "com.huawei.mytestapp.MainAbility",
    "icon": "$media:icon",
    "description": "$string:mainability_description",
    "label": "test",
    "type": "page",
    "launchType": "standard"
},
```

运行上述程序，首先页面中会出现 4 个 Button，效果如图 8.33 所示。

先不进行任何单击 Button 的操作，此时观察 HiLog 面板，如图 8.34 所示。

查看 HiLog 面板时，能够看到定义在 Data 中 onStart()函数已经被执行了，同时数据库已经构建完毕，其中应该有上述代码中定义的 3 个数据，接下来单击 query 按钮，查看数据是否存在，如图 8.35 所示。

可以观察到，在数据库中的 3 个数据 Zhang San、Li Si、Wang Wu，均已被查询完成，然后执行插入操作，并单击查询，即可看到新插入的数据，如图 8.36 所示。

数据库被成功改变，Zhao Liu 被成功添加，接下来测试更新功能，如图 8.37 所示。

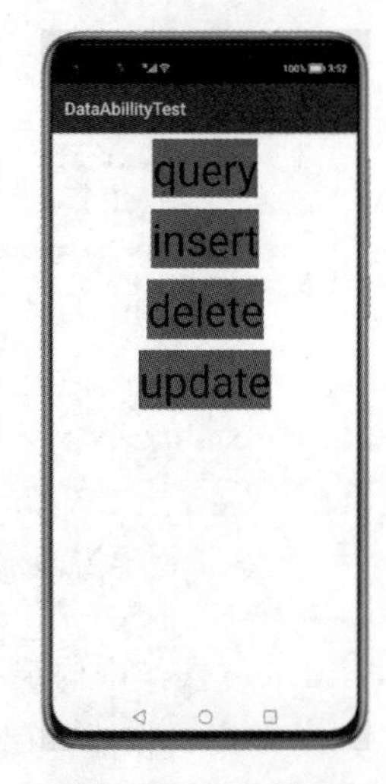

图 8.33　UI 界面

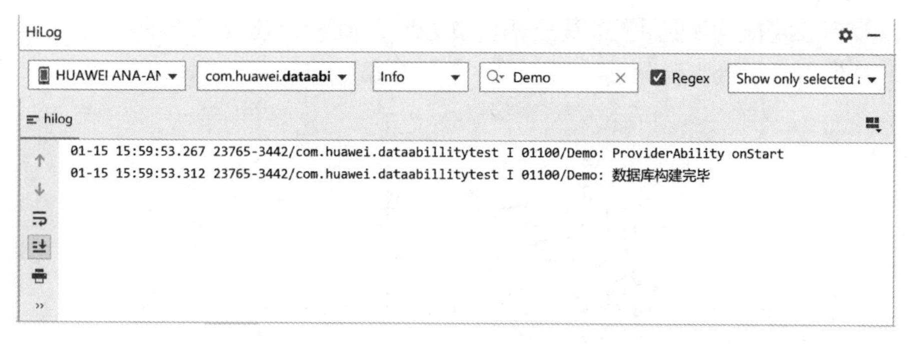

图 8.34 数据库构建完毕

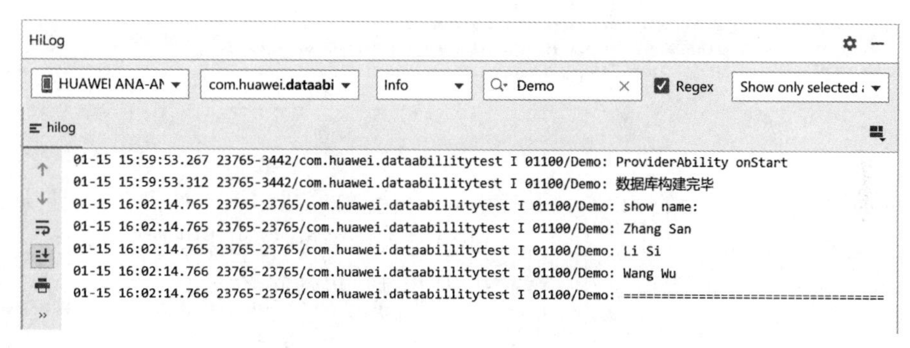

图 8.35 单击 query 按钮,查看到数据

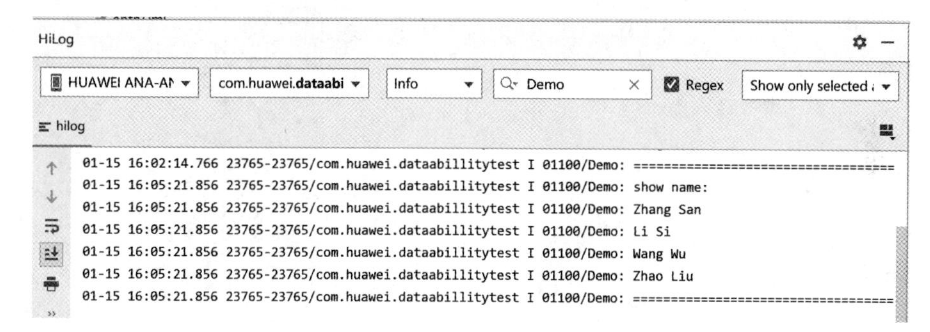

图 8.36 查询到新插入的数据

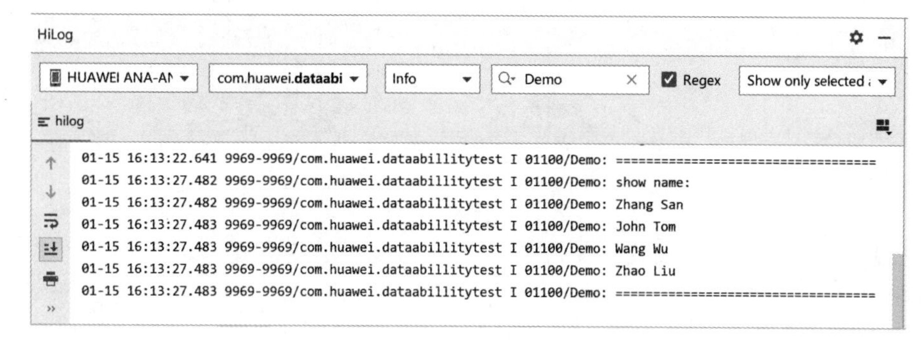

图 8.37 更新后的数据

可以看到 userId 为 1 的 Li Si 被成功替换为 John Tom。最后进行删除功能的测试，因为删除的条件谓语设置为删除 userId 为 1～2 的数据，所以会删除掉两个数据，如图 8.38 所示。

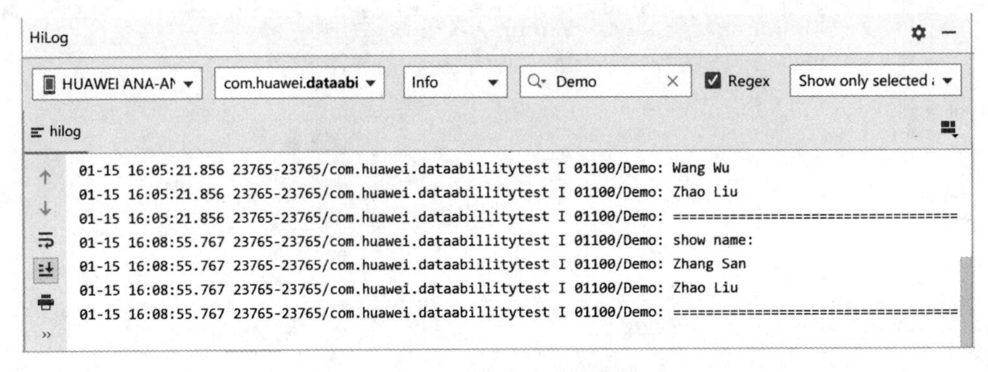

图 8.38　删除成功后的数据

可以看到，userId 为 0 的 Zhang San 和 userId 为 3 的 Zhao Liu 被留下来，而 userId 为 1～2 的数据被删除。通过以上的 HiLog 输出显示可以看出，跨程序共享数据功能已经基本实现，不止 Datatest 项目，其他项目只要拥有权限，都可以访问之前定义的 DataAbility。

第 9 章

多　媒　体

9.1　图像

9.1.1　图像场景概述

随着时代的发展，超清镜头和强大的照相算法使得拍摄的图像具有非常好的效果，但同时也伴随着图片数据量的暴增，5G 时代的高速网络也使传输高清多媒体数据成为家常便饭，因此手机要接收和处理的图像数据越来越多，且越来越大，人们对手机图像处理能力的要求也越来越高。

HarmonyOS 原生支持的图像模块支持对图像业务的开发，常见功能包括图像解码、图像编码、基本的位图操作、图像编辑等。当然，也支持通过接口组合实现更复杂的图像处理逻辑。

9.1.2　图像解码

HarmonyOS 的图像解码就是将系统所支持格式的存档图片，如 JPEG、PNG、JPG 等格式，解码成统一的位图格式的图像(在 HarmonyOS 中是 PixelMap 类)。

HarmonyOS 原生支持的图像解码类为 ImageSource 类，截至本书截稿为止支持格式包括 JPEG、PNG、GIF、HEIF、WebP 和 BMP。这个类的功能十分强大，可以在解码的同时对图像进行其他操作，例如旋转、裁剪和缩放。下面将通过一个实例来了解 HarmonyOS 的图像解码功能。

要想执行图像解码，首先需要一张图片。如果读者还没有 SD 卡读写权限，可以通过以下方法测试与图像相关的功能。首先随意下载一张图片，然后将其放置在项目目录下的 entry→src→main→resources→media 文件夹内，文件路径如图 9.1 所示，例如本范例中的 testPic.jpg。

然后在 IDE 界面最右侧找到 Gradle 工具，具体位置如图 9.2 所示。

打开 Gradle 工具，依照路径找到：项目名称→entry→ohos→compileDebugResources，双击 compileDebugResources，如图 9.3 所示。(compileDebugResources 用来打包调试用的资源文件，如果想要打包发布用的资源文件，则可以使用 compileReleaseResources

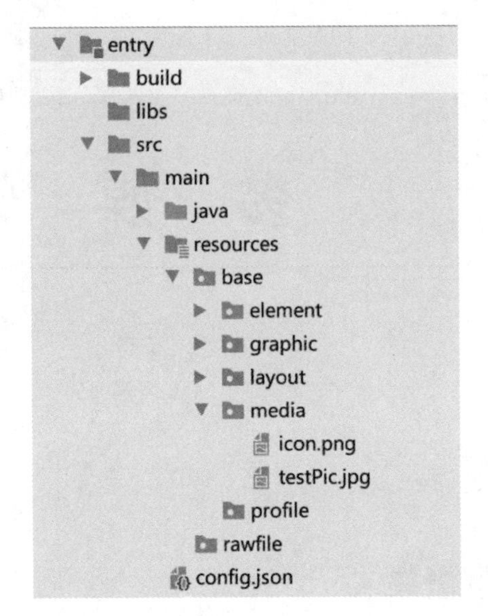

图 9.1　媒体文件的放置路径：media 文件夹

图 9.2　Gradle 工具在 IDE 界面上的位置

命令）。

Gradle 工具运行完毕后，在底边运行栏中会出现[compileDebugResources]：successful 字样，如图 9.4 所示，说明图像已经被打包成开发者可用的资源文件。

开发者可以在项目的 entry→build→generated→source→r→项目名称→ResourceTable 中找到它，如图 9.5 和图 9.6 所示。

至此，完成图像的准备工作，接下来可以使用 ImageSource 类来读取并解码这张图片。

将资源文件中的图片读取并解码需要两步，第一步通过资源文件的 Id（也就是图 9.6 中方框圈出的部分）获取资源文件的路径，代码如下：

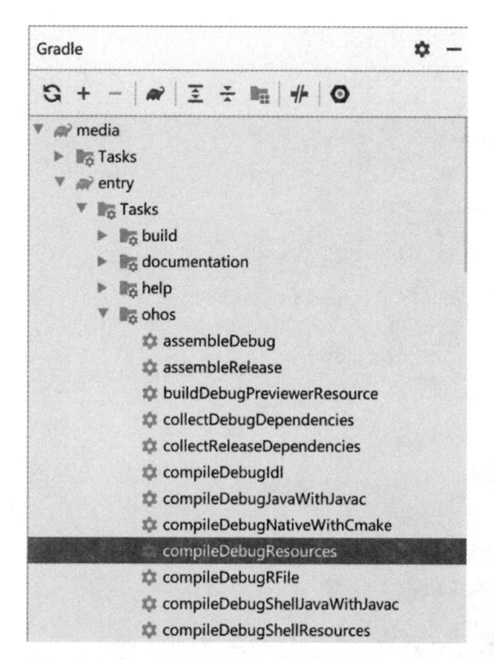

图 9.3 [compileDebugResources]在 Gradle 工具中的位置

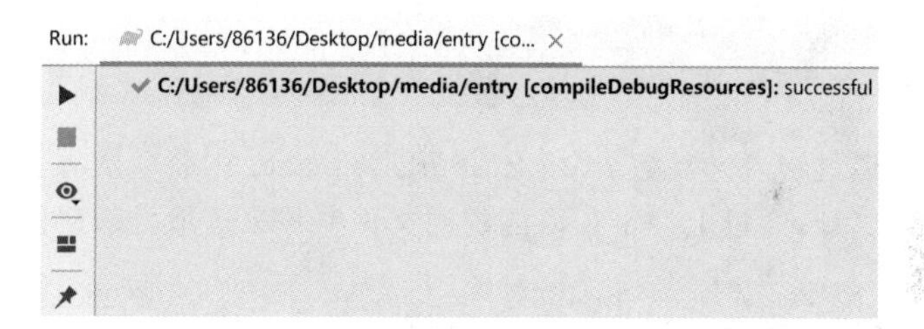

图 9.4 Gradle 工具[compileDebugResources]运行成功提示

```
entry
  build
    generated
      source
        r
          com
            harmony
              media
                ResourceTable.java
    intermediates
```

图 9.5 ResourceTable 的文件路径

```java
package com.harmony.media;

public final class ResourceTable {
    public static final int Graphic_background_ability_main = 0x1000004;

    public static final int Id_text_helloworld = 0x1000006;

    public static final int Layout_ability_main = 0x1000005;

    public static final int Media_icon = 0x1000002;
    public static final int Media_testPic = 0x1000003;

    public static final int String_app_name = 0x1000000;
    public static final int String_mainability_description = 0x1000001;
}
```

图 9.6 ResourceTable 文件内部的结构

```
//通过资源文件的 Id 获取资源文件的路径
public String getPathById(Context context, int id) {
    String path = "";
    if (context == null) {
        return path;
    }

    ResourceManager manager = context.getResourceManager();
    if (manager == null) {
        return path;
    }

    try {
        path = manager.getMediaPath(id);

    } catch (NotExistException e) {
        e.printStackTrace();
    } catch (WrongTypeException e) {
        e.printStackTrace();
    } catch (IOException e) {
        e.printStackTrace();
    }
    return path;
}
```

这个函数可以通过输入的参数 Id 在 ResourceManager 中获取资源文件的路径。

接下来实现第二步，通过上一步获取的资源文件路径来读取并解码图像，代码如下：

```
//读取并解码图像

public PixelMap getOriginalPixelMap(Context context, String path) {
    if (path == null || path.isEmpty()) {
        return null;
    }
    //通过资源文件路径获取 RawFileEntry 实例
    RawFileEntry assetManager = context.getResourceManager().getRawFileEntry(path);
    //图片源设置
    ImageSource.SourceOptions options = new ImageSource.SourceOptions();
    options.formatHint = "image/jpg";
    //图片解码设置
    ImageSource.DecodingOptions decodingOptions = new ImageSource.DecodingOptions();

    try {
        //通过 RawFileEntry 打开资源文件，也就是此处的图片
        Resource asset = assetManager.openRawFile();
```

```
            //使用 ImageSource 的 create()函数,将图片读取到 ImageSource 实例中
            ImageSource source = ImageSource.create(asset, options);

          //使用 ImageSource 的 createPixelMap()函数,将 ImageSource 实例解码为位图实例
            PixelMap pixelMap = source.createPixelmap(decodingOptions);
            return pixelMap;

        } catch (IOException e) {
            e.printStackTrace();
        }
        return null;
    }
```

在这种方法中,图片信息首先被传给 RawFileEntry 类,随后被转化为 Resource 类,然后才能被创建为 ImageSource 类,最后才能被解码为 PixelMap。看似非常复杂,但是实际上和图像解码相关的只有后两步,读者不需要详细了解前两步的细节,那是读取资源文件的通用步骤,只需知道这样写可以读取资源文件。

ImageSource 类在解码图片之前首先需要使用 create()函数读取图片。读取时需要一个 ImageSource.SourceOptions 类作为图片源设置,读者只需实例化 SourceOption,然后在读取图片的时候将其传入,没有必要额外添加任何信息。SourceOption 的 formatHint 仅作为读取时的格式提醒,在输入的格式提醒与要读取的文件格式匹配的情况下,读取速度会略微加快。

在完成读取之后,就可以对图像进行解码了。ImageSource 类使用 createPixelMap()函数执行解码操作。解码时需要一个 ImageSource.DecodingOptions 类作为解码设置,首先需要将其实例化,然后为其手动添加各项解码规则,常见的规则有旋转、缩放和裁剪。在这里并没有设置任何解码规则,所以 ImageSource 在解码时将采用默认值进行解码,不会对原图有任何更改。

图像解码完成后生成一个 PixelMap 类,其本身并不能够直接显示,为了展示一下效果,可以写一个简单的页面继承 AbilitySlice,然后重写其 onStart()方法,使用第 3 章讲到的 Image(ohos.agp.components.Image)组件将 PixelMap 显示在 UI 上,具体代码如下:

```
//展示图片解码效果

private Image image;
private PixelMap pixelMap;

public void onStart(Intent intent) {
    super.onStart(intent);
```

```
    this.setDisplayOrientation(AbilityInfo.DisplayOrientation.PORTRAIT);

    //从资源文件 Id 得到其路径
    String path = getPathById(this, ResourceTable.Media_testPic);
    //从路径解码图像
    pixelMap = getOriginalPixelMap(this, path);

    //创建容器组件 PositionLayout 实例
    PositionLayout myLayout = new PositionLayout(this);
    //容器组件 DirectionalLayout 的 LayoutConfig
     DirectionalLayout.LayoutConfig layoutConfig = new DirectionalLayout.LayoutConfig
(DirectionalLayout.LayoutConfig.MATCH_PARENT, DirectionalLayout.LayoutConfig.MATCH_
PARENT);
    myLayout.setLayoutConfig(layoutConfig);

    //将 PixelMap 放入 Image 组件中
    image = new Image(this);
    image.setPixelMap(pixelMap);
    image.setTop(50);
    image.setLeft(50);

    //将 Image 添加到容器中
    myLayout.addComponent(image);

    super.setUIContent(myLayout);
}
```

运行上述代码，效果如图 9.7 所示。

本书选择的图片是 1920×1080 像素的，在无缩放的情况下对于手机屏幕来讲较大，因此图像的右侧边缘超出了屏幕边界。解决方法是可以为 DecodingOptions 添加各种解码规则，以此改变图片的各种属性，使其适配屏幕。例如，在 getOriginalPixelMap()函数中添加如下代码：

```
decodingOptions.desiredSize = new Size(960,540);
```

代码将图片的边长缩小为原来的一半，这样便能够在屏幕中显示，效果如图 9.8 所示。

DecodingOptions 提供了丰富的解码规则供使用者选择。刚才使用的 desiredSize 可以对图片进行缩放和形变；desiredRegion 可以对图片进行裁剪；rotateDegrees 可以设置图片的旋转角度；sampleSize 可以设置图像的采样率，较低的采样率会降低清晰度，但是可以大幅节省占用的内存；desiredPixelFormat 可以设置位图的格式等。

图 9.7 将图片解码为位图后展示

图 9.8 缩小后的图像展示

9.1.3 位图操作

位图使用一个个像素来描述图像的格式，将图片解码为位图意味着我们可以在像素层级进行操作。HarmonyOS 的位图类 PixelMap 提供了强大的像素操作能力。下面将通过一个示例进行讲解。

9.1.2 节中将图片 testPic.jpg 解码成了 PixelMap 类实例 pixelMap，本节将使用 PixelMap 中的方法获取图片的一些信息。在刚才编写的简单展示页面文件中添加以下语句可以获取解码完成的 pixelMap 的信息，代码如下：

```
long capacity = pixelMap.getPixelBytesCapacity();
long BytesNumber = pixelMap.getPixelBytesNumber();
int rowBytes = pixelMap.getBytesNumberPerRow();
ImageInfo imageInfo = pixelMap.getImageInfo();
```

其中，capacity 表示这张位图所占用的空间，BytesNumber 代表这张位图的总像素大小，rowBytes 则代表每行像素的大小，imageInfo 则表示带有这张位图的一些基本信息，如像素格式(PixelFormat)和色彩空间(ColorSpace)等。读者可以利用第 3 章介绍的 Text 组件将上述属性值展示出来，效果如图 9.9 所示。

从图 9.9 可知，这张 960×540(边长缩为一半之后)的位图竟然要占用多于 2MB 的空间，可见位图格式占用的空间确实比 JPG 格式要大得多。其像素格式 ARGB_8888 代表其

图 9.9　位图像关信息

每个像素都是四通道，透明度、红色、绿色、蓝色按顺序各占一个通道，每个通道 1B，这也使其每行 960 像素共计 3840B。

还可以使用 PixelMap 类中的方法对位图的像素进行读写，实现对位图的修改。需要注意的是，为了进行位图像素的读写，需要在解码时将 decodingOptions 的参数 editable 设置为 true，代码如下：

```
//图片解码设置
ImageSource.DecodingOptions decodingOptions = newImageSource.DecodingOptions();
decodingOptions.desiredSize = new Size(960,540);
decodingOptions.editable = true;
```

在 onStart()中添加代码如下：

```
//从资源文件 Id 得到其路径
String path = getPathById(this, ResourceTable.Media_testPic);

//从路径解码图像
pixelMap = getOriginalPixelMap(this, path);

//创建容器组件 PositionLayout 实例
PositionLayout myLayout = new PositionLayout(this);
```

```
//容器组件 PositionLayout 的 LayoutConfig
    ...

//读取指定区域像素
int[] pixelArray = new int[40000];
Rect regionSrc = new Rect(100, 50, 200, 200);
pixelMap.readPixels(pixelArray, 0, 200, regionSrc);

//在指定区域写入像素
Rect regionDst = new Rect(500, 340, 200, 200);
pixelMap.writePixels(pixelArray, 0, 200, regionDst);

//将 PixelMap 放入 Image 组件中
image = new Image(this);
image.setPixelMap(pixelMap);
image.setTop(50);
image.setLeft(50);
    ...
```

在这里首先创建了一个 int 数组，用来存储像素数据，每个 int 格式的数据为 4 B，刚好能够存储一个像素的四通道数据。此处将这张图片的一个 200×200 区域的像素进行了复制，覆写在了另一个 200×200 的区域内。重新运行程序，效果如图 9.10 所示。与图 9.9 进行对比即可看出图片被进行了修改。

图 9.10　位图操作示例

9.1.4 图像编码

HarmonyOS 的 ImagePacker 类提供了方便的图像编码能力。开发者可以使用它将上述修改过的位图重新编码为适合存储用的图片格式。在之前写好的页面中，添加代码如下：

```
//将修改过的位图重新编码为适合存储的图片格式

//创建 ImagePacker 实例
ImagePacker imagePacker = ImagePacker.create();
//依据路径创建数据流
FileOutputStream outputStream = null;
try {
    outputStream = new FileOutputStream("/path/to/packed.file");
} catch (FileNotFoundException e) {
    e.printStackTrace();
}

//创建编码设置,并为其添加规则
ImagePacker.PackingOptions packingOptions = new
      ImagePacker.PackingOptions();
packingOptions.format = "image/jpeg";
packingOptions.quality = 90;

//预编码
boolean result = imagePacker.initializePacking(outputStream,packingOptions);

//将需要编码的位图放入 ImagePacker 中,并完成编码
result = imagePacker.addImage(pixelMap);
long dataSize = imagePacker.finalizePacking();
```

ImagePacker 兼具编码图像和存储的功能。为了进行编码工作，首先需要准备一个 PixelMap 实例作为编码的对象和一个 ImagePacker.PackingOptions 实例作为编码的设置，为了进行存储工作还需要一个由存储地址创建的 FileOutputStream 实例。在实际应用中，存储地址往往是在手机的 SD 卡上，因此还要申请 SD 卡读写的权限，否则运行时会报错。

ImagePacker 工作分为两步：initializePacking() 和 finalizePacking()。其中，initializePacking()可以根据输出流和编码设置来创建一个新的文件用于存储将要编码的位图，finalizePacking()则执行编码和写入文件的操作。

9.2 音视频

9.2.1 音视频场景概述

随着移动终端的普及和网络的提速，短而精的大流量传播内容逐渐获得各大平台、用户和资本的青睐。有调研结果显示，截至 2020 年 6 月，中国短视频用户规模已达 8.18 亿，日

均使用时长110分钟。微博、秒拍、快手、今日头条这些耳熟能详的企业纷纷入局短视频行业，募集了一批又一批优秀的内容制作团队入驻，视频已经不单是娱乐的工具，更成为许多人谋生的手段。

作为短视频输入和输出的主要平台，移动终端需要拥有强大的视频处理能力，包括视频的编码、解码、播放及录制等。

HarmonyOS视频模块支持视频业务的开发和生态开放，开发者可以通过已开放的接口很容易地实现视频媒体的播放、操作和新功能开发。视频媒体的常见操作有视频编解码、视频合成、视频提取、视频播放及视频录制等。另外，HarmonyOS作为一个跨平台的系统，其应用不仅可以在手机设备上正常工作，也可以在搭载了HarmonyOS的其他设备(如：平板、智慧屏、车机等)上运行。

虽然声频相比于视频缺少了画面，但是相比视频也确实"轻量化"了许多，让制作、传输和解析的成本大大降低，因此，迄今为止声频仍然是移动设备中不可缺少的一部分。依托于声频，开发者和相关公司发展出一系列新型媒体模式，如有声书、网络广播电台等。同时，声频还是手机上各种信息的载体之一，如来电铃声、通话语音等。

HarmonyOS提供了方便可靠的与声频相关的API，声频模块支持声频业务的开发，提供声频相关的功能，主要包括声频播放、声频采集、音量管理和短音播放等。

9.2.2 音视频编解码

从定义上讲，编码是信息从一种形式或格式转换为另一种形式的过程。用预先规定的方法将文字、数字或其他对象编成数码，或将信息、数据转换成规定的电脉冲信号。在音视频模块中，编码是指编码器将原始的音视频信息压缩为另一种格式的过程。

从意义上讲，编码的意义在于可以对音视频文件进行压缩，减小占用空间，增强网络传输效率等。

但是，压缩后的音视频格式往往不可以直接被播放，在播放之前还需要进行解码工作，所以，解码是一种用特定方法，把数码还原成它所代表的内容或将电脉冲信号、光信号、无线电波等转换成它所代表的信息、数据等的过程。在本模块中，解码是指解码器将接收的数据还原为音视频信息的过程，与编码过程相对应。

在HarmonyOS中，音视频的编解码工作都在Codec类中进行。Codec的编解码可分为普通模式和管道模式，在普通模式中Codec需要与Java的ByteBuffer类进行数据交互。在管道模式下开发者需要调用Codec类的setSource()方法，数据会自动解析并传输给Codec实例。这里主要介绍普通模式编解码。

首先需要使用Codec的构建函数来创建一个编码器，代码如下：

```
final Codec encoder = Codec.createEncoder();
```

Codec进行普通模式编码需要一个参数编码格式Format类实例，也就是要求Codec输出的是视频格式信息，Format类实例的创建代码如下，需要注意的是BITRATE_MODE必

须被设置,否则无法将 Format 设置到编码器上,代码如下:

```
//构造数据源格式,并设置 Codec 实例
Format fmt = new Format();
fmt.putStringValue(Format.MIME, Format.VIDEO_AVC);
fmt.putIntValue(Format.WIDTH, 640);
fmt.putIntValue(Format.HEIGHT, 480);
fmt.putIntValue(Format.BIT_RATE, 392000);
fmt.putIntValue(Format.COLOR_MODEL, 21);
fmt.putIntValue(Format.FRAME_RATE, 30);
fmt.putIntValue(Format.FRAME_INTERVAL, 1);
fmt.putIntValue(Format.BITRATE_MODE, 1);
```

然后将 Format 类示例设置到编码器中,函数会返回一个布尔值,如果 Format 设置成功,则返回值为 true,否则返回值为 false,代码如下:

```
boolean formatStatus = encoder.setCodecFormat(fmt);
HiLog.fatal(hiLogLabel,"Format Status:" + String.valueOf(formatStatus));
```

通过 start()函数启动编码器,同样地,启动成功则返回值为 true,否则返回值为 false,需要注意的是 Format 设置成功是编码器启动成功的必要条件,代码如下:

```
boolean encoderStatus = encoder.start();
HiLog.fatal(hiLogLabel,"Encoder Status:" + String.valueOf(encoderStatus));
```

如果 Format 设置正确并且启动成功,应该可以在 HiLog 中看到输出效果如图 9.11 所示。

```
02-05 10:36:37.600 27205-27205/com.huawei.mytestapp F 00000/onStart: Format Status:true
02-05 10:36:37.608 27205-27205/com.huawei.mytestapp F 00000/onStart: Encoder Status:true
```

图 9.11 HiLog 显示 Format 设置成功且编码器启动成功

编码器启动完毕,但是没有数据执行编码,所以下面主要讲如何给编码器输入数据,代码如下:

```
//需要编码的数据,可以是相机拍摄的画面等
Byte[] data = {0,1,2,3,4,5,6};
BufferInfo bufferInfo = new BufferInfo();
//得到 Codec 内部的 Buffer
ByteBuffer buffer = encoder.getAvailableBuffer(-1);
HiLog.fatal(hiLogLabel,"BufferGot");
//将数据装入 Buffer
buffer.put(data);
HiLog.fatal(hiLogLabel,"BufferLoaded");
```

```
//检查放入的数据
for(int i = 0; i < 6; i++){
    Byte showBuffer = buffer.get();
    HiLog.fatal(hiLogLabel,"第" + i + "个数据为" + buffer.get(i));
}
bufferInfo.setInfo(0, data.length, System.currentTimeMillis(), 0);
//将填充好数据的 Buffer 写入 Codec,执行编码
encoder.writeBuffer(buffer, bufferInfo);
HiLog.fatal(hiLogLabel,"BufferWriten");
```

上述代码首先创建了一个 Byte 数组并储存了一些数据，然后从编码器中取出 Buffer（此 Buffer 用于装载编码数据与编码器进行交互），再将数组的数据放入 Buffer 中，最后将 Buffer 送入编码器执行编码。运行上述代码，HiLog 输出信息如图 9.12 所示。

```
02-05 10:36:37.608 27205-27205/com.huawei.mytestapp F 00000/onStart: BufferGot
02-05 10:36:37.608 27205-27205/com.huawei.mytestapp F 00000/onStart: BufferLoaded
02-05 10:36:37.608 27205-27205/com.huawei.mytestapp F 00000/onStart: 第0个数据为0
02-05 10:36:37.608 27205-27205/com.huawei.mytestapp F 00000/onStart: 第1个数据为1
02-05 10:36:37.608 27205-27205/com.huawei.mytestapp F 00000/onStart: 第2个数据为2
02-05 10:36:37.608 27205-27205/com.huawei.mytestapp F 00000/onStart: 第3个数据为3
02-05 10:36:37.608 27205-27205/com.huawei.mytestapp F 00000/onStart: 第4个数据为4
02-05 10:36:37.608 27205-27205/com.huawei.mytestapp F 00000/onStart: 第5个数据为5
02-05 10:36:37.608 27205-27205/com.huawei.mytestapp F 00000/onStart: BufferWriten
```

图 9.12　编码器数据的装载与检查

可见 Byte 数组中的数据被成功装载到 Buffer 中。

在调用 writeBuffer()函数后，编码器会根据 Format 的参数对 Buffer 中的数据进行编码，如果想查询、修改编码后的数据，需要给编码器添加监听器 Codec.IcidecListener，代码如下：

```
//向编码器添加监听器 Codec.IcidecListener
Codec.ICodecListener listener = new Codec.ICodecListener() {
    @Override
    public void onReadBuffer(ByteBuffer ByteBuffer, BufferInfo bufferInfo, int trackId) {
        HiLogLabel hiLogLabel = new HiLogLabel(3,0,"listener");
        HiLog.fatal(hiLogLabel,"BufferReaded");
        for(int i = 0; i < 6; i++){
            HiLog.fatal(hiLogLabel,"第" + i + "个数据为" + ByteBuffer.get(i));
        }

    }
    @Override
    public void onError(int errorCode, int act, int trackId) {
        throw new RunTimeException();
    }
};
```

当编码器成功编码数据后，会回调 onReadBuffer()函数，其中的回调参数 ByteBuffer 就是编码后的数据，开发者可任意对其进行操作，例如在这里取出 Buffer 的前 6 个数据，打印在 HiLog 中，如图 9.13 所示。

```
02-05 11:11:32.772 29242-29443/com.huawei.mytestapp F 00000/listener: BufferRead
02-05 11:11:32.772 29242-29443/com.huawei.mytestapp F 00000/listener: 第0个数据为0
02-05 11:11:32.772 29242-29443/com.huawei.mytestapp F 00000/listener: 第1个数据为0
02-05 11:11:32.772 29242-29443/com.huawei.mytestapp F 00000/listener: 第2个数据为0
02-05 11:11:32.772 29242-29443/com.huawei.mytestapp F 00000/listener: 第3个数据为1
02-05 11:11:32.772 29242-29443/com.huawei.mytestapp F 00000/listener: 第4个数据为103
02-05 11:11:32.772 29242-29443/com.huawei.mytestapp F 00000/listener: 第5个数据为66
```

图 9.13 编码后数据的检查

可见，当前 Buffer 中的数据已经被改变了，这就是编码后的数据。

解码过程与编码类似，只不过需要调用 Codec 的 CreateDecoder()函数，代码如下：

```
final Codec decoder = Codec.createDecoder();
```

其余操作与编码过程一致，具体代码如下：

```
//创建解码器实例
final Codec decoder = Codec.createDecoder();
//设置格式
decoder.setCodecFormat(fmt);
//添加监听
decoder.registerCodecListener(listener2);
//启动解码器
boolean decoderStatus = decoder.start();
//得到 Codec 内部的 Buffer
ByteBuffer buffer2 = decoder.getAvailableBuffer(-1);
//放入数据
buffer2.put(data);
BufferInfo bufferInfo2 = new BufferInfo();
bufferInfo2.setInfo(0, data.length, System.currentTimeMillis(), 0);
//写入解码器,开始解码
decoder.writeBuffer(buffer2, bufferInfo2);
```

监听器与编码器相同，可以在 onReadBuffer 中对解码出的数据进行任意操作，代码如下：

```
Codec.ICodecListener listener2 = new Codec.ICodecListener() {
    @Override
    public void onReadBuffer(ByteBuffer ByteBuffer, BufferInfo bufferInfo, int trackId) {
        HiLogLabel hiLogLabel = new HiLogLabel(3,0,"listener2");
        HiLog.fatal(hiLogLabel,"BufferReaded");
```

```
        }
        @Override
        public void onError(int errorCode, int act, int trackId) {
            throw new RunTimeException();
        }
    };
```

上面的示例只是简单地讲解了 HarmonyOS 视频编解码器的用法，具体的视频编解码实例会在第 11 章的视频流直播实战部分进行详细讲解。

9.2.3 视频播放

HarmonyOS 提供了功能强大的播放器 Player 类，支持当前各种主流音视频格式的播放。Player 提供了方便的视频源接口，可以方便地读取并播放本地视频和远程网络视频。

首先介绍一下播放本地视频，与 9.2.2 节相同，首先需要准备一段视频，读者可以随意下载一段视频，并按照 9.1.2 节中的步骤将其放入到资源文件中。

对于本地视频，可以使用参数类型为 BaseFileDescriptor 类的 setSource() 函数来为 Player 设置视频源，所以首先要取得视频文件的 BaseFileDescriptor。与 9.1.2 节一样，在将视频文件打包进资源文件后，即可使用 ResourceID 获取这个视频的路径，代码如下：

```
//获取视频路径
public String getPathById(Context context, int id) {
    String path = "";
    if (context == null) {
        return path;
    }

    ResourceManager manager = context.getResourceManager();
    if (manager == null) {
        return path;
    }

    try {
        path = manager.getMediaPath(id);

    } catch (NotExistException e) {
        e.printStackTrace();
    } catch (WrongTypeException e) {
        e.printStackTrace();
    } catch (IOException e) {
        e.printStackTrace();
    }
    return path;
}
```

取得路径后，使用该路径获取 BaseFileDescriptor，代码如下：

```
public BaseFileDescriptor getBFD(Context context, String path) {
    if (path == null || path.isEmpty()) {
        return null;
    }
    //通过资源文件路径获取 RawFileEntry 实例
    RawFileEntry assetManager = context.getResourceManager().getRawFileEntry(path);

    try {
        //通过 RawFileEntry 打开资源文件，也就是我们的视频
        BaseFileDescriptor bfd = assetManager.openRawFileDescriptor();

        return bfd;

    } catch (IOException e) {
        e.printStackTrace();
    }
    return null;
}
```

有了 setSource()函数播放任务就变简单了。只要短短 6 行代码，即可对视频进行播放，代码如下：

```
String videoPath = getPathById(this, ResourceTable.Media_testVid3);
BaseFileDescriptor bfd = getBFD(this, videoPath);
Player mPlayer = new Player(this);
mPlayer.setSource(bfd);
mPlayer.prepare();
mPlayer.play();
```

按照之前学习的知识，写一个简单的页面(Ability 或 AbilitySlice)进行测试，但是会发现播放时只有声音而没有画面，这是因为 UI 是人机视觉交互的唯一途径，然而 Player 并不是一个 UI 组件，它只负责将音视频解码并播放出来。对于声频部分它可以自动连接到手机的声频播放组件，所以可以传出声音，而画面部分解码完成后却不知道将画面信息传递到哪里，因此必须手动创建一个 UI 组件，并将其和 Player 连接起来。在这里推荐使用 HarmonyOS 提供的 SurfaceProvider 组件，这个组件能够提供一个窗口，将画面信息显示于其上，下面介绍具体步骤：

(1) 创建一个类(刚才所写的页面也可以)，实现 SurfaceOps.Callback 接口。

(2) 重写 SurfaceOps.Callback 中的方法，将回调参数 SurfaceOps 添加到 Player 实例中。

(3) 为 SurfaceProvider 实例中的 SurfaceOps 添加回调。

完整代码如下：

```
//创建 SurfaceProvide 组件,将其与 Player 连接
public class VideoTest extends AbilitySlice implements SurfaceOps.Callback {

    private Player mPlayer;
    private SurfaceProvider surfaceProvider;
    private SurfaceOps surfaceOps;
    private DirectionalLayout myLayout;
    private DirectionalLayout.LayoutConfig layoutConfig;
    private String videoPath;
    private BaseFileDescriptor bfd;

    @Override
    public void onStart(Intent intent) {
        super.onStart(intent);

        this.setDisplayOrientation(AbilityInfo.DisplayOrientation.PORTRAIT);
        //读取视频
        videoPath = getPathById(this, ResourceTable.Media_testVid3);
        bfd = getBFD(this, videoPath);
        //实例化 Player 组件
        mPlayer = new Player(this);
        //为 Player 添加视频源
        mPlayer.setSource(bfd);
        //随便创建一个布局,用来承载 SurfaceProvider
        myLayout = new DirectionalLayout(this);
        //实例化 SurfaceProvider 组件,并为其设置属性
        surfaceProvider = new SurfaceProvider(this);
        layoutConfig = new DirectionalLayout.LayoutConfig(960,540);
        layoutConfig.setMargins(60,60,60,60);
        surfaceProvider.setLayoutConfig(layoutConfig);
        //为 SurfaceProvider 的 SurfaceOps 添加回调
        surfaceProvider.getSurfaceOps().get().addCallback(this);
        surfaceProvider.pinToZTop(true);    //这一行代码很重要,不添加则会没有画面

        myLayout.addComponent(surfaceProvider);
        super.setUIContent(myLayout);
    }

    public String getPathById(Context context, int id) {
        String path = "";
        if (context == null) {
            return path;
        }

        ResourceManager manager = context.getResourceManager();
        if (manager == null) {
            return path;
```

```
        }

        try {
            path = manager.getMediaPath(id);

        } catch (NotExistException e) {
            e.printStackTrace();
        } catch (WrongTypeException e) {
            e.printStackTrace();
        } catch (IOException e) {
            e.printStackTrace();
        }
        return path;
    }

    public BaseFileDescriptor getBFD(Context context, String path) {
        if (path == null || path.isEmpty()) {
            return null;
        }
        //通过资源文件路径获取 RawFileEntry 实例
        RawFileEntry assetManager = context.getResourceManager().getRawFileEntry(path);

        try {
            //通过 RawFileEntry 打开资源文件,也就是我们的视频
            BaseFileDescriptor bfd = assetManager.openRawFileDescriptor();

            return bfd;

        } catch (IOException e) {
            e.printStackTrace();
        }
        return null;
    }

    @Override
    public void surfaceCreated(SurfaceOps surfaceOps) {
        //将 Player 与 SurfaceOps 建立连接,也就是与 SurfaceProvider 建立连接
        mPlayer.setSurfaceOps(surfaceOps);
        //准备
        mPlayer.prepare();
        //播放
        mPlayer.play();
    }

    @Override
    public void surfaceChanged(SurfaceOps surfaceOps, int i, int i1, int i2) {
```

```
    }

    @Override
    public void surfaceDestroyed(SurfaceOps surfaceOps) {
        //不需要播放时,记得即时释放资源
        mPlayer.release();

    }
}
```

运行上述代码,效果如图 9.14 所示。

图 9.14 视频播放效果

读者还可以用 Button、ProcessBar 等组件为视频播放组件添加按钮和进度条等功能,使其变得更加像一个播放器界面。

下面再简单讲解一下如何从网络读取视频并播放的方法。其实在播放环节与本地视频播放完全一致,其差别仅在读取环节,Player 可以使用参数类型为 Source 类的 setSource()函数设置视频源,Source 类实例可以通过统一资源标志符(Uniform Resource Identifier,URI)来创建。如在上述例子中,只需将"为 Player 添加视频源"这一过程替换为如下两行代码:

```
//为 Player 添加视频源
Source source = new Source("https://stream7.iqilu.com/10339/upload_tran scode/202002/18/
20200218114723HDu3hhxqIT.mp4");
mPlayer.setSource(source);
```

括号中的Uri可以替换为自己的视频资源，播放的效果与本地视频完全一致。

9.2.4 声频资源的加载与播放

在9.2.3节中讲到Player类可以播放视频。事实上，Player也可以实现声频的播放，所以对于音乐(如mp3等格式)的播放可以参照9.2.3节中的视频加载和播放方法，但是不需要配置SurfaceProvider。

短音播放主要负责tone音的生成与播放及系统音的播放，这些任务主要由SoundPlayer类实现。Soundplayer的两个构造器分别对应tone音和系统音。

无参构造器用于构造播放tone音的SoundPlayer实例，调用createSound()函数来加载一些系统预设的tone音，还可以为其设置持续时间。至此tone音设置完毕，使用play()、pause()函数可以分别对其进行播放与暂停，在tone音完成使命后，可以调用release()函数释放资源。tone音进行播放、暂停和资源释放的代码如下：

```
//播放、暂停和资源释放 DTMF_0 格式的 tone 音

public void demo1(){
    //步骤 1: 实例化对象
    SoundPlayer soundPlayer = new SoundPlayer();
    //步骤 2: 创建 DTMF_0(高频 1336Hz,低频 941Hz)持续时间 1000ms 的 tone 音
    soundPlayer.createSound(ToneDescriptor.ToneType.DTMF_0, 1000);
    //步骤 3: tone 音播放、暂停和资源释放
    soundPlayer.play();
    soundPlayer.pause();
    soundPlayer.release();
}
```

除了DTMF_0以外，tone音还支持许多其他格式，读者可自行探索。

参数类型为String的构造器可以直接传入包含系统音的包名来构建用于播放系统音的SoundPlayer类实例，然后使用playSound()直接设置并播放想要播放的系统音，代码如下：

```
public void demo2(){
    //步骤 1: 实例化对象
    SoundPlayer soundPlayer = new SoundPlayer("packageName");
    //步骤 2: 播放键盘敲击音,音量为 1.0
    soundPlayer.playSound(SoundPlayer.SoundType.KEY_CLICK, 1.0f);
}
```

除键盘敲击声KEY_CLICK外，HarmonyOS还自带很多种类的系统音，如字符删除声KEYPRESS_DELETE、输入无效声KEYPRESS_INVALID等。直接调用以上demo方法即可实现tone音或系统音的播放。

9.3 相机

9.3.1 相机场景概述

随着时代的发展，相机已经是各大智能设备的必备组成部分，相机的能力也逐渐成为多媒体能力的一个重要组成部分。本节将讲解一下与相机相关的开发知识。

在正式进入开发讲解前，有必要先了解一下相机开发的基本流程。在调用相机之后一般进行拍照或者录像行为，为了完成这一系列的拍照或者录像的能力开发，需要进行5个关键步骤，如图9.15所示。

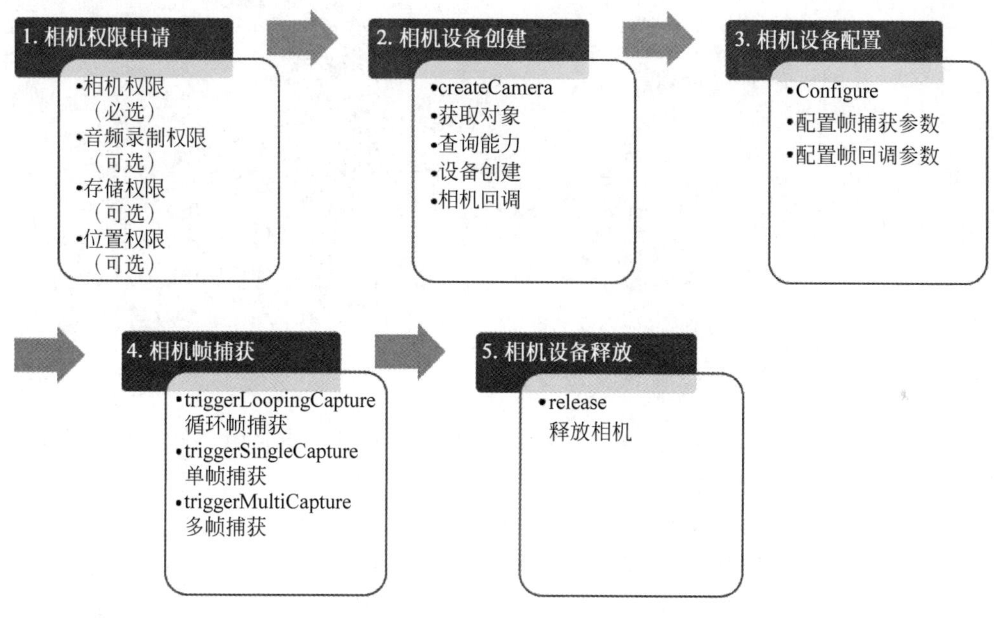

图9.15 相机开发关键步骤

基本上所有相机调用的场景都需要遵循这5个关键步骤来完成与相机相关的能力调用。

9.3.2 相机预览

相机预览是应用开发时最常用的核心功能之一，完整的相机预览功能可以通过以下4个核心步骤完成。

1. 添加相机权限

同第8章的多设备协同权限一样，首先需要在config.json文件中声明需要添加的权限。下面以相机能力及录音能力为例，展示如何添加这两个权限的信息，代码如下：

```
//添加相机权限和录音权限

"reqPermissions": [{
        "name": "ohos.permission.CAMERA",
        "grantMode": "user_grant",
        "reason": "$string:permreason_camera",
        "usedScene": {
            "ability": [
                "com.harmony.myapplication.MainAbility"
            ],
            "when": "always"
        }
    },
      {
        "name": "ohos.permission.MICROPHONE ",
        "grantMode": "user_grant",
        "reason": "",
        "usedScene": {
          "ability": [
            "com.harmony.myapplication.MainAbility"
          ],
          "when": "always"
        }
      }
]
```

除了这两个权限之外,如果开发者需要在使用拍照功能的同时用到外部存储能力及位置信息能力,还需要添加这两个能力所对应的权限。若需要保存图片或视频至设备的外部,则需要申请存储权限 ohos.permission.WRITE_USER_STORAGE,若需要保存图片或视频的位置信息,则需要申请位置权限 ohos.permission.LOCATION。

config.json 文件配置完成之后,就可以直接在设置中开启应用的对应权限或者直接让应用主动询问是否开启。若想要在应用内检测相机权限及录音能力权限是否开启,并在未开启状态下主动询问是否开启对应的权限,则可以在程序内添加代码如下:

```
//主动询问权限是否开启
String[] permission = {"ohos.permission.MICROPHONE","ohos.permission.CAMERA"};
for (int i = 0;i < permission.length;i++){
    if(verifyCallingOrSelfPermission(permission[i]) != 0){
        if(canRequestPermission(permission[i])){
            requestPermissionsFromUser(permission, 0);
        }
    }
}
```

2. 创建相机 UI

在 HarmonyOS 中实现相机的画面预览，可以通过创建 SurfaceProvider 实现。参照 9.2.3 节视频播放部分的知识，创建用于实现相机预览和照片展示的 SurfaceProvider 实例 surfaceProvider1 和 surfaceProvider2，随后建立一个用于拍照的 Button，将三者添加到布局中，具体代码如下：

```
//实现相机的画面预览

//定义 SurfaceProvider 的大小
public static final int VIDEO_WIDTH = 960;
public static final int VIDEO_HEIGHT = 960;
//删除导航栏等
Window window = this.getWindow();
window.setStatusBarVisibility(Component.INVISIBLE);
window.setNavigationBarColor(Color.TRANSPARENT.getValue());
//布局容器
myLayout = new DirectionalLayout(this);
LayoutConfig config = new LayoutConfig(LayoutConfig.MATCH_PARENT, LayoutConfig.MATCH_
PARENT);
myLayout.setLayoutConfig(config);
myLayout.setOrientation(Component.HORIZONTAL);
ShapeElement background = new ShapeElement();
background.setRgbColor(new RgbColor(255,255,255));
myLayout.setBackground(background);
//SurfaceProvider1,用于进行摄像头画面预览
config.width = VIDEO_WIDTH;
config.height = VIDEO_HEIGHT;
config.setMargins(0,0,0,0);
config.alignment = LayoutAlignment.VERTICAL_CENTER;
surfaceProvider1 = new SurfaceProvider(this);
surfaceProvider1.setLayoutConfig(config);
surfaceProvider1.getSurfaceOps().get().addCallback(callback1);
surfaceProvider1.pinToZTop(true);
//SurfaceProvider2,用于进行照片展示
config.width = VIDEO_WIDTH;
config.height = VIDEO_HEIGHT;
config.setMargins(0,0,0,0);
config.alignment = LayoutAlignment.VERTICAL_CENTER;
surfaceProvider2 = new SurfaceProvider(this);
surfaceProvider2.setLayoutConfig(config);
surfaceProvider2.getSurfaceOps().get().addCallback(callback2);
surfaceProvider2.pinToZTop(true);
//Button,单击拍摄照片
config.height = LayoutConfig.MATCH_CONTENT;
config.width = LayoutConfig.MATCH_CONTENT;
```

```
config.setMargins(0,0,0,0);
button = new Button(this);
button.setLayoutConfig(config);
button.setText("拍照");
button.setTextSize(50);
button.setMultipleLine(true);
ShapeElement buttonBackground = new ShapeElement();
buttonBackground.setRgbColor(new RgbColor(0xFF51A8DD));
buttonBackground.setCornerRadius(25);
button.setBackground(buttonBackground);
button.setClickedListener(new Component.ClickedListener() {
    @Override
    public void onClick(Component component) {
        //拍照
        mCamera.triggerSingleCapture(takePictureConfig);
    }
});
//将组件添加到布局容器中,并将布局容器作为 UI 的根布局
myLayout.addComponent(surfaceProvider1);
myLayout.addComponent(surfaceProvider2);
myLayout.addComponent(button);
super.setUIContent(myLayout);
```

在这段代码中定义了两个宽 960、高 960 的预览窗口,并且增加了描述 surfaceProvider1 的回调 callback1 和描述 surfaceProvider2 的回调 callback2,该回调将定义该组件在其生命周期不同阶段的执行内容,callback1 和 callback2 的具体代码如下:

```
//callback1 和 callback2 的具体代码
private SurfaceOps.Callback callback1 = new SurfaceOps.Callback() {
    @Override
    public void surfaceCreated(SurfaceOps surfaceOps) {
        //将 SurfaceProvider 中的 Surface 与 previewsurface 建立连接
        previewSurface = surfaceOps.getSurface();
    }

    @Override
    public void surfaceChanged(SurfaceOps surfaceOps, int i, int i1, int i2) {
    }

    @Override
    public void surfaceDestroyed(SurfaceOps surfaceOps) {
    }
};

private SurfaceOps.Callback callback2 = new SurfaceOps.Callback() {
```

```
    @Override
    public void surfaceCreated(SurfaceOps surfaceOps) {
        //将 SurfaceProvider 中的 Surface 与 datasurface 建立连接
        dataSurface = surfaceOps.getSurface();
        //初始化相机
        openCamera();
    }

    @Override
    public void surfaceChanged(SurfaceOps surfaceOps, int i, int i1, int i2) {
    }

    @Override
    public void surfaceDestroyed(SurfaceOps surfaceOps) {
    }
};
```

在这段代码中，使用 surfaceOps.getSurface()方法获取 surfaceProvider1 和 surfaceProvider2 用于预览显示 Surface 的对象 previewSruface 和 dataSurface，并且用 openCamera()方法来创建相机对象。openCamera()方法中具体创建相机对象的方法参考下一节“创建相机对象”的步骤。

图 9.16 用于展示相机能力的 UI 界面

以上代码创建的相机 UI 效果如图 9.16 所示。

从左至右分别是用于预览相机画面 surfaceProvider1、用于展示照片的 surfaceProvider2 和拍照按钮 Button。

3. 创建相机对象

在 HarmonyOS 上实现一个相机应用，无论将来想应用到哪个或者哪些设备上，都必须先创建一个独立的相机设备，然后才能继续相机的其他操作。在 HarmonyOS 中，相机设备的创建首先需要创建一个 CameraKit 实例并通过 CameraKit.getInstance(Context context)方法获取唯一的 CameraKit 对象，实现代码如下：

```
private void openCamera(){                    //获取 CameraKit 对象
    cameraKit = CameraKit.getInstance(this);
    if (cameraKit == null) {
        //处理 cameraKit 获取失败的情况
        return;
    }
}
```

如上述代码所示，在这里可以使用 openCamera()方法获取对象。如果此步骤操作失败，相机可能已经被占用或无法使用。如果被占用，则必须等到相机被释放后才能重新获取 CameraKit 对象。

接下来需要获取设备支持的逻辑相机列表，并基于某个逻辑相机创建相机对象，具体实现这一方法需要 4 个步骤。

（1）通过 getCameraIds()方法，获取当前使用的设备所支持的逻辑相机列表。逻辑相机列表中存储了当前设备所拥有的所有逻辑相机 ID，如果列表不为空，则说明列表中的每个 ID 都支持独立创建相机对象，否则，说明正在使用的设备无可用的相机。当无可用相机时则不能继续后续的操作。

（2）通过 CameraStateCallbackImpl()方法创建用于接收有关摄像机设备的状态更新的回调对象，必须提供此回调实例才能打开摄像机设备。

（3）创建执行回调的 EventHandler。

（4）通过 createCamera（String cameraId，CameraStateCallback callback，EventHandler handler）方法来创建一个可用的相机对象，如果此步骤执行成功，则表示相机系统的硬件已经完成了上电。当通过该方法完成相机创建之后，会在 CameraStateCallback 中触发 onCreated（Camera camera）回调。在进入相机设备的配置步骤前，需确保相机设备已经创建成功。否则会触发相机设备创建失败的回调，并返回错误码，进行错误处理后，重新执行相机设备的创建。

以上 4 个步骤可以通过下面的代码实现，将该部分的代码放在获取 CameraKit 对象之后运行即可完成完整的相机创建流程，代码如下：

```
//创建相机对象
try {
    String[] cameraIds = cameraKit.getCameraIds();          //获取当前设备的逻辑相机列表
    if (cameraIds.length <= 0) {
        System.out.println("cameraIds size is 0");
    }
    //相机创建和相机运行时的回调
    CameraStateCallbackImpl cameraStateCallback = newCameraStateCallbackImpl();
    if(cameraStateCallback == null) {
        System.out.println("cameraStateCallback is null");
    }
    //执行回调的 EventHandler
    EventHandler eventHandler = new EventHandler(EventRunner.create("CameraCb"));
    if(eventHandler == null) {
        System.out.println("eventHandler is null");
    }
    //创建一个可用的相机对象
    cameraKit.createCamera(cameraIds[0], cameraStateCallback, eventHandler);
    } catch (IllegalStateException e) {
        System.out.println("getCameraIds fail");
}
```

在相机创建的同时可以获取与相机相关的信息，例如调用 getDeviceLinkType(String physicalId)方法可以获取物理相机的连接方式信息；调用 getCameraInfo(String cameraId)方法可以查询相机硬件朝向等信息；通过调用 getCameraAbility(String cameraId)方法可以查询相机能力信息(例如支持的分辨率列表等)。更多的详细方法和调用信息可以通过官方 API 文档进行查询。

前文讲到相机创建后会在 CameraStateCallback 中触发 onCreated(Camera camera)回调，下面介绍 CameraStateCallback 中的函数，代码如下：

```
//CameraStateCallbackImpl
private final class CameraStateCallbackImpl extends CameraStateCallback {
    //相机回调
    @Override
    public void onCreated(Camera camera) {
        //相机创建时回调
        CameraConfig.Builder cameraConfigBuilder = camera.getCameraConfigBuilder();
        if (cameraConfigBuilder == null) {
            System.out.println("onCreated cameraConfigBuilder is null");
            return;
        }
        //配置预览的 Surface
        cameraConfigBuilder.addSurface(previewSurface);
        try {
            //相机设备配置
            camera.configure(cameraConfigBuilder.build());
        } catch (IllegalArgumentException e) {
            System.out.println("Argument Exception");
        } catch (IllegalStateException e) {
            System.out.println("State Exception");
        }
    }
    @Override
    public void onConfigured(Camera camera) {
        //相机配置
         FrameConfig.Builder frameConfigBuilder1 = camera.getFrameConfigBuilder(FRAME_
CONFIG_PREVIEW);
        //配置预览 Surface
        frameConfigBuilder1.addSurface(previewSurface);
        try {
            //启动循环帧捕获
            int triggerId = camera.triggerLoopingCapture(frameConfigBuilder1.build());
            //传出拍照设置,用于在按钮中实现拍照
            mCamera = camera;
            takePictureConfig = frameConfigBuilder2.build();
        } catch (IllegalArgumentException e) {
```

```
                System.out.println("Argument Exception");
            } catch (IllegalStateException e) {
                System.out.println("State Exception");
            }
        }
        @Override
        public void onReleased(Camera camera) {
            //释放相机设备
            if (camera != null) {
                camera.release();
                camera = null;
            }
        }
    }
```

可以看到，CameraStateCallback 有 3 个函数，分别是 onCreated(Camera)、onConfigured(Camera)和 onReleased(Camera)，这 3 个函数分别在相机被创建时、相机被配置时和相机被释放时回调，其参数 Camera 就是创建完成的相机对象，可对相机对象进行操作。在这里主要为相机配置预览和拍照的 Surface，可以看到在 onCreated(Camera)函数中使用 CameraConfig 类为 Camera 添加 previewSurface，用于相机画面的预览。

有关 onConfigure(Camera)和 onRelease(Camera)函数，会在下一节讲解。

4. 相机设备配置及释放

相机创建成功后会调用 configured(CameraConfig)方法实现相机的相关配置。随后会回调 onConfigure(Camera)函数，代码如下：

```
//与相机相关的配置
@Override
public void onConfigured(Camera camera) {
    //相机配置
    FrameConfig.Builder frameConfigBuilder1 = camera.getFrameConfigBuilder(FRAME_CONFIG_
PREVIEW);
    //配置预览 Surface
    frameConfigBuilder1.addSurface(previewSurface);
    try {
        //启动循环帧捕获
        int triggerId = camera.triggerLoopingCapture(frameConfigBuilder1.build());
        //传出拍照设置,用于在按钮中实现拍照
        mCamera = camera;
        takePictureConfig = frameConfigBuilder2.build();
    } catch (IllegalArgumentException e) {
        System.out.println("Argument Exception");
    } catch (IllegalStateException e) {
        System.out.println("State Exception");
    }
}
```

在 onConfigured(Camera)中，创建了 FrameConfig 类实例，选用了预览模式并配置了预览 Surface 实例 previewSurface。随后对创建好的 Camera 使用 triggerLoopingCapture(FrameConfig)启动循环帧捕获，相机拍摄的每一帧画面都会传输到 previewSurface 上，从而实现了相机画面的预览。

在 onReleased(Camera)函数中主要实现了对相机对象的置空操作。

此时运行代码，预览效果如图 9.17 所示。

由图 9.17 可以看出，左侧用于预览的 SurfaceProvider 使用的 previewSurface 接收到了数据，显示了摄像头的画面，右侧的 SurfaceProvider 中的 Surface 没有被配置到相机中，所以没有画面。

图 9.17 实现预览后的 UI 界面

9.3.3 相机拍照

实现拍照与实现相机预览类似，只不过需要在回调类 CameraStateCallback 中对相机额外配置拍照 Surface，具体代码修改如下：

```
//对相机额外配置拍照 Surface

private final class CameraStateCallbackImpl extends CameraStateCallback {
    //相机回调
    @Override
    public void onCreated(Camera camera) {
        //相机创建时回调
        CameraConfig.Builder cameraConfigBuilder = camera.getCameraConfigBuilder();
        if (cameraConfigBuilder == null) {
            System.out.println("onCreated cameraConfigBuilder is null");
            return;
        }
        //配置预览的 Surface
        cameraConfigBuilder.addSurface(previewSurface);
        //配置拍照的 Surface
        cameraConfigBuilder.addSurface(dataSurface);
        try {
            //相机设备配置
            camera.configure(cameraConfigBuilder.build());
        } catch (IllegalArgumentException e) {
            System.out.println("Argument Exception");
        } catch (IllegalStateException e) {
            System.out.println("State Exception");
        }
    }
    @Override
```

```
    public void onConfigured(Camera camera) {
        //相机配置
        FrameConfig.Builder frameConfigBuilder1 = camera.getFrameConfigBuilder(FRAME_
CONFIG_PREVIEW);
        FrameConfig.Builder frameConfigBuilder2 = camera.getFrameConfigBuilder(FRAME_
CONFIG_PICTURE);
        //配置预览 Surface
        frameConfigBuilder1.addSurface(previewSurface);
        //配置拍照的 Surface
        frameConfigBuilder2.addSurface(dataSurface);
        try {
            //启动循环帧捕获
            int triggerId = camera.triggerLoopingCapture(frameConfigBuilder1.build());
            //传出拍照设置,用于在按钮中实现拍照
            mCamera = camera;
            takePictureConfig = frameConfigBuilder2.build();
        } catch (IllegalArgumentException e) {
            System.out.println("Argument Exception");
        } catch (IllegalStateException e) {
            System.out.println("State Exception");
        }
    }
    @Override
    public void onReleased(Camera camera) {
        //释放相机设备
        if (camera != null) {
            camera.release();
            camera = null;
        }
    }
}
```

注意到在 onCreate(Camera)方法中为 CameraConfig 实例额外配置了一个 dataSurface，在 onConfigured(Camera)函数中创建了两个 FrameConfig 实例，其中前者选用了预览模式并配置了预览 Surface 实例 previewSurface，后者采用了拍照模式并配置了拍照 Surface 实例 dataSurface。随后使用前者启动循环帧捕获，相机拍摄的每一帧画面都会传输到 previewSurface 上，从而实现了相机画面的预览。将后者和相机实例设置为全局变量 takePictureConfig 和 mCamera，方便在拍照按钮中调用，每按一次按钮就取一帧画面到 dataSurface，从而实现拍照画面的展示效果，按钮的逻辑代码如下：

```
button.setClickedListener(new Component.ClickedListener() {
    @Override
    public void onClick(Component component) {
        //拍照
```

```
            mCamera.triggerSingleCapture(takePictureConfig);
        }
    });
```

每当按钮被触发就会回调 onClick(Component)，在其中使用 takePicture 作为参数对 mCamera 调用 triggerSingleCapture(FrameConfig)函数，从而实现单次拍照，效果如图 9.18 所示。

图 9.18　实现预览和照相后的 UI 界面

此时左侧的 surfaceProvider1 仍实时展示摄像头的画面，而右侧的 surfaceProvider2 展示了按下拍照按钮时相机捕捉到的一帧画面，此画面为静态的。

9.3.4　连拍与录像

连拍功能方便用户一次拍照获取多张照片，用于捕捉精彩瞬间。同普通拍照的实现流程一致，不同点仅在于调用的函数不同，单次拍摄调用的是 triggerSingleCapture (frameConfig)方法，而连拍需要使用 triggerMultiCapture(List < FrameConfig > frameConfigs)方法。具体代码在此处就不进行详细介绍了，感兴趣的读者可以查阅官方 API 进行了解。

录像对于手机用户而言也是非常重要的功能之一，毕竟比起静态的图片，生动的视频更适合作为生活的记录载体。在 HarmonyOS 中，启动录像和启动预览类似，但需要另外创建并配置一个录像 Surface 才能使用，开发者需要在相机的 onConfigure()回调函数中配置录像的 Surface，然后使用 triggerLoopingCapture()方法启动循环帧捕获(与预览功能使用的函数一样)，也就是录像功能。感兴趣的读者可以自行进行开发尝试。

第10章 应用实战：第三方组件的使用——弹幕

15min

现如今用户在观看视频时，弹幕已经成了不可或缺的功能，通过随时随地发送弹幕的方式，极大地提高了用户交互体验的乐趣。甚至在很多的年会、演出等具有大屏互动的活动现场，都有弹幕的身影，因此弹幕开发也成为各类多媒体应用开发中必不可少的功能组成部分。

本章将基于鸿蒙开源的弹幕库第三方组件，讲解如何在鸿蒙应用开发中使用第三方组件，并完成一个应用实例，实现弹幕的隐藏、显示、暂停、继续、发送、定时发送弹幕等一系列第三方组件提供的主要功能。

所谓第三方组件，是指除本地类库、系统类库以外的类库组件，它们功能强大，种类繁多，是由来自世界各地的开发者进行开发并不停地维护和升级。当需要实现一个特定的功能，而自身的组件无法满足，必须进行大量开发时，若有开发者已经实现该功能并以第三方组件的形式开源，则开发者只需引用该第三方组件，并直接调用功能接口，就能实现一些想要的功能。一个优秀的第三方组件可以完成特定的复杂的功能，从而节省大量研发成本。

这里，我们为开发者提供了弹幕的开源第三方组件，请扫描前言处二维码下载，进入danmaku-flame-master-ohos工程下的entry\libs文件，可以看到DanmakuFlameMaster.jar包，如图10.1所示，新建工程引用jar包，在应用开发中通过调用第三方组件中的功能接口，实现弹幕效果。

下载完组件的jar包后，启动DevEco Studio，打开已创建的某个Java工程，将刚下载的jar包导入工程目录entry→libs下，如图10.2所示。

打开module级别下的build.gradle文件，在dependences标签中增加对libs目录下jar包的引用，代码如下：

```
dependencies {
    implementation fileTree(dir: 'libs', include: ['*.jar'])
    …
}
```

并在导入的jar包上右击，选择Add as Library进行引用，如图10.3所示。

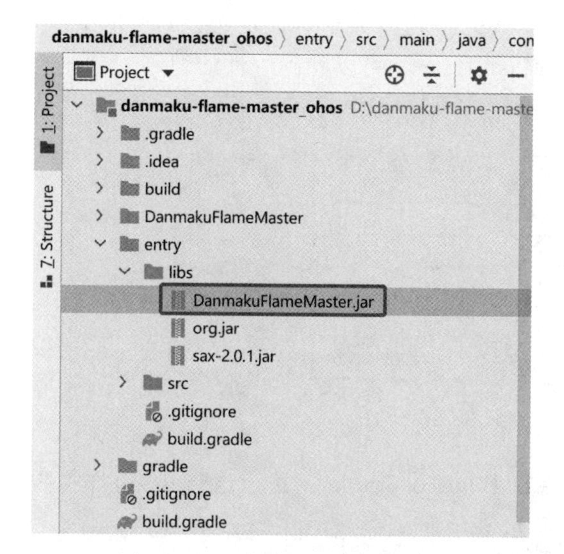

图 10.1 下载弹幕库组件 jar 包

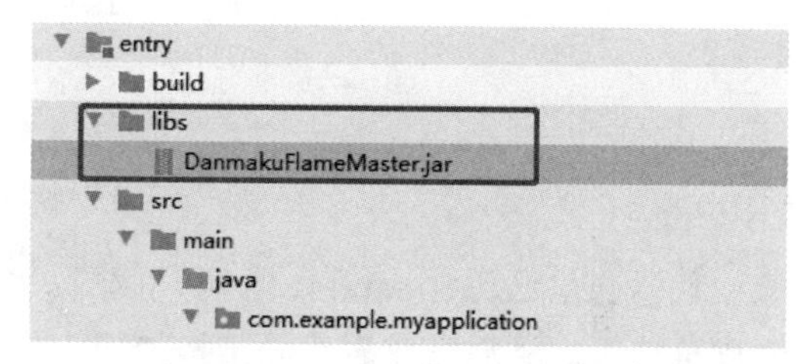

图 10.2 将第三方组件 jar 包导入 libs 目录下

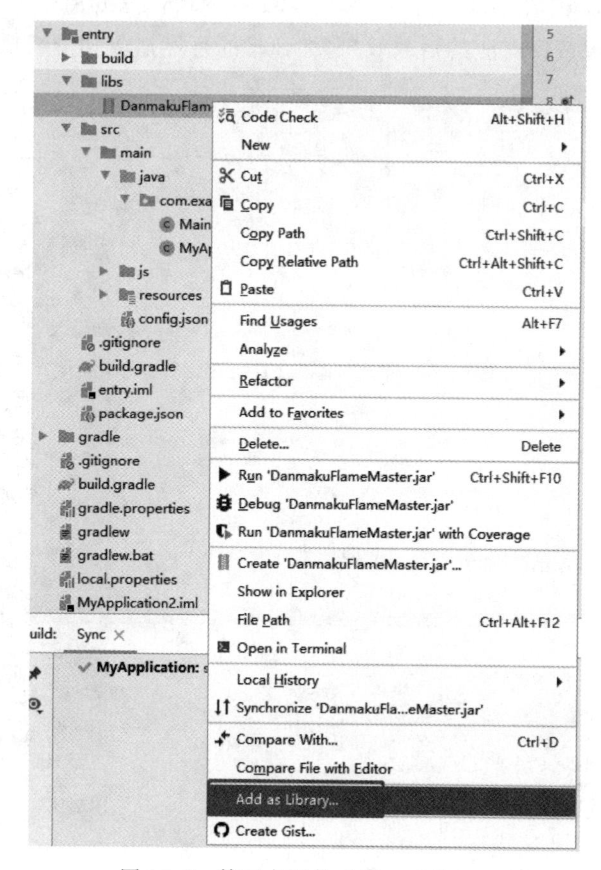

图 10.3 第三方组件 Add as Library

这里 Level 选择 Project Library，Add to module 中选择需要引用的模块，并单击 OK 按钮，如图 10.4 所示。此时在 DanmakuFlameMaster. jar 下可以看到组件的源码。至此即完成了组件的引用工作，接下来即可使用组件提供的功能进行弹幕开发。

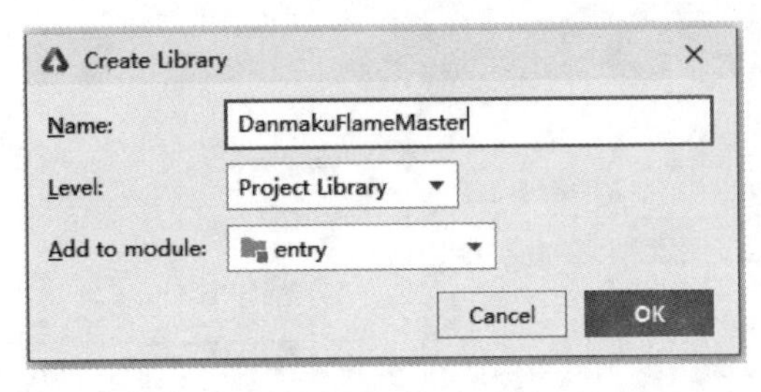

图 10.4　第三方组件 Create Library

代码仓里已包含 sample 案例，下载源码后在 DevEco Studio 中可以直接运行，即可看到弹幕效果。下载完成后代码结构如图 10.5 所示。

其中，DanmakuFlameMaster 模块是弹幕库组件的具体源码，entry 模块为应用案例。通过模拟器运行 entry 模块，可以直接看到弹幕效果，如图 10.6 所示。单击界面下方的功能按钮，弹幕也会实现按钮对应的播放效果，如弹幕的暂停播放和继续播放等。

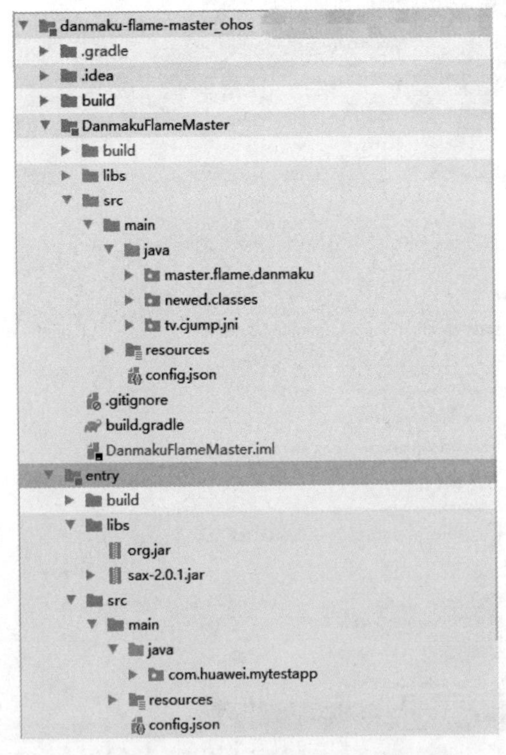

图 10.5　Gitee 弹幕库工程结构

图 10.6　弹幕运行效果图

这里对 entry 模块进行一些简单讲解，看一看是如何实现弹幕效果的。首先需要设计一个界面布局，进入 entry→src→main→resources→base→layout，打开 ability_main. xml 布局文件，可以看到整个界面的布局代码。其中，弹幕库组件为开发者提供了 DanmakuView UI 组件，用一个全透明的遮罩层进行弹幕信息的显示。弹幕组件还提供了一些核心功能，包括弹幕的隐藏、显示、暂停、继续、发送、定时发送等，这里通过一系列 button 来接入这些功能。ability_main. xml 中的代码如下：

```
<!-- ability_main.xml -->
<?xml version = "1.0" encoding = "utf-8"?>
<StackLayout
    xmlns:ohos = "http://schemas.huawei.com/res/ohos"
    ohos:id = "$+id:all"
    ohos:height = "match_parent"
    ohos:width = "match_parent"
    ohos:bottom_padding = "0"
    ohos:left_padding = "@dimen/activity_horizontal_margin"
    ohos:right_padding = "@dimen/activity_horizontal_margin"
    ohos:top_padding = "@dimen/activity_vertical_margin"
    ohos:background_element = "#000000"
    ohos:alpha = "1.0fp" >

<ohos.agp.components.surfaceprovider.SurfaceProvider
        ohos:id = "$+id:sf_video"
        ohos:height = "match_parent"
        ohos:width = "match_parent"/>

<master.flame.danmaku.ui.widget.DanmakuView
        ohos:id = "$+id:sv_danmaku"
        ohos:width = "match_parent"
        ohos:height = "match_parent" />

<DependentLayout
        ohos:id = "$+id:media_controller"
        ohos:width = "match_parent"
        ohos:height = "match_parent"
        ohos:orientation = "vertical"
        ohos:alignment = "left"
>

<DirectionalLayout
            ohos:width = "match_parent"
            ohos:height = "match_content"
            ohos:layout_alignment = "bottom"
            ohos:alignment = "bottom"
            ohos:orientation = "horizontal"
```

```
            ohos:bottom_margin = "0px"
            ohos:align_parent_bottom = "true"
            ohos:background_element = "#614534"
>

<Button
                ohos:margin = "5px"
                ohos:weight = "1"
                ohos:id = "$ + id:btn_rotate"
                ohos:width = "match_content"
                ohos:height = "match_content"
                ohos:text = "旋转屏幕"
                ohos:text_size = "20fp"
                ohos:background_element = "#7C9CAC"/>
            …
</DirectionalLayout>
</DependentLayout>
</StackLayout>
```

在上述代码中，最外层的 StackLayout 布局结构，可以直接在屏幕上开辟出一块空白的区域，添加到这个布局中的视图都是以层叠的方式显示，第一个添加到布局中的视图显示在最底层，最后一个被显示在最顶层。上一层的视图会覆盖下一层的视图。StackLayout 布局的示意图如图 10.7 所示。

通常在弹幕功能的实现中，UI 的布局结构示意图如图 10.8 所示。

图 10.7　StackLayout 布局结构示意图

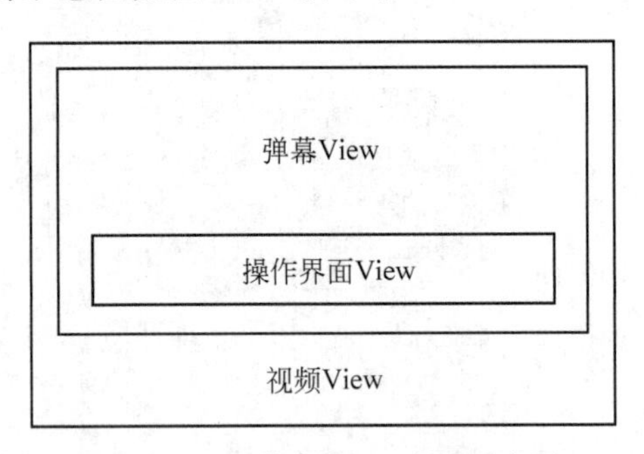

图 10.8　弹幕功能 UI 布局示意图

由图 10.8 可以看出，视频 View 位于最底层，只用来播放视频，不涉及任何交互操作，弹幕 View 位于视频 View 上方，用于显示弹幕，操作界面 View 位于弹幕 View 上方，用于对弹幕进行操作，如显示、隐藏、暂停等，因此在 ability_main. xml 布局文件中，采用 StackLayout 对上述 View 进行整体布局，其中，SurfaceProvider 用于视频播放，第三方组件提供的 DanmakuView 用于弹幕显示，DirectionalLayout 用于将一组功能操作 button 进行水平方向的线性布局。完成 XML 布局文件后，进入 entry→src→main→java→包名→slice，打开 MainAbilitySlice. java 文件，可以看到以下代码。其中，通过 super. setUIContent() 对

ability_main. xml 布局文件进行引用。

```
public class MainAbilitySlice extends AbilitySlice {
    @Override
    public void onStart(Intent intent) {
        super.onStart(intent);
super.setUIContent(ResourceTable.Layout_ability_main);
    }
    ...
}
```

完成界面布局后，接下来看一看功能逻辑实现。首先在 MainAbilitySlice. java 文件中，找到整个页面的 onStart()方法，在 onStart()方法中，设置窗口管理服务 WindowManager，并调用实现 findView()方法，这部分代码如下：

```
//MainAbilitySlice.java
public class MainAbilitySlice extends AbilitySlice implements Component.ClickedListener{
    .......
    @Override
    public void onStart(Intent intent) {
        super.onStart(intent);
        super.setUIContent(ResourceTable.Layout_ability_main);
        WindowManager windowManager = WindowManager.getInstance();
        Window window = windowManager.getTopWindow().get();
        WindowManager.LayoutConfig layoutConfig = new WindowManager.LayoutConfig();
        layoutConfig.flags = WindowManager.LayoutConfig.MARK_ALLOW_LAYOUT_COVER_SCREEN;
        layoutConfig.flags = layoutConfig.flags | WindowManager.LayoutConfig.MARK_FULL_
SCREEN;
        layoutConfig.windowBrightness = 1.0f;
        window.setLayoutConfig(layoutConfig);
window.addFlags(WindowManager.LayoutConfig.MARK_LAYOUT_ATTACHED_IN_DECOR);
       window.addFlags(WindowManager.LayoutConfig.MARK_LAYOUT_INSET_DECOR);
        window.setBackgroundColor(new RgbColor(0xB7B208));
        DanmakuView danmakuView = (DanmakuView) findComponentById(ResourceTable.Id_sv_
danmaku);
        danmakuView.addDrawTask(null);
        danmakuView.setLayoutRefreshedListener(null);
        try {
            findViews();
        } catch (IOException e) {
            e.printStackTrace();
        }
    }
}
```

其中，WindowManager 主要用来管理窗口的一些状态、属性、消息的收集和处理等，通过 WindowManager. getInstance ( ) 可以获取 WindowManager 的实例，并通过 WindowManager 中的静态类 LayoutConfig 可以获取和设置当前窗口的一些属性，如上述代码中，MARK_ALLOW_LAYOUT_COVER_SCREEN 设置窗口扩展以覆盖整个屏幕，同时保持如状态栏的正确显示，layoutConfig. windowBrightness 用于设置窗口亮度。

在 onStart()方法最后，通过调用 findViews()方法实现对页面布局控件的初始化，对每个控件注册一个监听器，并实现对弹幕的显示。findViews()方法代码如下：

```
//MainAbilitySlice.java 中的 findViews()方法
private void findViews() throws IOException {
        mAll = findComponentById(ResourceTable.Id_all);
        mSurfaceProvider = (SurfaceProvider) findComponentById(ResourceTable.Id_sf_video);
        mMediaController = findComponentById(ResourceTable.Id_media_controller);
        mBtnRotate = (Button) findComponentById(ResourceTable.Id_btn_rotate);
        mBtnHideDanmaku = (Button) findComponentById(ResourceTable.Id_btn_hide);
        mBtnShowDanmaku = (Button) findComponentById(ResourceTable.Id_btn_show);
        mBtnPauseDanmaku = (Button) findComponentById(ResourceTable.Id_btn_pause);
        mBtnResumeDanmaku = (Button) findComponentById(ResourceTable.Id_btn_resume);
        mBtnSendDanmaku = (Button) findComponentById(ResourceTable.Id_btn_send);
        mBtnSendDanmakus = (Button) findComponentById(ResourceTable.Id_btn_send_danmakus);

        mAll.setClickedListener(this);
        mSurfaceProvider.setClickedListener(this);
        mBtnRotate.setClickedListener(this);
        mBtnHideDanmaku.setClickedListener(this);
        mMediaController.setClickedListener(this);
        mBtnShowDanmaku.setClickedListener(this);
        mBtnPauseDanmaku.setClickedListener(this);
        mBtnResumeDanmaku.setClickedListener(this);
        mBtnSendDanmaku.setClickedListener(this);
        mBtnSendDanmakus.setClickedListener(this);

        //设置最大显示行数
        HashMap<Integer, Integer> maxLinesPair = new HashMap<Integer, Integer>();
        maxLinesPair.put(BaseDanmaku.TYPE_SCROLL_RL, 5); //滚动弹幕最大显示 5 行

        //设置是否禁止重叠
        HashMap<Integer, Boolean> overlappingEnablePair = new HashMap<Integer, Boolean>();
        overlappingEnablePair.put(BaseDanmaku.TYPE_SCROLL_RL, true);
        overlappingEnablePair.put(BaseDanmaku.TYPE_FIX_TOP, true);

        mDanmakuView = (IDanmakuView) findComponentById(ResourceTable.Id_sv_danmaku);
        mContext = DanmakuContext.create();
        mContext.setDanmakuStyle(IDisplayer.DANMAKU_STYLE_STROKEN, 3)   //设置描边样式
                .setDuplicateMergingEnabled(false)      //设置不合并相同内容弹幕
```

```
                .setScrollSpeedFactor(0.9f)                    //设置弹幕滚动速度及缩放比例
                .setScaleTextSize(1.2f)                        //设置字体缩放比例
                .setCacheStuffer(new SimpleTextCacheStuffer(), mCacheStufferAdapter)
                .setMaximumLines(maxLinesPair)                 //设置最大行数策略
                .preventOverlapping(overlappingEnablePair) //设置禁止重叠策略
                .setDanmakuMargin(40);
        if (mDanmakuView != null) {
            InputStream stream = this.getResourceManager().getRawFileEntry("resources/
rawfile/comments.xml").openRawFile();
            mParser = createParser( stream );
            mDanmakuView.setCallback(new master.flame.danmaku.controller.DrawHandler.Callback() {
                @Override
                public void updateTimer(DanmakuTimer timer) {
                }
                @Override
                public void drawingFinished() {
                }
                @Override
                public void danmakuShown(BaseDanmaku danmaku) {
                }
                @Override
                public void prepared() {
                    mDanmakuView.start();
                }
            });
            mDanmakuView.setOnDanmakuClickListener(new IDanmakuView.OnDanmakuClickListener() {
                @Override
                public boolean onDanmakuClick(IDanmakus danmakus) {
                    BaseDanmaku latest = danmakus.last();
                    if (null != latest) {
                        return true;
                    }
                    return false;
                }

                @Override
                public boolean onDanmakuLongClick(IDanmakus danmakus) {
                    return false;
                }

                @Override
                public boolean onViewClick(IDanmakuView view) {
                    mMediaController.setVisibility(Component.VISIBLE);
                    return false;
                }
```

```
            });
            mDanmakuView.prepare(mParser, mContext);
            mDanmakuView.showFPS(true);
            mDanmakuView.enableDanmakuDrawingCache(true);
        }
    }
```

在 findViews()方法中，首先获取了控件对象，并为每个控件注册了监听器，在 MainAbilitySlice 的 onClick()单击事件方法中，实现具体的单击逻辑，然后通过 DanmakuContext.create()创建了 DanmakuContext 的实例 mContext，DanmakuContext 用于设置弹幕的各种全局配置，如设置字体、最大显示行数等。例如在上述代码中，通过 mContext.setDanmakuStyle()设置了弹幕的描边样式，通过 mContext.setScrollSpeedFactor()设置弹幕滚动速度的系数，设置的数值越大则滚动速度越慢，通过 mContext.setMaximumLines()设置弹幕的最大滚动行数，这个属性只对滚动弹幕有效，其中 setMaximumLines()方法需要传入一个哈希表的数据类型，这里传入定义的 HashMap <Integer, Integer> maxLinesPair，其中 maxLinesPair 的 key 值为基础弹幕库类 BaseDanmaku 的静态成员变量 TYPE_SCROLL_RL，即从右向左滚动，value 值为设置的最大行数值 5。

mParser 为弹幕解析器 BaseDanmakuParser 的全局实例对象，这里通过 createParser()方法解析数据流，为解析器添加数据源，其中数据源 InputStream 为从资源文件目录 resources/rawfile 下解析的 comments.xml 文件。在 createParser()方法中，首先通过 DanmakuLoaderFactory.create()创建 BiliBili 网站的弹幕加载器，并通过 loader.load(stream)将数据流载入弹幕加载器中。随后新建一个弹幕解析器 parser，通过 loader.getDataSource()方法取出数据源，并通过 parser.load(dataSource)将数据源放入解析器，最终返回解析器 parser。createParser()的实现代码如下：

```
//设置数据源
private BaseDanmakuParser createParser(InputStream stream) {
    if (stream == null) {
        return new BaseDanmakuParser() {
            @Override
            protected Danmakus parse() {
                return new Danmakus();
            }
        };
    }
    ILoader loader = DanmakuLoaderFactory.create(DanmakuLoaderFactory.TAG_BILI);
    try {
        loader.load(stream);
    } catch (IllegalDataException e) {
```

```
            e.printStackTrace();
        }
        BaseDanmakuParser parser = new BiliDanmukuParser();
        IDataSource<?> dataSource = loader.getDataSource();
        parser.load(dataSource);
        return parser;
    }
```

随后在 findViews()方法中，通过 setCallback()方法来为 mDanmakuView 设置回调函数，并通过调用 setOnDanmakuClickListener()方法为 mDanmakuView 设置了单击事件。此时 mParser 和 mContext 均已完成初始化，下面开始进行弹幕的绘制。核心代码即通过调用 mDanmakuView.prepare()方法，传入 mParser 和 mContext 变量，实现弹幕的准备和启动。这里在源码执行过程中，执行上述回调函数中的 prepared()方法，从而执行 mDanmakuView.start()方法，开启弹幕。感兴趣的读者可以通过阅读源码，进行深入理解。mDanmakuView.showFPS(true)用于设置是否显示 FPS，mDanmakuView.enableDanmakuDrawingCache(true)用于提升屏幕的绘制效率。

下面看一下功能操作 button 的单击事件逻辑。onClick()方法的实现代码如下：

```
    //各类 Button 的 onClick()方法
    @Override
    public void onClick(Component v) {
        if (v == mAll || v == mMediaController || v == mSurfaceProvider) {
            mMediaController.setVisibility(mMediaController.getVisibility() == Component.
INVISIBLE ? Component.VISIBLE:Component.INVISIBLE);
        }
        if (mDanmakuView == null || !mDanmakuView.isPrepared())
            return;
        if (v == mBtnRotate) {                          //旋转屏幕
            int orientation = this.getDisplayOrientation();
            if (orientation == 0){
                    this.setDisplayOrientation(AbilityInfo.DisplayOrientation.PORTRAIT);
            }else{
                    this.setDisplayOrientation(AbilityInfo.DisplayOrientation.LANDSCAPE);
            }
        } else if (v == mBtnHideDanmaku) {              //隐藏弹幕
            mDanmakuView.hide();
        } else if (v == mBtnShowDanmaku) {              //显示弹幕
            mDanmakuView.show();
        } else if (v == mBtnPauseDanmaku) {             //暂停弹幕
            mDanmakuView.pause();
        } else if (v == mBtnResumeDanmaku) {            //继续弹幕
            mDanmakuView.resume();
        } else if (v == mBtnSendDanmaku) {
```

```
            addDanmaku(false);
        } else if (v == mBtnSendDanmakus) {
            Boolean b = (Boolean) mBtnSendDanmakus.getTag();
            timer.cancel();
            if (b == null || !b) {
                mBtnSendDanmakus.setText("取消定时");
                timer = new Timer();
                timer.schedule(new AsyncAddTask(), 0, 1000);
                mBtnSendDanmakus.setTag(true);
            } else {
                mBtnSendDanmakus.setText("定时发送");
                mBtnSendDanmakus.setTag(false);
            }
        }
    }
```

在onClick()方法中，当用户的单击焦点在最外层的StackLayout、mMediaController或者SurfaceProvider时，则设置mMediaController的隐藏或显示。当单击“旋转屏幕”按钮，即mBtnRotate时，通过getDisplayOrientation()获取当前Ability的显示方向，并通过setDisplayOrientation()设置横竖屏显示，其中AbilityInfo.DisplayOrientation.PORTRAIT表示竖屏，AbilityInfo.DisplayOrientation.LANDSCAPE表示横屏。当单击“隐藏弹幕”“显示弹幕”“暂停弹幕”和“继续弹幕”4个按钮时，第三方组件分别提供了hide()、show()、pause()和resume()方法，直接调用相应方法即可实现对应功能。当单击“发送弹幕”按钮，即mBtnSendDanmaku时，执行addDanmaku()方法，用来新增一条文本弹幕信息，addDanmaku()方法的实现代码如下：

```
//addDanmaku()增加弹幕信息
private void addDanmaku(boolean islive) {
        BaseDanmaku danmaku = mContext.mDanmakuFactory.createDanmaku(BaseDanmaku.TYPE_
SCROLL_RL);
        if (danmaku == null || mDanmakuView == null) {
            return;
        }
        danmaku.text = "这是你自己发的" + System.nanoTime();
        danmaku.padding = 5;
        danmaku.priority = 0;            //可能会被各种过滤器过滤并隐藏显示
        danmaku.isLive = islive;
        danmaku.setTime(mDanmakuView.getCurrentTime() + 1200);
        danmaku.textSize = 25f * (mParser.getDisplayer().getDensity() - 0.6f);
        danmaku.textColor = Color.RED.getValue();
        danmaku.textShadowColor = Color.WHITE.getValue();
        danmaku.borderColor = Color.GREEN.getValue();
        mDanmakuView.addDanmaku(danmaku);
    }
```

在 addDanmaku()方法中，首先通过 mContext. mDanmakuFactory. createDanmaku()创建弹幕对象，TYPE_SCROLL_RL 表示从右向左滚动弹幕，随后对所创建的弹幕对象设置属性。danmaku. text 用于设置弹幕的具体文本内容，danmaku. priority 设置为 0，表示此条弹幕可能会被各种过滤器过滤并隐藏显示，若设置为 1 则表示一定会显示。danmaku. textShadowColor 用于设置描边颜色，danmaku. borderColor 用于设置弹幕的边框颜色，若设置为 0 则表示无边框。完成设置后，通过 mDanmakuView. addDanmaku(danmaku)将这条弹幕信息添加至 mDanmakuView 中，即可显示新增弹幕。

至此就实现了弹幕的核心功能，运行 entry，模拟器中弹幕会自动开始播放，单击界面下方的操作按钮，也会触发相应的功能。这里单击“发送弹幕”按钮，会在弹幕中显示文本内容为“这是你自己发的”的弹幕信息，并用绿色文本框包裹，与上述代码的属性设置相同，效果如图 10.9 所示。

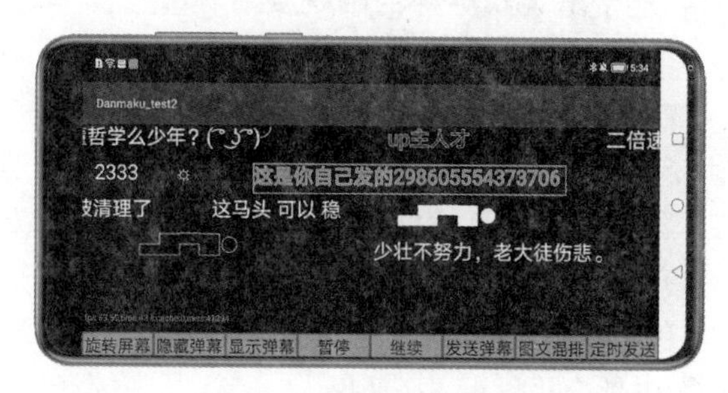

图 10.9　发送弹幕效果图

弹幕库的第三方组件还为开发者提供了非常多的功能，这里只简单地讲解了其中几个基本功能。感兴趣的读者可以通过阅读源码，尝试开发弹幕的更多功能。

也欢迎广大开发者自身开发并开源更多的第三方组件，供 HarmonyOS 应用开发者使用，共同助力 HarmonyOS 的生态建设与发展。

第11章

应用实战：视频流直播

随着网络通信技术的不断提高，实时传输高清视频数据成为可能，而且与录播相比，直播的实时互动性要强得多，能带给观众更好的临场观看体验。截至2020年3月，中国网络直播用户规模达5.60亿，达到了中国总网民人数的62 %，并且这一数据仍在不断增长。

可见，网络直播已经成为人们日常生活中很重要的一部分，其内容不但涉及游戏、语音聊天等娱乐领域，也催生了诸如直播带货这样的商业活动，甚至还促进了以教育为首的专业领域知识付费行为。基于音视频流的直播已经深深改变了生活的方方面面。

为了提升观众的直播观看体验，现在互联网公司在直播的各个中间环节都开发了很多相关技术，去解决各种各样的问题。一整套直播的技术流程是十分复杂的，不过如果只需实现最基本的视频流直播功能，可以将其抽象为一个发送端（Sender）和一个接收端（Receiver）的协同运作。其中，Sender负责视频信号的采集、编码和发送，Receiver则负责编码信号的接收、解码和播放。下面将基于HarmonyOS的基本API和UDP传输协议，开发一套属于自己的视频流直播App，大致思路如图11.1所示。

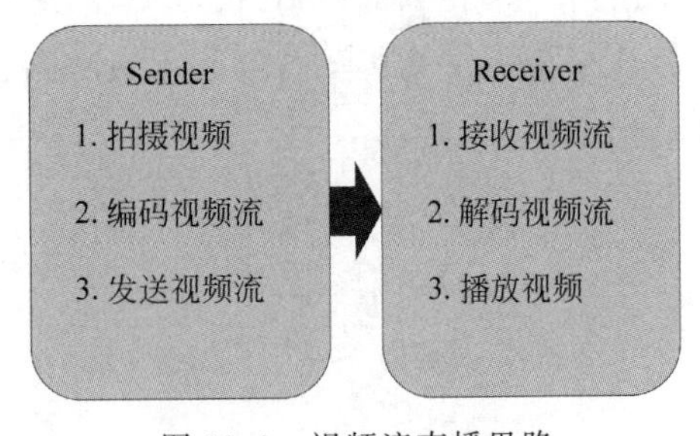

图11.1　视频流直播思路

遵循这个思路，可以创建两个简单的工程，分别实现以上发送端和接收端功能。工程源代码请扫描前言二维码下载（发送端为Video_Streaming_Sender，接收端为Video_Streaming_Receiver）。

14min

11.1　发送端

11.1.1　发送端工程结构

本工程的发送端主要需要实现对相机能力的调用、相机帧提取、相机帧处理、帧数据编码及编码后数据的发送。具体工程结构如图11.2所示，整个工程主要分为4个部分：

(1) rtp包。这个开源rtp包实现一个基于UDP协议的发送，在本工程中不提供对它的详细讲解，只需知道如何使用它进行数据发送（rtp包可以在上述Gitee本项目的开源地址中下载）。

（2）slice包。这个包是工程实现的核心，对其他包进行调用，从而实现视频的拍摄、编码、发送。其中，Sender是在发送端上运行的PageAbility。

（3）utils包。实现视频编码、线程调用并提供其他的API，以供Sender和Receiver调用。

（4）最外层的MainAbility。作为应用程序的入口，实现了对发送端PageAbility的路由。

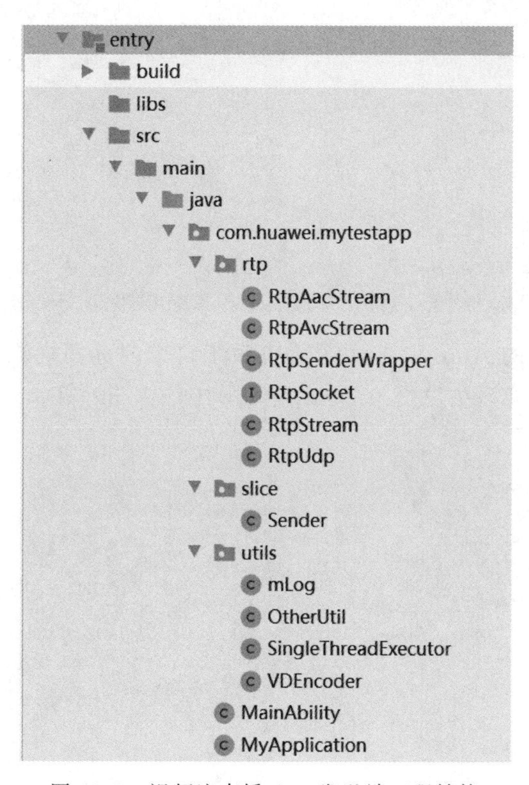

图11.2　视频流直播App发送端工程结构

由于发送端需要使用相机能力，并且需要执行数据的发送，所以需要申请一些权限，例如网络使用权限ohos.permission.CAMERA等。权限申请需要在entry模块的config.json中进行，具体需要在config.json的module字段内部添加reqPermissions字段，代码如下：

```
/* config.json */
"reqPermissions": [
  {
        "name": "ohos.permission.INTERNET",
        "reason": "reason",
        "usedScene": {
     "ability": [
     "com.huawei.mytestapp"
     ],
```

```
        "when": "always"
        }
        },
        {
            "name": "ohos.permission.CAMERA",
            "reason": "reason",
            "usedScene": {
        "ability": [
        "com.huawei.mytestapp"
        ],
        "when": "always"
        }
        }
        ]
```

对于涉及视频播放的 App 而言，由于手机系统自带的标题栏比较影响观看体验，所以应当将其去掉。具体的实现方法是在 config.json 中加入 metaData 字段，实现标题栏的摘除，代码如下：

```
"metaData":{
        "customizeData":[
        {
    "name": "hwc-theme",
            "value": "androidhwext:style/Theme.Emui.Light.NoTitleBar",
    "extra": ""
        }
    ]
}
```

本工程的 config.json 结构代码如下，可供参考：

```
{
    "app": {
        "bundleName": "com.huawei.mytestapp",
        "vendor": "huawei",
        "version": {
        "code": 1,
        "name": "1.0"
        },
        "apiVersion": {
    "compatible": 3,
    "target": 4
        }
    },
    "deviceConfig": {},
```

```
"module": {
    "package": "com.huawei.mytestapp",
    "name": ".MyApplication",
    "reqCapabilities": [
    "video_support"
    ],
    "deviceType": [
    "phone"
    ],
    "distro": {
    "deliveryWithInstall": true,
    "moduleName": "entry",
    "moduleType": "entry"
    },
     "abilities": [
    {
     "skills": [
         {
            "entities": [
             "entity.system.home"
         ],
         "actions": [
             "action.system.home"
             ]
         }
     ],
     "orientation": "landscape",
     "formEnabled": false,
     "name": "com.huawei.mytestapp.MainAbility",
     "icon": "$media:icon",
     "description": "$string:mainability_description",
     "label": "MyApplication",
     "type": "page",
     "launchType": "standard",
     "metaData":{
        "customizeData":[
        {
         "name": "hwc-theme",
         "value": "androidhwext:style/Theme.Emui.Light.NoTitleBar",
         "extra": ""
        }
        ]
     }
    }
     ],
    "reqPermissions": [
         {
    "name": "ohos.permission.INTERNET",
```

```
            "reason": "reason",
            "usedScene": {
                "ability": [
                "com.huawei.mytestapp"
                ],
                "when": "always"
            }
        },
        {
            "name": "ohos.permission.DISTRIBUTED_DATASYNC",
            "reason": "reason",
            "usedScene": {
                "ability": [
                "com.huawei.mytestapp"
                ],
            "when": "always"
            }
        }
            ],
            "network": {
            "uses-cleartext": true
            }
        }
}
```

下面将正式开始讲解发送端的开发流程。

11.1.2 发送端核心实现——Sender

发送端工程 slice 包下只有一个 Java 文件，也就是 Sender.java，此文件内含视频采集、编码、发送这一流程的逻辑框架，也包含了与相机相关能力的调用实现。其实现的主要功能如图 11.3 所示。

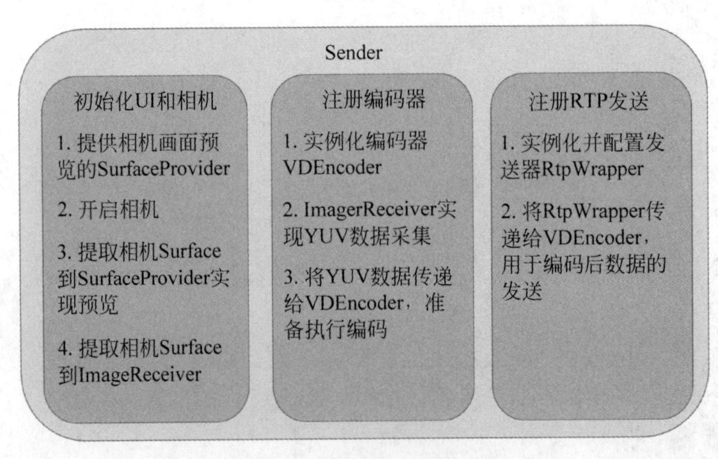

图 11.3　Sender 的主要功能

Sender 继承自 AbilitySlice 类，所以是一个 PageAbility，一旦这个 AbilitySlice 被创建，会自动调用其 onStart(Intent)方法，所以 onStart(Intent)一般可以视为主函数。Sender 的 onStart(Intent)函数代码如下：

```
@Override
public void onStart(Intent intent) {
    super.onStart(intent);
    //初始化 UI 和相机,实现视频帧的获取
    initUIandCamera();
    //注册编码器,实现视频帧的编码
    registerEncoder();
    //注册发送,实现编码数据的发送
    registerRTP();
}
```

除去调用父类的 super.onStart(Intent)方法，其实只有 3 个步骤，事实上，这 3 行代码调用了 3 个封装好的函数，分别实现了视频帧的获取、编码器的注册及 RTP 发送机的注册。

先来看主函数运行的第一个函数 initUIandCamera()，其功能就是初始化了一个 UI 界面和相机，并且实现了二者之间的联动。其代码如下：

```
//初始化 UI 界面和相机
private void initUIandCamera(){

        //删除导航栏等
        Window window = this.getWindow();
        window.setStatusBarVisibility(Component.INVISIBLE);
        window.setNavigationBarColor(Color.TRANSPARENT.getValue());

        //布局容器
        myLayout = new DirectionalLayout(this);
        LayoutConfig config = new LayoutConfig(LayoutConfig.MATCH_PARENT, LayoutConfig.
MATCH_PARENT);
        myLayout.setLayoutConfig(config);
        myLayout.setOrientation(Component.HORIZONTAL);
        ShapeElement background = new ShapeElement();
        background.setRgbColor(new RgbColor(150,150,150));
        myLayout.setBackground(background);

        //SurfaceProvider,用于播放摄像头画面
        config.width = VIDEO_WIDTH;
        config.height = VIDEO_HEIGHT;
        config.setMargins(50,0,0,0);
        config.alignment = LayoutAlignment.VERTICAL_CENTER;
        surfaceProvider = new SurfaceProvider(this);
```

```
        surfaceProvider.setLayoutConfig(config);
        surfaceProvider.getSurfaceOps().get().addCallback(callback);
        surfaceProvider.pinToZTop(true);

        //Button,单击后开始编码和发送
        config.height = LayoutConfig.MATCH_CONTENT;
        config.width = LayoutConfig.MATCH_CONTENT;
        config.setMargins(300,0,0,0);
        button = new Button(this);
        button.setLayoutConfig(config);
        button.setText("编码并发送");
        button.setTextSize(50);
        button.setMultipleLine(true);
        ShapeElement buttonBackground = new ShapeElement();
        buttonBackground.setRgbColor(new RgbColor(0xFF51A8DD));
        buttonBackground.setCornerRadius(25);
        button.setBackground(buttonBackground);
        button.setClickedListener(new Component.ClickedListener() {
            @Override
            public void onClick(Component component) {
                //为按钮添加功能,实现单击后开始编码并发送
                vdEncoder.start();
                //实现单击后改变状态显示
                text.setText("状态: 已启动发送");
                text.setTextColor(Color.GREEN);
            }
        });

        //Text,用于显示状态
        config.setMargins(-300,500,0,0);
        text = new Text(this);
        text.setText("状态: 未启动发送");
        text.setTextColor(Color.RED);
        text.setLayoutConfig(config);
        text.setTextSize(50);
        text.setMultipleLine(true);

        //将组件添加到布局容器中,并将布局容器作为UI的根布局
        myLayout.addComponent(surfaceProvider);
        myLayout.addComponent(button);
        myLayout.addComponent(text);
        super.setUIContent(myLayout);
    }
```

其中,首先利用第3章JavaUI的知识,创建了一个线性布局容器DirectionalLayout,然后设置一个用于视频播放的SurfaceProvider和一个用于启动发送的按钮Button,此外还有

一个用于显示发送状态的 Text 组件。为了美观，还将系统自带的导航栏和状态栏进行了调整。

下面重点讲述两个问题，SurfaceProvider 如何获取播放的数据源，以及相机的初始化。其中一行代码如下：

```
surfaceProvider.getSurfaceOps().get().addCallback(callback);
```

上述代码为 surfaceProvider 内含的 SurfaceOps 添加了一个回调类实例 callback，这个回调类是 HarmonyOS 的原生类 SurfaceOps. Callback。

若要将 SurfaceOps. Callback 实例化，则需要重写其中的 surfaceCreated(SurfaceOps) 方法、surfaceChanged(SurfaceOps,int,int,int)方法和 surfaceDestroyed(SurfaceOps)方法，这 3 种方法都是回调函数，顾名思义 surfaceCreated(SurfaceOps)就是当 Surface 被创建时会被调用，surfaceChanged(SurfaceOps,int,int,int)是当 Surface 被改变(例如大小发生了变化)时被调用，而 surfaceDestroyed(SurfaceOps)则是当 Surface 被删除时才会被调用。工程中自定义回调类的代码如下：

```
private SurfaceOps.Callback callback = new SurfaceOps.Callback() {
    @Override
    public void surfaceCreated(SurfaceOps surfaceOps) {
        //将 SurfaceProvider 中的 Surface 与 previewsurface 建立连接
    previewsurface = surfaceOps.getSurface();
    //初始化相机
    openCamera();
    }

    @Override
    public void surfaceChanged(SurfaceOps surfaceOps, int i, int i1, int i2) {
        }

    @Override
    public void surfaceDestroyed(SurfaceOps surfaceOps) {
    }
};
```

简单起见，不考虑 Surface 发生变化和需要删除的情况，所以只需重写 surfaceCreated (SurfaceOps)函数。在 surfaceCreated(SurfaceOps)中一共只有两行代码，在第一行代码中，SurfaceOps 的 Surface 实例与 previewSurface 指向同一存储地址，后续 previewSurface 从相机获取帧数据时，SurfaceOps 中的 Surface 也可以得到同样的数据，这样相机所拍摄到的画面就可以显示在 SurfaceProvider 上。surfaceCreated(SurfaceOps)函数的第二行调用了 openCamera()函数，用于初始化相机，下面看一下 openCamera()的代码，其代码如下：

```
//openCamera()初始化相机
private void openCamera(){
    //获取 CameraKit 对象
    cameraKit = CameraKit.getInstance(this);
    if (cameraKit == null) {
        return;
    }
    try {
    //获取当前设备的逻辑相机列表 cameraIds
    String[] cameraIds = cameraKit.getCameraIds();
    if (cameraIds.length <= 0) {
                System.out.println("cameraIds size is 0");
    }
    //用于相机创建和相机运行的回调
    CameraStateCallbackImpl cameraStateCallback = new CameraStateCallbackImpl();
    if(cameraStateCallback == null) {
                System.out.println("cameraStateCallback is null");
        }
    //创建用于运行相机回调的线程
    EventHandler eventHandler = new EventHandler(EventRunner.create("CameraCb"));
    if(eventHandler == null) {
                System.out.println("eventHandler is null");
        }
        //初始化相机
        cameraKit.createCamera(cameraIds[0], cameraStateCallback, eventHandler);
    } catch (IllegalStateException e) {
        System.out.println("getCameraIds fail");
    }
}
```

其实上述代码即为第 9 章中的创建相机的过程，不过这里注意 CameraStateCallbackImpl 这个类，该类继承自 CameraStateCallback，重写了其中的方法来达到提取帧数据的目的。CameraStateCallbackImpl 的实现代码如下：

```
//CameraStateCallbackImpl 提取帧数据
private final class CameraStateCallbackImpl extends CameraStateCallback {
    //相机回调
    @Override
    public void onCreated(Camera camera) {
    //相机创建时回调
    CameraConfig.Builder cameraConfigBuilder = camera.getCameraConfigBuilder();
    if (cameraConfigBuilder == null) {
System.out.println("onCreated cameraConfigBuilder is null");
return;
    }
```

```
    //配置预览的 Surface
    cameraConfigBuilder.addSurface(previewSurface);
    //配置拍照的 Surface
    dataSurface = imageReceiver.getRecevingSurface();
    cameraConfigBuilder.addSurface(dataSurface);
    try {
    //相机设备配置
    camera.configure(cameraConfigBuilder.build());
    } catch (IllegalArgumentException e) {
    System.out.println("Argument Exception");
    } catch (IllegalStateException e) {
    System.out.println("State Exception");
    }
    }
    @Override
    public void onConfigured(Camera camera) {
    //相机配置
     FrameConfig.Builder frameConfigBuilder = camera.getFrameConfigBuilder(FRAME_CONFIG_
PREVIEW);
    //配置预览 Surface
    frameConfigBuilder.addSurface(previewSurface);
    //配置拍照的 Surface
    frameConfigBuilder.addSurface(dataSurface);
    try {
    //启动循环帧捕获
int triggerId = camera.triggerLoopingCapture(frameConfigBuilder.build());
    } catch (IllegalArgumentException e) {
    System.out.println("Argument Exception");
    } catch (IllegalStateException e) {
    System.out.println("State Exception");
    }
    }
    @Override
    public void onReleased(Camera camera) {
     //释放相机设备
     if (camera != null) {
    camera.release();
    camera = null;
        }
    }
}
```

总体上与第 9 章中的相机回调一样，为相机配置了一个预览 Surface 和一个拍照 Surface，预览 Surface 就是上文讲到的 previewSurface，不同的是在这里拍照 Surface 被设置为 ImageReceiver 内含的 Surface，这样利用 ImageReceiver 的能力就可以很方便地对

Surface 中的内容进行提取和处理，为后续将帧数据发送给编码器处理做准备，相机数据流向如图 11.4 所示，具体如何进行数据处理在后文中会讲到。

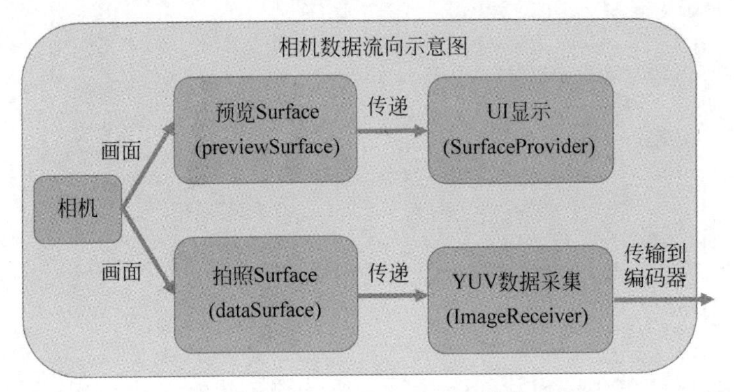

图 11.4　相机数据流向示意图

至此，主函数中运行的第 1 个函数 initUIandCamera()就运行完毕了，UI 和相机也完成了初始化。实现的效果为手机上显示一个含有 SurfaceProvider 和 Button 的 UI 界面，相机会源源不断地将拍摄到的画面帧数据传输到 previewSurface 和 ImageReceiver 内含的 Surface 上，而由于 previewSurface 和 SurfaceProvider 已经建立了联系，所以在 SurfaceProvider 上会显示相机拍摄到的画面。目前基本完成了一个相机预览效果，要实现对视频数据的编码，就需要利用 ImageReceiver 的功能。

下面来讲述主函数中的第 2 个函数 registerEncoder()，这个函数实例化了编码器，实现了 ImageReceiver 对图像数据的处理和传输至编码器，代码如下：

```
//实例化编码器
private void registerEncoder() {
    //实例化编码器
    vdEncoder = new VDEncoder(15);
    //创建 ImageReceiver 并配置监听器
    imageReceiver = ImageReceiver.create(VIDEO_WIDTH, VIDEO_HEIGHT, ImageFormat.YUV420_
888, 10);
    IImageArrivalListenerImpl listener = new IImageArrivalListenerImpl();
    imageReceiver.setImageArrivalListener(listener);
    }
```

首先在 registerEncoder()方法中实例化了一个 VDEncoder 类，这个类归属于 utils 包，里面包含了对视频进行编码的方法，在后续的章节中会讲到它的具体实现。后续会有关于 ImageReceiver 的创建与配置，通过 create(int, int, int, int)方法可以接收图像的宽度、高度、格式和最大可接收的图像数目来创建 ImageReceiver 实例，随后需要为 ImageReceiver 实例配置一个监听器 ImageArrivalListener，这个监听器会在 ImageReceiver 接收到图像数据时触发 onImageArrival(ImageReceiver)方法，重写 onImageArrival(ImageReceiver)即可实现对相机帧数据的处理和后续将其传输到编码器。重写的方法代码如下：

```
//监听器,当有数据传入 ImageReceiver 时触发
private class IImageArrivalListenerImpl implements
        ImageReceiver.IImageArrivalListener {
    //对监听事件的响应逻辑,实现对图像数据处理和向编码器的传输
    @Override
    public void onImageArrival(ImageReceiver imageReceiver) {
        mLog.log("imagearival", "arrival");
        Image mImage = imageReceiver.readNextImage();
        if (mImage != null) {
            BufferInfo bufferInfo = new BufferInfo();
            ByteBuffer mBuffer;
            Byte[] YUV_DATA = new Byte[VIDEO_HEIGHT * VIDEO_WIDTH * 3 / 2];
            int i;
            //采集 YUV 格式数据
            mBuffer = mImage.getComponent(ImageFormat.ComponentType.YUV_Y).getBuffer();
            for (i = 0; i < VIDEO_WIDTH * VIDEO_HEIGHT; i++) {
                YUV_DATA[i] = mBuffer.get(i);
            }
            mBuffer = mImage.getComponent(ImageFormat.ComponentType.YUV_V).getBuffer();
            for (i = 0; i < VIDEO_WIDTH * VIDEO_HEIGHT / 4; i++) {
                YUV_DATA[(VIDEO_WIDTH * VIDEO_HEIGHT) + i * 2] = mBuffer.get(i * 2);
            }
            mBuffer = mImage.getComponent(ImageFormat.ComponentType.YUV_U).getBuffer();
            for (i = 0; i < VIDEO_WIDTH * VIDEO_HEIGHT / 4; i++) {
                YUV_DATA[(VIDEO_WIDTH * VIDEO_HEIGHT) + i * 2 + 1] = mBuffer.get(i * 2);
    }
            bufferInfo.setInfo(0, VIDEO_WIDTH * VIDEO_HEIGHT * 3 / 2, mImage.getTimestamp(), 0);
            //将 YUV 数据传入 Encoder
            vdEncoder.addFrame(YUV_DATA);
            mImage.release();
            return;
        }
    }
}
```

ImageReceiver 提供了 readNextImage()方法，此方法可以将其接收的一个 Surface 整体读取为一个 Image。Image 可以被采集为各种各样的格式，例如 YUV、RAW10、JPEG 等，在这里选择采样为 YUV 色彩格式，所以使用 Image 的 getComponent(ComponentType)方法，设置参数为 ImageFormat. ComponentType. YUV_Y 用于采集每个像素的明亮度、ImageFormat. ComponentType. YUV_U 用于采集每个像素的色度、ImageFormat. ComponentType. YUV_V 用于采集每个像素的浓度。为了减少数据量从而方便传输，采用 YUV420 的色彩格式，即对每 4 个像素的色度和浓度保留一次采样数据。完成了对相机帧画面的 YUV 采集后，将其

数据存入数组中，使用编码器的 addFrame(Bytes[])方法将其传输到编码器中，准备进行编码，有关编码器相关的方法，将在后续的章节中介绍。

至此，主函数中的第 2 个函数 registerEncoder()就运行完毕了。实现的效果为实例化了编码器 VDEncoder，将相机每一帧的画面进行了 YUV420 色彩格式的采集，并将其传输到编码器中，准备进行编码处理。

最后讲解一下主函数中的第 3 个函数 registerRTP()。这个函数调用了 rtp 包内的相关功能，实现了对 RTP 发送机的注册，这一步实现了较好的自动化封装，代码如下：

```
private void registerRTP(){
    mRtpSenderWrapper = new RtpSenderWrapper("192.168.31.185", 5004, false);
    vdEncoder.setRTPSender(mRtpSenderWrapper);
}
```

在这里首先使用局域网 IP 地址、端口号和是否广播 3 个参数构造 rtp 包内的 RtpSenderWrapper 类，IP 地址为直播信号发送目标的机器 IP 地址，也就是接收端的 IP，端口号要与接收端开放的端口号一致。这一步实质上是告诉 RTP 发送机要将数据发送到的 IP 和端口。接下来调用 VDEncoder 中的 setRTPSender(RtpSenderWrapper)方法，将 RTP 发送机传递给编码器，用于对每一帧数据编码完成后的传输。

至此，Sender 主方法全部讲解完毕。实现的效果为初始化了 UI 和相机，实现了相机帧的预览和 YUV 格式采集，实例化了 RTP 发送机，并将采集到的 YUV 帧数据和 RTP 发送机传递给了编码器 VDEncoder。后续的编码和传输工作将在 VDEncoder 类中进行讲解。

下面附上 Sender.java 的全部代码，其代码如下：

```
package com.huawei.mytestapp.slice;
//本工程类导入
import com.huawei.mytestapp.rtp.RtpSenderWrapper;
import com.huawei.mytestapp.utils.VDEncoder;
import com.huawei.mytestapp.utils.mLog;
//HarmonyOS 类导入
import ohos.aafwk.ability.AbilitySlice;
import ohos.aafwk.content.Intent;
import ohos.agp.colors.RgbColor;
import ohos.agp.components.Button;
import ohos.agp.components.Component;
import ohos.agp.components.DirectionalLayout;
import ohos.agp.components.DirectionalLayout.LayoutConfig;
import ohos.agp.components.element.ShapeElement;
import ohos.agp.components.surfaceprovider.SurfaceProvider;
import ohos.agp.graphics.Surface;
import ohos.agp.graphics.SurfaceOps;
import ohos.agp.utils.LayoutAlignment;
```

```
import ohos.eventhandler.EventHandler;
import ohos.eventhandler.EventRunner;
import ohos.media.camera.CameraKit;
import ohos.media.camera.device.*;
import ohos.media.common.BufferInfo;
import ohos.media.image.Image;
import ohos.media.image.ImageReceiver;
import ohos.media.image.common.ImageFormat;
import static ohos.media.camera.device.Camera.FrameConfigType.FRAME_CONFIG_PREVIEW;
//Java 类导入
import java.nio.ByteBuffer;

public class Sender extends AbilitySlice {
    private DirectionalLayout myLayout;
    private ImageReceiver imageReceiver;
    private CameraKit cameraKit;
    private SurfaceProvider surfaceProvider;
    private Surface previewSurface;
    private Surface dataSurface;

    public static final int VIDEO_WIDTH = 640;
    public static final int VIDEO_HEIGHT = 480;
    private VDEncoder vdEncoder;
    public RtpSenderWrapper mRtpSenderWrapper;
    private Button button;

    @Override
    public void onStart(Intent intent) {
        super.onStart(intent);
        //初始化 UI 和相机,实现视频帧的获取
        initUIandCamera();
        //注册编码器,实现视频帧的编码
        registerEncoder();
        //注册发送,实现编码数据的发送
        registerRTP();
    }
    private void initUIandCamera(){
        //布局容器
        myLayout = new DirectionalLayout(this);
        LayoutConfig config = new LayoutConfig(LayoutConfig.MATCH_PARENT, LayoutConfig.
MATCH_PARENT);
        myLayout.setLayoutConfig(config);
        myLayout.setOrientation(Component.VERTICAL);
        myLayout.setPadding(32,32,32,32);
        //SurfaceProvider,用于播放摄像头画面
        config.width = VIDEO_WIDTH;
```

```
        config.height = VIDEO_HEIGHT;
        config.alignment = LayoutAlignment.HORIZONTAL_CENTER;
        surfaceProvider = new SurfaceProvider(this);
        surfaceProvider.setLayoutConfig(config);
        surfaceProvider.getSurfaceOps().get().addCallback(callback);
        surfaceProvider.pinToZTop(true);
        //Button,单击后开始编码和发送
        config.height = LayoutConfig.MATCH_CONTENT;
        config.width = LayoutConfig.MATCH_CONTENT;
        config.setMargins(0,50,0,0);
        button = new Button(this);
        button.setLayoutConfig(config);
        button.setText("编码并发送");
        button.setTextSize(50);
        ShapeElement background = new ShapeElement();
        background.setRgbColor(new RgbColor(0xFF51A8DD));
        background.setCornerRadius(25);
        button.setBackground(background);
        button.setClickedListener(new Component.ClickedListener() {
            @Override
            public void onClick(Component component) {
                //为按钮添加功能,实现单击后开始编码并发送
                vdEncoder.start();
            }
        });
        //将组件添加到布局容器中,并将布局容器作为 UI 的根布局
        myLayout.addComponent(surfaceProvider);
        myLayout.addComponent(button);
        super.setUIContent(myLayout);
    }

    private void registerRTP(){
        mRtpSenderWrapper = new RtpSenderWrapper("192.168.31.185", 5004, false);
        vdEncoder.setRTPSender(mRtpSenderWrapper);
    }

    private void registerEncoder() {
        //实例化编码器
        vdEncoder = new VDEncoder(15);
        //创建 ImageReceiver 并配置监听器
        imageReceiver = ImageReceiver.create(VIDEO_WIDTH, VIDEO_HEIGHT, ImageFormat.YUV420_888,
10);
        IImageArrivalListenerImpl listener = new IImageArrivalListenerImpl();
        imageReceiver.setImageArrivalListener(listener);
    }
    //监听器,当有数据传入 ImageReceiver 时触发
```

```
    private class IImageArrivalListenerImpl implements ImageReceiver.IImageArrivalListener {
        //对监听事件的响应逻辑,实现对图像数据处理和向编码器的传输
        @Override
        public void onImageArrival(ImageReceiver imageReceiver) {
            mLog.log("imagearival", "arrival");
            if (mImage != null) {
                BufferInfo bufferInfo = new BufferInfo();
                ByteBuffer mBuffer;
                Byte[] YUV_DATA = new Byte[VIDEO_HEIGHT * VIDEO_WIDTH * 3 / 2];
                int i;
                //采集 YUV 格式数据
                mBuffer = mImage.getComponent(ImageFormat.ComponentType.YUV_Y).getBuffer();
                for (i = 0; i < VIDEO_WIDTH * VIDEO_HEIGHT; i++) {
                    YUV_DATA[i] = mBuffer.get(i);
                }
                mBuffer = mImage.getComponent(ImageFormat.ComponentType.YUV_V).getBuffer();
                for (i = 0; i < VIDEO_WIDTH * VIDEO_HEIGHT / 4; i++) {
                    YUV_DATA[(VIDEO_WIDTH * VIDEO_HEIGHT) + i * 2] = mBuffer.get(i * 2);
                }
                mBuffer = mImage.getComponent(ImageFormat.ComponentType.YUV_U).getBuffer();
                for (i = 0; i < VIDEO_WIDTH * VIDEO_HEIGHT / 4; i++) {
                YUV_DATA[(VIDEO_WIDTH * VIDEO_HEIGHT) + i * 2 + 1] = mBuffer.get(i * 2);
                }
                bufferInfo.setInfo(0, VIDEO_WIDTH * VIDEO_HEIGHT * 3 / 2, mImage.
getTimestamp(), 0);
                //将 YUV 数据传入 Encoder
                vdEncoder.addFrame(YUV_DATA);
                mImage.release();
                return;
            }
        }
    }

    private void openCamera(){
        //获取 CameraKit 对象
        cameraKit = CameraKit.getInstance(this);
        if (cameraKit == null) {
            return;
        }
        try {
            //获取当前设备的逻辑相机列表 cameraIds
            String[] cameraIds = cameraKit.getCameraIds();
            if (cameraIds.length <= 0) {
                System.out.println("cameraIds size is 0");
            }
            //用于相机创建和相机运行的回调
```

```
            CameraStateCallbackImpl cameraStateCallback = new CameraStateCallbackImpl();
            if(cameraStateCallback == null) {
                System.out.println("cameraStateCallback is null");
            }
            //创建用于运行相机的线程
            EventHandler eventHandler = new EventHandler(EventRunner.create("CameraCb"));
            if(eventHandler == null) {
                System.out.println("eventHandler is null");
            }
            //创建相机
            cameraKit.createCamera(cameraIds[0], cameraStateCallback, eventHandler);
        } catch (IllegalStateException e) {
            System.out.println("getCameraIds fail");
        }
    }

    private final class CameraStateCallbackImpl extends CameraStateCallback {
        //相机回调
        @Override
        public void onCreated(Camera camera) {
            //相机创建时回调
            CameraConfig.Builder cameraConfigBuilder = camera.getCameraConfigBuilder();
            if (cameraConfigBuilder == null) {
                System.out.println("onCreated cameraConfigBuilder is null");
                return;
            }
            //配置预览的 Surface
            cameraConfigBuilder.addSurface(previewSurface);
            //配置拍照的 Surface
            dataSurface = imageReceiver.getRecevingSurface();
            cameraConfigBuilder.addSurface(dataSurface);
            try {
                //相机设备配置
                camera.configure(cameraConfigBuilder.build());
            } catch (IllegalArgumentException e) {
                System.out.println("Argument Exception");
            } catch (IllegalStateException e) {
                System.out.println("State Exception");
            }
        }

        @Override
        public void onConfigured(Camera camera) {
        //相机配置
         FrameConfig.Builder frameConfigBuilder = camera.getFrameConfigBuilder(FRAME_CONFIG_
PREVIEW);
```

```
        //配置预览 Surface
        frameConfigBuilder.addSurface(previewSurface);
        //配置拍照的 Surface
        frameConfigBuilder.addSurface(dataSurface);
        try {
            //启动循环帧捕获
            int triggerId = camera.triggerLoopingCapture(frameConfigBuilder.build());
        } catch (IllegalArgumentException e) {
            System.out.println("Argument Exception");
        } catch (IllegalStateException e) {
            System.out.println("State Exception");
        }
    }

    @Override
    public void onReleased(Camera camera) {
        //释放相机设备
        if (camera != null) {
            camera.release();
            camera = null;
        }
    }

    private SurfaceOps.Callback callback = new SurfaceOps.Callback() {
        @Override
        public void surfaceCreated(SurfaceOps surfaceOps) {
            //将 SurfaceProvider 中的 Surface 与 previewsurface 建立连接
            previewSurface = surfaceOps.getSurface();
            //初始化相机
            openCamera();
        }

        @Override
        public void surfaceChanged(SurfaceOps surfaceOps, int i, int i1, int i2) {
        }

        @Override
        public void surfaceDestroyed(SurfaceOps surfaceOps) {
        }
    };

    @Override
    public void onActive() {
        super.onActive();
    }
```

```
    @Override
    public void onForeground(Intent intent) {
        super.onForeground(intent);
    }
}
```

11.1.3 发送端核心工具——VDEncoder

VDEncoder 实现了对 Sender.java 传输进来的 YUV 格式数据进行 H264 编码，并使用 Sender.java 传输进来的 RTPSenderWrapper 将 H264 格式数据进行发送。VDEncoder 的工作原理如图 11.5 所示。

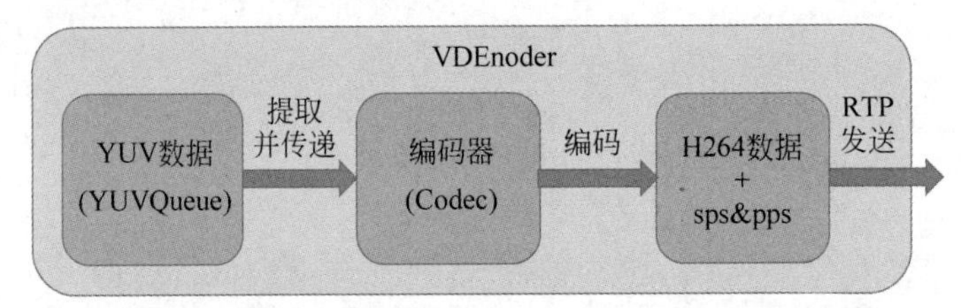

图 11.5 工具类 VDEncoder 工作原理示意图

正如 9.2.2 节所提到的，编码器的核心功能是由类 Codec 实现的，此处 VDEncoder 对 Codec 进行了个性化封装，额外实现了数据发送等功能。VDEncoder 构造器代码如下：

```
//VDEncoder 构造器
public VDEncoder(int framerate){
    //建立编码格式
    Format fmt = new Format();
    fmt.putStringValue("mime", "video/avc");
    fmt.putIntValue("width", Sender.VIDEO_WIDTH);
    fmt.putIntValue("height", Sender.VIDEO_HEIGHT);
    fmt.putIntValue("bitrate", 392000);
    fmt.putIntValue("color-format", 21);
    fmt.putIntValue("frame-rate", framerate);
    fmt.putIntValue("i-frame-interval", 1);
    fmt.putIntValue("bitrate-mode", 1);
    //实例化 Codec
    mCodec = Codec.createEncoder();
    //为 Codec 添加编码格式
    mCodec.setCodecFormat(fmt);
    //为 Codec 添加监听
    mCodec.registerCodecListener(encoderlistener);
    //Codec 启动
    mCodec.start();
    //实例化单线程池，用于执行编码工作，避免线程的频繁创建和销毁
    singleThreadExecutor = new SingleThreadExecutor();
}
```

可以看到，主要使用了9.2.2节中视频编码器Codec的构造方式，选择编码的格式为avc（H264），并设置编码窗口的宽和高及帧率等信息，在这里考虑到手机的性能和传输的带宽，帧率及宽和高设置需要低一点。

相比9.2.2节的代码，此处添加了两点，其一使用registerCodecListener（Codec.ICodecListener）为Codec添加了编码监听器，一旦有编码完成的数据进入Codec的ByteBuffer中，就会回调onReadBuffer（ByteBuffer，BufferInfo，int）函数，从回调参数中可以提取ByteBuffer中的数据，从而进行自定义处理。由于本工程实现视频流直播，这一步的作用就是将编码好的每一帧数据都进行即时发送。Codec.ICodecListener的代码如下：

```
//发送编码后的帧数据
private Codec.ICodecListener encoderlistener = new Codec.ICodecListener() {
    @Overrid
    public void onReadBuffer(ByteBuffer ByteBuffer, BufferInfo bufferInfo, int i) {
        int pos = 0;
        boolean nothing = false;
        Byte[] outdata = new Byte[bufferInfo.size];
        ByteBuffer.get(outdata);
        Byte[] output = new Byte[bufferInfo.size * 2];
        mLog.log("endata", "encoded data:" + outdata.length);

        //手动为H264数据添加pps和sps头
        if(m_info != null){
            System.arraycopy(outdata, 0, output, pos, outdata.length);
            pos += outdata.length;
        }else{
            //保存pps和sps,只有第一个帧里有,保存起来供后面使用
            ByteBuffer spsPpsBuffer = ByteBuffer.wrap(outdata);
            mLog.log("xmc", "swapYV12toI420:outData:" + outdata);
            mLog.log("xmc", "swapYV12toI420:spsPpsBuffer:" + spsPpsBuffer);
            for(int ii = 0;ii < outdata.length;ii++){
                mLog.log("xmc333", "run: get data rtpData[i] = " + ii + ":" + outdata[ii]);
                                                          //输出sps和pps循环
        }

        if (spsPpsBuffer.getInt() == 0x00000001) {
            m_info = new Byte[outdata.length];
            System.arraycopy(outdata, 0, m_info, 0, outdata.length);
        } else{
            nothing = true; }
        }
        //key frame 编码器生成关键帧时只有 00 00 00 01,65 没有 pps 和 sps,所以要加上
        if(output[4] == 0x65) {
            System.arraycopy(m_info, 0, output, 0, m_info.length);
            System.arraycopy(outdata, 0, output, m_info.length, outdata.length);
```

```
        }

        mLog.log("endata", "pos:" + pos);
        //实现 RTP 发送
        if (!nothing && output!= null) {
            mRtpSenderWrapper.sendAvcPacket(output, 0, pos, 0);
        }
    }

    @Override
    public void onError(int i, int i1, int i2) {
        throw new RunTimeException();
    }
};
```

这个函数的大部分代码都在为编码好的每一帧 H264 数据添加 pps 和 sps 信息，用于解码器的识别。其实本段最重要的功能是将编码好的 H264 数据通过 RtpSenderWrapper 的 sendAvcPacket(Byte[], int, int, int)方法进行发送，这种方法的 4 个参数分别为要发送的数据、偏移量(为了保证帧数据的完整性，置零即可)、发送数据的大小和使用时间(置零即可)。使用的 RtpSenderWrapper 为 11.1.2 节中 Sender 传过来的实例，setRTPSender(RtpSenderWrapper)方法代码如下：

```
public void setRTPSender(RtpSenderWrapper rtpSenderWrapper) {
    this.mRtpSenderWrapper = rtpSenderWrapper;
}
```

至此，实例化并设置好了编码器的编码参数，并且完成了编码后数据的传输逻辑。实现的效果：来自 Sender 的每一帧 YUV 数据传入编码器的 Codec 中被编码为 H264 格式数据，然后被 RtpSenderWrapper 发送出去。

11.1.2 节中给 Sender 配置了一个按钮 Button，这个按钮监听到单击事件后就会调用 VDEncoder 的 start()函数开始编码工作，start()函数的代码如下：

```
public void start() {
    //YUV 数据队列数据清除
    YUVQueue.clear();
    //解码器工作状态
    isRuning = true;
    //调用线程池开始进行编码
    startEncoderThread();
}
```

其中，YUVQueue 是一个 ArrayBlockingQueue < Byte[]>队列，Sender 通过 addFrame(Byte[])函数将 YUV 数据传输到这个队列中，等待编码。VDEncoder 工作时将从这个队

列中提取数据以便执行编码。由于 addFrame(Byte[])函数在相机完成初始化时就被执行了，而编码工作需要在 Button 被按下时才会开始，这个时间差会导致 YUVQueue 中积累了很多帧的数据，如果不在开始解码之前清除 YUVQueue 中的数据，会导致直播延迟，所以首先需要使用 clear()方法清除 YUV 队列的数据，再开始编码工作。

由于需要实现视频流的编码工作，所以必须创建一个线程用于不断地从 YUVQueue 中提取数据，这是比较占用资源的，而且为了避免线程的反复创建和销毁消耗系统资源，选择了线程池的办法为其进行线程配置。这也是相比 9.2.2 节添加的第二点，构造器中额外实例化了一个单线程池 SingleThreadExcutor，用于提取 YUVQueue 中的数据并发送给 Codec。单线程池 SingleThreadExcutor 归属 utils 包，将在后面进行简单介绍。start()函数中的 startEncoderThread()方法将调用这个单线程池执行数据的提取与发送，startEncoderThread()方法的代码如下：

```
//调用单线程池执行数据的提取与发送
private void startEncoderThread() {
    singleThreadExecutor.execute(new Runnable() {
        @Override
        public void run() {
            Byte[] outData;
            Byte[] data;
            //是否需要缩放
            ByteBuffer outputBuffer;
            int outputBufferIndex;
            BufferInfo bufferInfo = new BufferInfo();
            while (isRuning) {
                try {
                    data = YUVQueue.take();
                    System.out.println("aaa " + data.length);
                } catch (InterruptedException e) {
                    e.printStackTrace();
                    break;
                }
                ByteBuffer buffer = mCodec.getAvailableBuffer(-1);
                buffer.put(data);
                bufferInfo.setInfo(0, data.length, System.currentTimeMillis(), 0);
                mCodec.writeBuffer(buffer, bufferInfo);
            }
        }
    });
}
```

主要功能实现了在单线程池中对 YUV 队列使用 take()方法获取 YUV 数据，将其装入 Codec 的 Buffer 中，最后将 Buffer 发送给 Codec 执行编码。

至此，VDEncoder 的全部内容就完成了。实现的效果：从 YUVQueue 中获取 YUV 数据，将其传递给 Codec 进行 H264 编码，将编码完成后的数据使用 RtpSenderWrapper 进行发送。

另附 VDEncoder 的全部代码如下：

```
//VDEncoder.java
package com.huawei.mytestapp.utils;

import com.huawei.mytestapp.rtp.RtpSenderWrapper;
import com.huawei.mytestapp.slice.Sender;
import ohos.media.codec.Codec;
import ohos.media.common.BufferInfo;
import ohos.media.common.Format;

import java.nio.ByteBuffer;
import java.util.concurrent.ArrayBlockingQueue;

public class VDEncoder {
    Codec mCodec;
    public boolean isRuning = false;
    private ArrayBlockingQueue<Byte[]> YUVQueue = newArrayBlockingQueue<>(OtherUtil.
QueueNum);
    private SingleThreadExecutor singleThreadExecutor;
    Byte[] m_info = null;
    public RtpSenderWrapper mRtpSenderWrapper;
    public VDEncoder(int framerate){
        //建立编码格式
        Format fmt = new Format();
        fmt.putStringValue("mime", "video/avc");
        fmt.putIntValue("width", Sender.VIDEO_WIDTH);
        fmt.putIntValue("height", Sender.VIDEO_HEIGHT);
        fmt.putIntValue("bitrate", 392000);
        fmt.putIntValue("color-format", 21);
        fmt.putIntValue("frame-rate", framerate);
        fmt.putIntValue("i-frame-interval", 1);
        fmt.putIntValue("bitrate-mode", 1);
        //实例化 Codec
        mCodec = Codec.createEncoder();
        //为 Codec 添加编码格式
        mCodec.setCodecFormat(fmt);
        //为 Codec 添加监听
        mCodec.registerCodecListener(encoderlistener);
        //Codec 启动
        mCodec.start();
        //实例化单线程池,用于执行编码工作,避免线程的频繁创建和销毁
        singleThreadExecutor = new SingleThreadExecutor();
```

```
    }
    public void setRTPSender(RtpSenderWrapper rtpSenderWrapper) {
        this.mRtpSenderWrapper = rtpSenderWrapper;
    }

    private Codec.ICodecListener encoderlistener = new Codec.ICodecListener() {
        @Override
        public void onReadBuffer(ByteBuffer ByteBuffer, BufferInfo bufferInfo, int i) {
            int pos = 0;
            boolean nothing = false;
            Byte[] outdata = new Byte[bufferInfo.size];
            ByteBuffer.get(outdata);
            Byte[] output = new Byte[bufferInfo.size * 2];
            mLog.log("endata", "encoded data:" + outdata.length);

            //手动为 H264 数据添加 pps 和 sps 头
            if(m_info != null){
                System.arraycopy(outdata, 0, output, pos, outdata.length);
                pos += outdata.length;
            }else{
                //保存 pps 和 sps,只有第一个帧里有,保存起来供后面使用
                ByteBuffer spsPpsBuffer = ByteBuffer.wrap(outdata);
                mLog.log("xmc", "swapYV12toI420:outData:" + outdata);
                mLog.log("xmc", "swapYV12toI420:spsPpsBuffer:" + spsPpsBuffer);
                for(int ii = 0;ii < outdata.length;ii++){
                     mLog.log("xmc333", "run: get data rtpData[i] = " + ii + ":" + outdata
[ii]); //输出 pps 和 sps 循环
                }
                if (spsPpsBuffer.getInt() == 0x00000001) {
                    m_info = new Byte[outdata.length];
                    System.arraycopy(outdata, 0, m_info, 0, outdata.length);
                } else{
                    nothing = true;
                }
            }
            //key frame 编码器生成关键帧时只有 00 00 00 01,65 没有 pps 和 sps,所以要加上
            if(output[4] == 0x65) {
                System.arraycopy(m_info, 0, output, 0, m_info.length);
                System.arraycopy(outdata, 0, output, m_info.length, outdata.length);
            }
            mLog.log("endata", "pos:" + pos);

            //实现 RTP 发送
            if (!nothing && output!= null) {
                mRtpSenderWrapper.sendAvcPacket(output, 0, pos, 0);
            }
```

```
        }

        @Override
        public void onError(int i, int i1, int i2) {
            throw new RunTimeException();
        }
    };
    public void destroy() {
        isRuning = false;
        mCodec.release();
        mCodec = null;
        singleThreadExecutor.shutdownNow();
    }

    public void addFrame(Byte[] Bytes) {
        if (isRuning) {
            OtherUtil.addQueue(YUVQueue, Bytes);
        }
    }

    public void start() {
        //YUV 数据队列数据清除
        YUVQueue.clear();
        //解码器工作状态
        isRuning = true;
        //调用线程池开始进行编码
        startEncoderThread();
    }

    private void startEncoderThread() {
        singleThreadExecutor.execute(new Runnable() {
            @Override
            public void run() {
                Byte[] outData;
                Byte[] data;
                //是否需要缩放
                ByteBuffer outputBuffer;
                int outputBufferIndex;
                BufferInfo bufferInfo = new BufferInfo();
                while (isRuning) {
                    try {
                        data = YUVQueue.take();
                        System.out.println("aaa " + data.length);
                    } catch (InterruptedException e) {
                        e.printStackTrace();
                        break;
```

```
                    }
                    ByteBuffer buffer = mCodec.getAvailableBuffer(-1);
                    buffer.put(data);
                    bufferInfo.setInfo(0, data.length, System.currentTimeMillis(), 0);
                    mCodec.writeBuffer(buffer, bufferInfo);
                }
            }
        });
    }
}
```

11.1.4　发送端其他工具类

utils 包中还有 3 个工具类协助 Sender 和 VDEncoder 完成工作，分别是 mLog.java、OtherUtil.java 和 SingleThreadExecutor.java，下面简单对其进行介绍。

首先是 mLog.java 文件，代码如下：

```
//mLog.java
package com.huawei.mytestapp.utils;
import ohos.hiviewdfx.HiLog;
import ohos.hiviewdfx.HiLogLabel;

public class mLog {
    public static void log(String TAG, String content) {
        HiLogLabel LABEL_LOG = new HiLogLabel(3, 0xD001100, TAG);
        HiLog.info(LABEL_LOG,content);
    }
}
```

实质上就是一个对 HiLog 类进行的自动化封装，使得日志的打印更加方便。

其次是 OtherUtil.java 文件，代码如下：

```
//OtherUtil.java
package com.huawei.mytestapp.utils;
import java.util.concurrent.ArrayBlockingQueue;

public class OtherUtil {
    public static final int QueueNum = 300;
    public static <T> void addQueue(ArrayBlockingQueue<T> queue, T t) {
        if (queue.size() >= QueueNum) {
            queue.poll();
        }
        queue.offer(t);
    }
}
```

这个类提供了对 ArrayBlockingQueue 的支持，用于 YUV 数据队列，设置了 YUV 队列的最大长度，并且提供了向队列中添加数据的方法 addQueue(ArrayBlockingQueue<T>, T)。

最后是 SingleThreadExecutor.java 文件，代码如下：

```
//SingleThreadExecutor.java
package com.huawei.mytestapp.utils;
import java.util.concurrent.*;

/**
 * 自定义单线程池
 */
public class SingleThreadExecutor {
    private BlockingQueue<Runnable> queue;
    private ThreadPoolExecutor threadPoolExecutor;

    public SingleThreadExecutor() {
        queue = new ArrayBlockingQueue<>(2);
        threadPoolExecutor = new ThreadPoolExecutor(
            1,
            1,
            0,
            TimeUnit.SECONDS,
            queue,
            Executors.defaultThreadFactory(),
            new ThreadPoolExecutor.DiscardPolicy());
        }
    public void execute(Runnable runnable) {
        threadPoolExecutor.execute(runnable);
    }

    public void shutdownNow() {
        threadPoolExecutor.shutdownNow();
    }
    public int getQueueNum() {
        return queue.size();
    }

    public void clearQueue() {
        queue.clear();
    }
}
```

SingleThreadExecutor 类本质上是对 Java 线程池方法的封装，用于支持从 YUV 队列中提取数据的相关操作。

11.2　接收端

11.2.1　接收端工程结构

本工程的接收端主要需要实现对编码数据的接收、数据的解码及解码后视频的播放。具体工程结构如图 11.6 所示，由于对于 UDP 传输数据的接收只需原生 Java 的 DatagramSocket 类便可以完成，所以不需要 rtp 包的支持，整个工程主要分为 3 个部分：

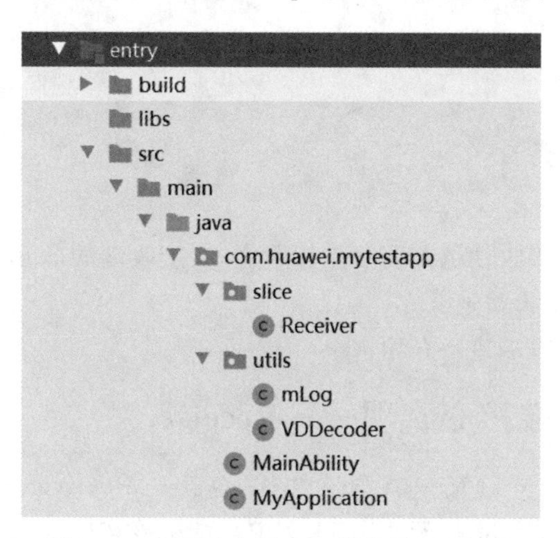

图 11.6　视频流直播 App 接收端工程结构

(1) slice 包。这个包是工程实现的核心，可以对其他包进行调用从而实现接收、解码、播放。其中 Receiver 是在接收端上运行的 PageAbility。

(2) utils 包。实现了视频解码并提供其他的 API，以供 Receiver 调用。

(3) 最外层的 MainAbility。作为应用程序的入口，实现了对接收端 PageAbility 的路由。

由于接收端需要使用 DatagramSocket 方法执行数据的接收，所以需要申请一些权限，例如网络使用权限 ohos. permission. INTERNET 等。权限申请需要在 entry 模块的 config. json 中进行，具体需要在 config. json 的 module 字段内部添加 reqPermissions 字段，代码如下：

```
"reqPermissions": [
    {
    "name": "ohos.permission.INTERNET",
    "reason": "reason",
        "usedScene": {
            "ability": [
                "com.huawei.mytestapp"
```

```
            ],
            "when": "always"
        }
    },
    {
        "name": "ohos.permission.DISTRIBUTED_DATASYNC",
        "reason": "reason",
        "usedScene": {
            "ability": [
                "com.huawei.mytestapp"
            ],
            "when": "always"
        }
    }
]
```

相同地，为了美观，接收端也需要摘除标题栏，在接收端的 config. json 中的相同位置加入同一字段即可，这里不再赘述。

下面正式开始对接收端进行介绍。

11.2.2 接收端核心实现——Receiver

发送端工程 slice 包下只有一个 Java 文件，也就是 Receiver. java，此文件内含数据接收、解码、播放这一流程的逻辑框架。图 11.7 中总结了该文件实现的主要功能。

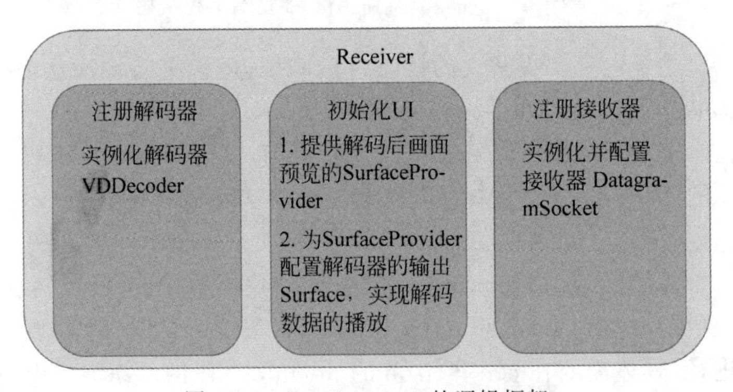

图 11.7　Receiver. java 的逻辑框架

Receiver 继承自 AbilitySlice 类，所以是一个 PageAbility，一旦这个 AbilitySlice 被创建，会自动调用其 onStart（Intent）方法，所以 onStart（Intent）一般可以视为主函数。Receiver 的 onStart(Intent)函数的代码如下：

```
public void onStart(Intent intent) {
    super.onStart(intent);
    //初始化 UI
```

```
        initUI();
        //实例化解码器
        registerDecoder();
        //在编码器中创建接收器
        registerSocket();
    }
```

和 11.1.2 节中的 Sender 类似，Receiver 的 onStart()函数只有 3 个步骤，分别是初始化 UI、实例化解码器和创建数据接收器。不过由于不涉及与相机和 RTP 发送相关的能力，这 3 个步骤相对较为简单，下面先讲解用于初始化 UI 的函数 initUI()，代码如下：

```
    //initUI()初始化 UI
    private void initUI(){
        //删除导航栏等
        Window window = this.getWindow();
        window.setStatusBarVisibility(Component.INVISIBLE);
        window.setNavigationBarColor(Color.TRANSPARENT.getValue());
        //布局容器
        myLayout = new DirectionalLayout(this);
        DirectionalLayout.LayoutConfig config = newDirectionalLayout.LayoutConfig(DirectionalLayout.
    LayoutConfig.MATCH_PARENT, DirectionalLayout.LayoutConfig.MATCH_PARENT);
        myLayout.setLayoutConfig(config);
        myLayout.setOrientation(Component.HORIZONTAL);
        ShapeElement background = new ShapeElement();
        background.setRgbColor(new RgbColor(150,150,150));
        myLayout.setBackground(background);
        //SurfaceProvider,用于播放摄像头画面
        config.width = VIDEO_WIDTH;
        config.height = VIDEO_HEIGHT;
        config.setMargins(50,0,0,0);
        config.alignment = LayoutAlignment.VERTICAL_CENTER;
        surfaceProvider = new SurfaceProvider(this);
        surfaceProvider.setLayoutConfig(config);
        surfaceProvider.getSurfaceOps().get().addCallback(callback);
        surfaceProvider.pinToZTop(true);
        //Button,单击后开始编码和发送
        config.height = DirectionalLayout.LayoutConfig.MATCH_CONTENT;
        config.width = DirectionalLayout.LayoutConfig.MATCH_CONTENT;
        config.setMargins(300,0,0,0);
        button = new Button(this);
        button.setLayoutConfig(config);
        button.setText("接收并解码");
        button.setTextSize(50);
        ShapeElement buttonBackground = new ShapeElement();
```

```
        buttonBackground.setRgbColor(new RgbColor(0xFF51A8DD));
        buttonBackground.setCornerRadius(25);
        button.setBackground(buttonBackground);
        button.setClickedListener(new Component.ClickedListener() {
            @Override
            public void onClick(Component component) {
                //为按钮添加功能,实现单击后开始编码并发送
                vdDecoder.start();
                //实现单击后改变状态显示
                text.setText("状态: 已启动接收");
                text.setTextColor(Color.GREEN);
            }
        });
        //Text,用于显示状态
        config.setMargins(-300,500,0,0);
        text = new Text(this);
        text.setText("状态: 未启动接收");
        text.setTextColor(Color.RED);
        text.setLayoutConfig(config);
        text.setTextSize(50);
        text.setMultipleLine(true);
        //将组件添加到布局容器中,并将布局容器作为UI的根布局
        myLayout.addComponent(surfaceProvider);
        myLayout.addComponent(button);
        myLayout.addComponent(text);
        super.setUIContent(myLayout);
    }
```

UI方面的整体结构与Sender的整体结构一致,首先创建了一个线性布局容器DirectionalLayout,然后设置了一个用于视频播放的SurfaceProvider和一个用于启动接收的按钮Button,此外还有一个用于显示状态的Text组件。系统自带的导航栏和状态栏仿照发送端进行处理。SurfaceProvider中的Surface通过回调函数与解码器的输出Surface建立联系,实现对解码后数据的播放。回调类SurfaceOps.Callback的代码如下:

```
//回调类SurfaceOps.Callback
private SurfaceOps.Callback callback = new SurfaceOps.Callback(){
    @Override
    public void surfaceCreated(SurfaceOps holder) {
        vdDecoder.isSurfaceCreated = true;
        vdDecoder.mSurface = holder.getSurface();
        vdDecoder.beginCodec();
        mLog.log("surfacecreated", "surfaceCreated!!!!!!");
    }

    @Override
```

```
    public void surfaceChanged(SurfaceOps holder, int format, int width, int height) {
    }

    @Override
    public void surfaceDestroyed(SurfaceOps holder) {
        vdDecoder.isSurfaceCreated = false;
        if (vdDecoder.isMediaCodecInit) {
            vdDecoder.mCodec.stop();
            vdDecoder.isMediaCodecInit = false;
        }
    }
};
```

应当注意到在回调函数 surfaceCreated(SurfaceOps)中，首先将解码器 VDDncoder 的布尔值 isSurfaceCreated 设置为 true，然后调用 beginCodec()函数以便启动解码器，这是一个判断机制，可以使编码器在 SurfaceProvider 初始化完毕后再向 Surface 传输数据，否则可能会导致空指针错误。关于 beginCodec()的逻辑，将在下一节中讲解。

至此，UI 初始化完毕，在 UI 初始化完毕后通过回调实现了解码器的激活。接下来介绍用于注册解码器的函数 registerDecoder()，代码如下：

```
private void registerDecoder(){
    //实例化解码器
    vdDecoder = new VDDecoder();
}
```

实际上调用了 VDDecoder 的构造器，实例化了一个解码器 VDDcoder。最后一个函数 registerSocket()的代码如下：

```
private void registerSocket(){
    //在编码器中创建接收器
    try {
        //创建接收器
        vdDecoder.socket = new DatagramSocket(5004);                    //端口号
        vdDecoder.socket.setReuseAddress(true);
        vdDecoder.socket.setBroadcast(true);
    } catch (SocketException e) {
        e.printStackTrace();
    }
}
```

这个函数在编码器中实例化了一个 DatagramSocket，用于发送端数据的接收，需要注意的是构造器参数为端口号，需要和发送端填写的端口号一致。

至此，完成了解码器的实例化并在解码器中建立了接收器，结合上述的 UI 部分，已经可以实现对数据的接收、解码和播放。有关接收和解码的实现，将在下一节讲述。

Receiver 的全部代码如下：

```
//Receiver.java
package com.huawei.mytestapp.slice;

import com.huawei.mytestapp.utils.VDDecoder;
import com.huawei.mytestapp.utils.mLog;
import ohos.aafwk.ability.AbilitySlice;
import ohos.aafwk.content.Intent;
import ohos.agp.colors.RgbColor;
import ohos.agp.components.Button;
import ohos.agp.components.Component;
import ohos.agp.components.DirectionalLayout;
import ohos.agp.components.element.ShapeElement;
import ohos.agp.components.surfaceprovider.SurfaceProvider;
import ohos.agp.graphics.SurfaceOps;
import ohos.agp.utils.LayoutAlignment;

import java.net.DatagramSocket;
import java.net.SocketException;

public class Receiver extends AbilitySlice {
    private DirectionalLayout myLayout = new DirectionalLayout(this);
    private SurfaceProvider surfaceProvider;
    private VDDecoder vdDecoder;
    private Button button;
    public static final int VIDEO_WIDTH = 640;
    public static final int VIDEO_HEIGHT = 480;

    @Override
    public void onStart(Intent intent) {
        super.onStart(intent);
        String[] permission = {"ohos.permission.INTERNET", "ohos.permission.DISTRIBUTED_
DATASYNC"};
        for (int i = 0;i < permission.length;i++){
            if(verifyCallingOrSelfPermission(permission[i]) != 0){
                if(canRequestPermission(permission[i])){
                    requestPermissionsFromUser(permission, 0);
                }
            }
        }
        //实例化解码器
        registerDecoder();
        //初始化 UI
```

```
        initUI();
        //在编码器中创建接收器
        registerSocket();
    }

    private void registerSocket(){
        //在编码器中创建接收器
        try {
            //创建接收器
            vdDecoder.socket = new DatagramSocket(5004); //端口号
            vdDecoder.socket.setReuseAddress(true);
            vdDecoder.socket.setBroadcast(true);
        } catch (SocketException e) {
            e.printStackTrace();
        }
    }

    private void registerDecoder(){
        //实例化解码器
        vdDecoder = new VDDecoder();
    }

    @Override
    public void onActive() {
        super.onActive();
    }

    @Override
    public void onForeground(Intent intent) {
        super.onForeground(intent);
    }

    private void initUI(){
        //布局容器
        DirectionalLayout.LayoutConfig config = newDirectionalLayout.LayoutConfig
(DirectionalLayout.LayoutConfig.MATCH_PARENT, DirectionalLayout.LayoutConfig.MATCH_
PARENT);
        myLayout.setLayoutConfig(config);
        myLayout.setOrientation(Component.VERTICAL);
        myLayout.setPadding(32,32,32,32);
        //SurfaceProvider,用于播放摄像头画面
        config.width = VIDEO_WIDTH;
        config.height = VIDEO_HEIGHT;
        config.alignment = LayoutAlignment.HORIZONTAL_CENTER;
        surfaceProvider = new SurfaceProvider(this);
        surfaceProvider.setLayoutConfig(config);
```

```
        surfaceProvider.getSurfaceOps().get().addCallback(callback);
        surfaceProvider.pinToZTop(true);
        //Button,单击开始后编码和发送
        config.height = DirectionalLayout.LayoutConfig.MATCH_CONTENT;
        config.width = DirectionalLayout.LayoutConfig.MATCH_CONTENT;
        config.setMargins(0,50,0,0);
        button = new Button(this);
        button.setLayoutConfig(config);
        button.setText("解码并显示");
        button.setTextSize(50);
        ShapeElement background = new ShapeElement();
        background.setRgbColor(new RgbColor(0xFF51A8DD));
        background.setCornerRadius(25);
        button.setBackground(background);
        button.setClickedListener(new Component.ClickedListener() {
            @Override
            public void onClick(Component component) {
                //为按钮添加功能,单击按钮后开始接收信号并解码
                vdDecoder.start();
            }
        });
        //将组件添加到布局容器中,并将布局容器作为 UI 的根布局
        myLayout.addComponent(surfaceProvider);
        myLayout.addComponent(button);
        super.setUIContent(myLayout);
    }

    private SurfaceOps.Callback callback = new SurfaceOps.Callback(){
        @Override
        public void surfaceCreated(SurfaceOps holder) {
            vdDecoder.isSurfaceCreated = true;
            vdDecoder.mSurface = holder.getSurface();
            vdDecoder.beginCodec();
            mLog.log("surfacecreated", "surfaceCreated!!!!!!");
        }

        @Override
        public void surfaceChanged(SurfaceOps holder, int format, int width, int height) {
        }

        @Override
        public void surfaceDestroyed(SurfaceOps holder) {
            vdDecoder.isSurfaceCreated = false;
            if (vdDecoder.isMediaCodecStart) {
                vdDecoder.mCodec.stop();
                vdDecoder.isMediaCodecStart = false;
            }
        }
    };
}
```

11.2.3 接收端核心工具——VDDecoder

VDDecoder 实现了对 Sender 发送的 H264 数据进行接收和解码，并将解码完成的数据传给 Receiver 的 SurfaceProvider 进行播放。与编码器一样，解码工作也是基于类 Codec 实现的，VDDecoder 对 Codec 进行了个性化封装，额外实现了数据接收等功能。VDDecoder 的工作原理如图 11.8 所示。

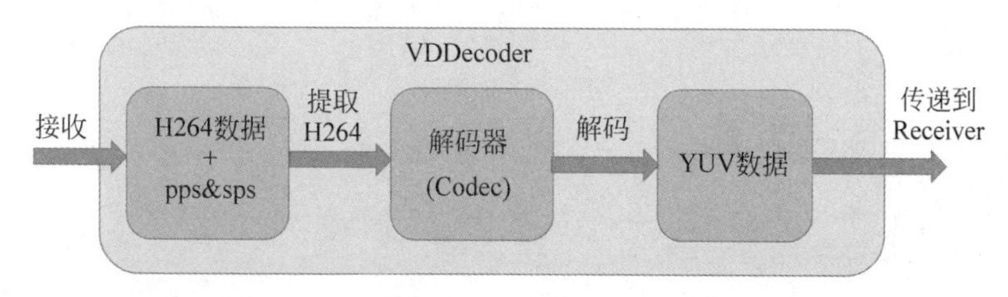

图 11.8 工具类 VDDecoder 工作原理示意图

与 VDEncoder 类似，VDDecoder 的构造器实例化了 Codec 并设置了解码格式（虽然对于解码器来讲不是必要的），构造器的代码如下：

```
//VDDecoder 构造器
public VDDecoder(){
    //解码格式
    Format fmt = new Format();
    fmt.putStringValue("mime", "video/avc");
    fmt.putIntValue("width", Receiver.VIDEO_WIDTH);
    fmt.putIntValue("height", Receiver.VIDEO_HEIGHT);
    fmt.putIntValue("bitrate", 392000);
    fmt.putIntValue("color-format", 21);
    fmt.putIntValue("frame-rate", 30);
    fmt.putIntValue("i-frame-interval", -1);
    fmt.putIntValue("bitrate-mode", 1);
    //Codec
    mCodec = Codec.createDecoder();
    mCodec.setCodecFormat(fmt);
    mCodec.registerCodecListener(decoderlistener);
}
```

值得注意的是，与编码器不同，没有直接调用 Codec 的 start()方法启动解码器，这是为了等待 SurfaceProvider 完成初始化。

在构造器中还为 Codec 添加了解码监听器，一旦有完成解码的数据进入 ByteBuffer，监听器就会回调 onReadBuffer(ByteBuffer, BufferInfo, int)函数，在回调函数中主要通过回调参数 ByteBuffer 获取解码后的数据，然后将这些数据传递给 Receiver 中的 SurfaceProvider。监听器的代码如下：

```
//解码监听器
private Codec.ICodecListener decoderlistener = new Codec.ICodecListener() {
    @Override
    public void onReadBuffer(ByteBuffer ByteBuffer, BufferInfo bufferInfo, int i) {
        Byte[] Bytes = new Byte[bufferInfo.size];
        Byte[] rotate_Bytes = new Byte[bufferInfo.size];
        Byte[] NV21_Bytes = new Byte[bufferInfo.size];
        Byte[] U_data = new Byte[bufferInfo.size / 6];
        Byte[] V_data = new Byte[bufferInfo.size / 6];
        int j;
        ByteBuffer.get(Bytes);
        mLog.log("Bytes", "Byte:" + Bytes);
        mSurface.showRawImage(Bytes, Surface.PixelFormat.PIXEL_FORMAT_YCRCB_420_SP, Receiver.
VIDEO_WIDTH, Receiver.VIDEO_HEIGHT);
    }

    @Override
    public void onError(int i, int i1, int i2) {
    }
};
```

在 Recevier 的 SurfaceProvider 完成初始化后，会回调 VDDecoder 中的 beginCodec() 方法，执行解码器的启动，beginCodec()的代码如下：

```
public synchronized void beginCodec() {
    System.out.println("isSurfaceCreated = " + Boolean.toString(isSurfaceCreated));
    if (isSurfaceCreated) {
    mCodec.start();
    isMediaCodecStart = true;
    }
}
```

可以看到，需要先对 Receiver 中 SurfaceProvider 的状态做出一个判断，初始化成功才会调用 Codec 的 start() 方法执行解码器的启动，解码器启动后还会将布尔值 isMediaCodecStart 置为 true，说明此时解码器已经可以输入数据并执行解码了。

当 Receiver 的按钮 Button 被按下时，会调用 VDDecoder 中的 start()方法，开始接收发送端数据，然后将数据传递给 Codec 执行解码。对于数据的接收，这里使用的是 Java 自带的 DatagramSocket 类，对于它的配置方法已在 11.2.1 节中进行了说明，此处不再赘述。下面主要介绍接收和解码的逻辑，start()的代码如下：

```
//接收发送端数据,并传递给 Codec 解码
public void start(){
    //一个新的线程
    new Thread(new Runnable() {
```

```
            @Override
            public void run() {
                Byte[] data = new Byte[80000];
                int h264Length = 0;
                //循环接收数据
                while (true){
                    mLog.log(TAG, "run: 正在接收数据!");
                    DatagramPacket datagramPacket = newDatagramPacket(data,data.length);
                    if (socket != null){
                        try {
                            datagramPacket = new DatagramPacket(data,data.length);
                            socket.receive(datagramPacket);   //用来接收数据
                        } catch (IOException e) {
                            e.printStackTrace();
                        }
                    }
                    rtpData = datagramPacket.getData();        //获取(已经发送的)数据
                    if (rtpData != null ){
                        mLog.log(TAG, "run: receiving message !!!");
                        if (rtpData[0] == -128 && rtpData[1] == 96){
                            int l1 = (rtpData[12]<< 24)& 0xff000000;
                            int l2 = (rtpData[13]<< 16)& 0x00ff0000;
                            int l3 = (rtpData[14]<< 8) & 0x0000ff00;
                            int l4 = rtpData[15]&0x000000FF;
                            mLog.log(TAG, "run:l1 == " + l1 + "l2 = " + l2 + "l3 == " + l3 + "
l4 == " + l4);
                            h264Length = l1 + l2 + l3 + l4;
                            mLog.log(TAG, "run: h264Length == " + h264Length);
                            System.arraycopy(rtpData,16, h264Data,0,h264Length);
                            //打印 sps、pps
                            mLog.log(TAG,"run:h264Data[0] = " + h264Data[0] + "," + h264Data[1] + ",
" + h264Data[2] + "," + h264Data[3] + "," + h264Data[4] + "," + h264Data[5] + "," + h264Data[6] + "," +
h264Data[7] + "," + h264Data[8] + "," + h264Data[9] + "," + h264Data[10] + "," + h264Data[11] + "," +
h264Data[12] + "," + h264Data[13] + "," + h264Data[14] + "," + h264Data[15] + "," + h264Data[16] + "," +
h264Data[17] + "," + h264Data[18] + "," + h264Data[19] + "," + h264Data[20] + "," + h264Data[21] + "," +
h264Data[22]);
                            if (isMediaCodecStart) {
                                //解码数据
                                decoding(h264Data);
                                mLog.log("receiveStatus","length:" + h264Data.length);
                            }
                        }
                    }else{
                        mLog.log(TAG, "run: null receiving message !!!");
                    }
                }
            }
        }).start();
    }
```

由于直播需要源源不断地提供视频信号，所以在这里需要新增一个线程用于数据的接收和解码，因此创建了一个新的 Thread，用于运行 Runnable，在 Runnable 中重写 run()方法，用于实现所需功能。

在 run()方法中，通过 DatagramSocket 使 DatagramPacket 接收数据，为了便于解码工作，还需要将 DatagramPacket 中的数据传入 Byte[]中。如上代码所示 Byte[]实例 rtpData 就是接收的 RTP 数据，由于在发送端给每一帧的 RTP 数据添加了 pps 和 sps 信息，在这里需要对 pps 和 sps 信息进行解析，获取 H264 视频画面的字节流长度 h264Length，然后根据 h264Length 截取 RTP 数据中真正用于解码的 H264 数据，这部分数据在本段代码中是一个 Byte[]，被称为 h264Data。

获取 h264Data 后，调用函数 decoding(Byte[])对其进行解码，decoding(Byte[])函数代码如下：

```
private void decoding(Byte[ ] video) {
    ByteBuffer mBuffer = mCodec.getAvailableBuffer( - 1);
    BufferInfo info = new BufferInfo();
    info.setInfo(0, video.length, 0, 0);
    mBuffer.put(video);
    mCodec.writeBuffer(mBuffer, info);
}
```

具体就是将 h264Data 写入 Codec 的 ByteBuffer 中，调用 writeBuffer(ByteBuffer, BufferInfo)执行解码。

至此，VDDecoder 实现了接收发送端数据并对其进行解码，将解码后的数据传送给 Receiver 的 SurfaceProvider 进行播放。

VDDecoder 的全部代码如下：

```
//VDDecoder.java
package com.huawei.mytestapp.utils;

import com.huawei.mytestapp.slice.Receiver;
import ohos.agp.graphics.Surface;
import ohos.media.codec.Codec;
import ohos.media.common.BufferInfo;
import ohos.media.common.Format;

import java.io.IOException;
import java.net.DatagramPacket;
import java.net.DatagramSocket;
import java.nio.ByteBuffer;

public class VDDecoder {
```

```
    //解码分辨率
    public Surface mSurface;
    //解码器
    public Codec mCodec;
    //是否播放
    public boolean isMediaCodecStart = false;
    public boolean isSurfaceCreated = false;
    //设置 TAG
    public static final String TAG = "VDDecoder" ;
    //存储数据
    Byte[] rtpData = new Byte[80000];        //长
    Byte[] h264Data = new Byte[80000];       //短
    //设置接收用 Socket
    public DatagramSocket socket;
    private Codec.ICodecListener decoderlistener = new Codec.ICodecListener() {
        @Override
        public void onReadBuffer(ByteBuffer ByteBuffer, BufferInfo bufferInfo, int i) {
            Byte[] Bytes = new Byte[bufferInfo.size];
            Byte[] rotate_Bytes = new Byte[bufferInfo.size];
            Byte[] NV21_Bytes = new Byte[bufferInfo.size];
            Byte[] U_data = new Byte[bufferInfo.size / 6];
            Byte[] V_data = new Byte[bufferInfo.size / 6];
            int j;
            ByteBuffer.get(Bytes);
            mLog.log("Bytes1", "Byte:" + Bytes);
            mSurface.showRawImage(Bytes, Surface.PixelFormat.PIXEL_FORMAT_YCRCB_420_SP,
Receiver.VIDEO_WIDTH, Receiver.VIDEO_HEIGHT);
        }

        @Override
        public void onError(int i, int i1, int i2) {
        }
    };

    public VDDecoder(){
        //解码格式
        Format fmt = new Format();
        fmt.putStringValue("mime", "video/avc");
        fmt.putIntValue("width", Receiver.VIDEO_WIDTH);
        fmt.putIntValue("height", Receiver.VIDEO_HEIGHT);
        fmt.putIntValue("bitrate", 2646000);
        fmt.putIntValue("color-format", 21);
        fmt.putIntValue("frame-rate", 30);
        fmt.putIntValue("i-frame-interval", -1);
        fmt.putIntValue("bitrate-mode", 1);
        //Codec
```

```
        mCodec = Codec.createDecoder();
        mCodec.setCodecFormat(fmt);
        mCodec.registerCodecListener(decoderlistener);
    }

    public synchronized void beginCodec() {
        System.out.println("isSurfaceCreated = " + Boolean.toString(isSurfaceCreated));
        if (isSurfaceCreated) {
            mCodec.start();
            isMediaCodecStart = true;
        }
    }

    public void start(){
        //一个新的线程
        new Thread(new Runnable() {
            @Override
            public void run() {
                Byte[] data = new Byte[80000];
                int h264Length = 0;
                //循环接收数据
                while (true){
                    mLog.log(TAG, "run: 正在接收数据!");
                    DatagramPacket datagramPacket = newDatagramPacket(data,data.length);
                    if (socket != null){
                        try {
                            datagramPacket = newDatagramPacket(data,data.length);
                            socket.receive(datagramPacket);   //用来接收数据
                        } catch (IOException e) {
                            e.printStackTrace();
                        }
                    }
                    rtpData = datagramPacket.getData();       //获取(已经发送的)数据
                    if (rtpData != null ){
                        mLog.log(TAG, "run: receiving message !!!");
                        if (rtpData[0] == -128 && rtpData[1] == 96){
                            int l1 = (rtpData[12]<< 24)& 0xff000000;
                            int l2 = (rtpData[13]<< 16)& 0x00ff0000;
                            int l3 = (rtpData[14]<< 8) & 0x0000ff00;
                            int l4 = rtpData[15]&0x000000FF;
                            mLog.log(TAG, "run:l1 == "+l1+"l2 = "+l2+"l3 == " + l3 + "
l4 == "+ l4);

                            h264Length = l1+l2+l3+l4;
                            mLog.log(TAG, "run: h264Length == "+h264Length);
                            System.arraycopy(rtpData,16,h264Data,0,h264Length);
                            //打印 sps、pps
```

```
                            mLog.log(TAG,"run:h264Data[0] = " + h264Data[0] + "," +
h264Data[1] + "," + h264Data[2] + "," + h264Data[3] + "," + h264Data[4] + "," + h264Data[5]
+ "," + h264Data[6] + "," + h264Data[7] + "," + h264Data[8] + "," + h264Data[9] + "," + h264Data
[10] + "," + h264Data[11] + "," + h264Data[12] + "," + h264Data[13] + "," + h264Data[14] + "," +
h264Data[15] + "," + h264Data[16] + "," + h264Data[17] + "," + h264Data[18] + "," + h264Data
[19] + "," + h264Data[20] + "," + h264Data[21] + "," + h264Data[22]);
                        if (isMediaCodecStart) {
                            //解码数据
                            decoding(h264Data);
                            mLog.log("receiveStatus", "length:" + h264Data.length);
                        }
                    }
                }else{
                    mLog.log(TAG, "run: null receiving message !!!");
                }
            }
        }
    }).start();
}

private void decoding(Byte[] video) {
    ByteBuffer mBuffer = mCodec.getAvailableBuffer(-1);
    BufferInfo info = new BufferInfo();
    info.setInfo(0, video.length, 0, 0);
    mBuffer.put(video);
    mCodec.writeBuffer(mBuffer, info);
}
}
```

11.2.4 接收端其他工具类

utils包中还有1个工具类用于协助Receiver和VDDecoder完成它们的工作，即mLog.java，具体实现代码如下：

```
package com.huawei.mytestapp.utils;
import ohos.hiviewdfx.HiLog;
import ohos.hiviewdfx.HiLogLabel;

public class mLog {
    public static void log(String TAG, String content) {
        HiLogLabel LABEL_LOG = new HiLogLabel(3, 0xD001100, TAG);
        HiLog.info(LABEL_LOG,content);
    }
}
```

与发送端相同，实质上就是一个对 HiLog 类进行的自动化封装，使得日志的打印更加方便，便于程序调试。

11.3 运行与效果

11.3.1 发送端运行

9.1 节中提到在 config.json 中添加了发送端所需的权限，不过这些权限需要用户在手机的设置→隐私→权限管理中逐项开启，较为麻烦。为方便开发者的调试和用户的使用，下面先介绍一种 App 自动申请权限的方法。

在发送端的 MainAblility 中，找到 onStart(Intent)函数，将其内容进行修改，代码如下：

```
@Override
public void onStart(Intent intent) {
    super.onStart(intent);
    //权限申请
    String[] permission = {"ohos.permission.INTERNET","ohos.permission.DISTRIBUTED_
DATASYNC","ohos.permission.CAMERA"};
    for (int i = 0;i < permission.length;i++){
        if(verifyCallingOrSelfPermission(permission[i]) != 0){
            if(canRequestPermission(permission[i])){
                requestPermissionsFromUser(permission, 0);
            }
        }
    }
    super.setMainRoute(Sender.class.getName());
}
```

首先定义了一个由三项权限名称组成的字符串数组，然后使用 requestPermissionsFromUser (String[], int)方法即可实现 App 在运行过程中自动申请这三项权限，用户仅需单击“始终允许”按钮即可，效果如图 11.9 所示。

由于视频流直播需要调用摄像头和网络传输等能力，所以需要在真机上进行测试。在运行之前需按照 1.4.3 节的内容进行证书的申请和配置，方能正常运行。

完成了证书配置后，将真机与计算机通过 USB 进行连接，再使用 DevEco 的 Run 工具，选择真机运行本工程的 entry，等待 gradle 配置完毕后即可在真机上看到效果。图 11.10 为 UI 初始化完毕后的效果，此时摄像头还未打开，所以 SurfaceProvider 没有画面显示，为黑色。

在完成了摄像头权限的申请后，打开安装的发送端 App，即可看到摄像头拍摄到的画面，如图 11.11 所示。

此时单击“编码并发送”按钮即可将摄像头显示的画面数据进行 H264 编码和 RTP 发送，效果如图 11.12 所示。

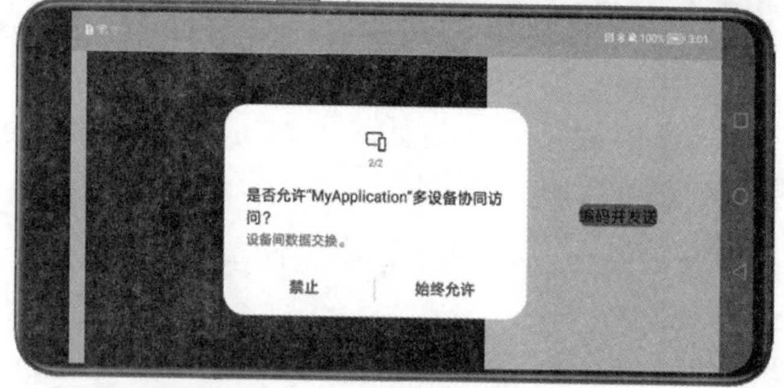

图 11.9 发送端 App 运行时自动申请权限效果图

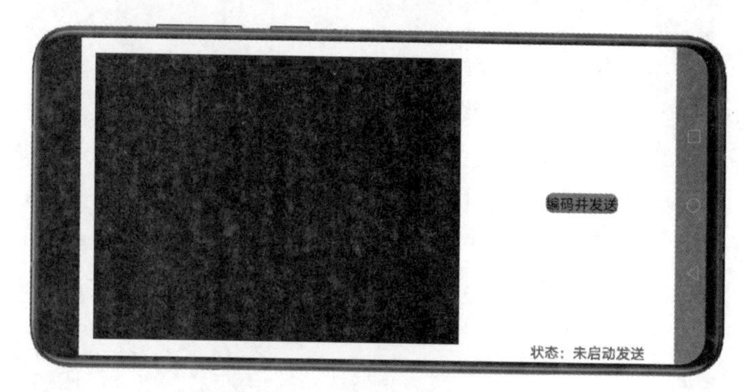

图 11.10 发送端的 UI 布局

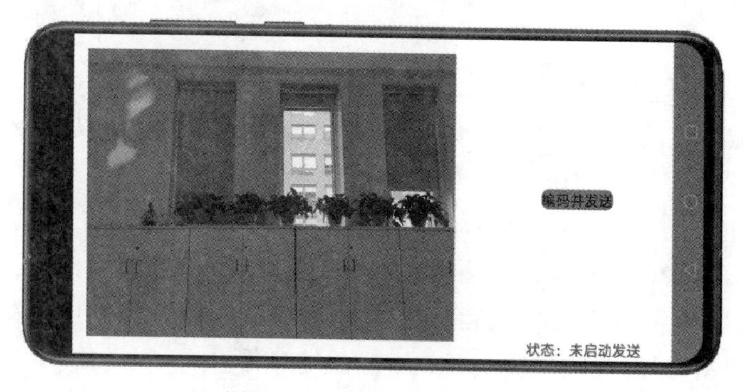

图 11.11 发送端效果展示

图 11.12 发送端进行编码发送效果展示

11.3.2 接收端运行

接收端的运行与发送端类似。首先为了方便可以使用相同的方法执行接收端的自动权限申请，代码如下：

```
@Override
public void onStart(Intent intent) {
    super.onStart(intent);
    //权限申请
    String[] permission = {"ohos.permission.INTERNET","ohos.permission.DISTRIBUTED_
DATASYNC"};
    for (int i = 0;i < permission.length;i++){
        if(verifyCallingOrSelfPermission(permission[i]) != 0){
            if(canRequestPermission(permission[i])){
                requestPermissionsFromUser(permission, 0);
            }
        }
    }
    super.setMainRoute(Receiver.class.getName());
}
```

运行后的权限申请画面如图 11.13 所示。

UI 初始化完毕后的效果，此时并没有开始接收并解码数据，所以 SurfaceProvider 是没有画面显示的，此时为黑色，如图 11.14 所示。

单击“接收并解码”按钮后，接收端会开始接收并解码数据，将解码后的视频传送到 SurfaceProvider 中，实现了视频流的播放，效果如图 11.15 所示。

需要注意的是，发送端与接收端运行的设备应处于统一局域网中，在发送端应当填写接收端的 IP 地址和端口号，接收端应当开放对应的端口号。

至此，本书关于 HarmonyOS 应用开发的内容就全部结束了。期待读者能够将本书的

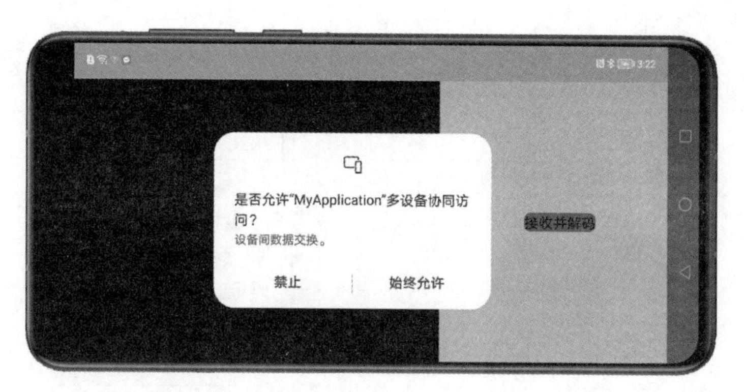

图 11.13　接收端 App 运行时自动权限申请

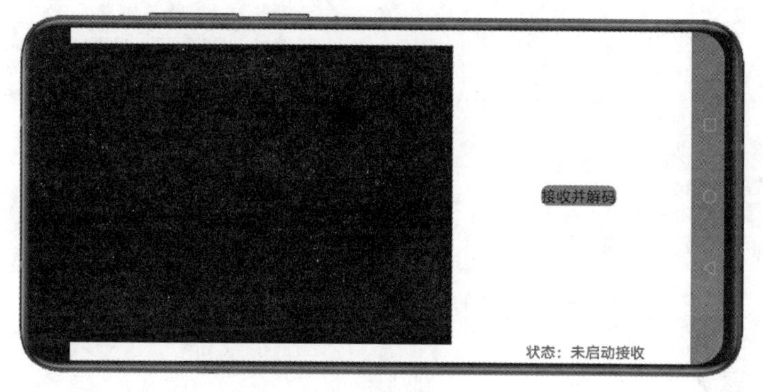

图 11.14　接收端 UI 展示

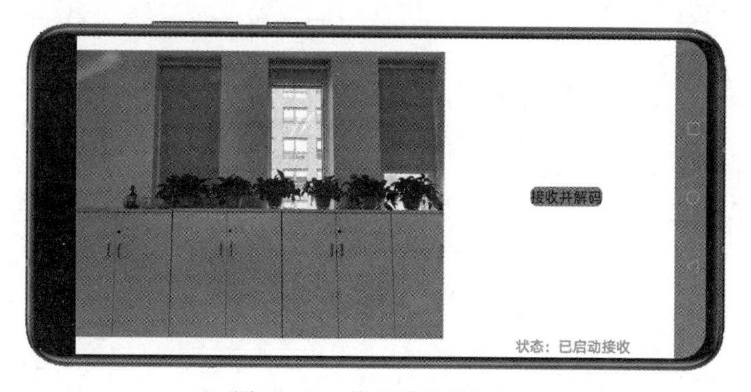

图 11.15　接收端效果展示

基础知识应用于更多全场景下的项目实战中，随着 HarmonyOS 的不断发展，期待越来越多的开发者加入 HarmonyOS！

图书推荐

书　　名	作　　者
鸿蒙应用程序开发	董昱
鸿蒙操作系统开发入门经典	徐礼文
华为方舟编译器之美——基于开源代码的架构分析与实现	史宁宁
鲲鹏架构入门与实战	张磊
华为 HCIA 路由与交换技术实战	江礼教
Flutter 组件精讲与实战	赵龙
Flutter 实战指南	李楠
Dart 语言实战——基于 Angular 框架的 Web 开发	刘仕文
Dart 语言实战——基于 Flutter 框架的程序开发	亢少军
IntelliJ IDEA 软件开发与应用	乔国辉
Vue＋Spring Boot 前后端分离开发实战	贾志杰
Vue. js 企业开发实战	千锋教育高教产品研发部
Python 人工智能——原理、实践及应用	杨博雄 主编，于营、肖衡、潘玉霞、高华玲、梁志勇 副主编
Python 深度学习	王志立
Python 异步编程实战——基于 AIO 的全栈开发技术	陈少佳
物联网——嵌入式开发实战	连志安
智慧建造——物联网在建筑设计与管理中的实践	[美]周晨光(Timothy Chou)著；段晨东、柯吉译
TensorFlow 计算机视觉原理与实战	欧阳鹏程、任浩然
分布式机器学习实战	陈敬雷
计算机视觉——基于 OpenCV 与 TensorFlow 的深度学习方法	余海林、翟中华
深度学习——理论、方法与 PyTorch 实践	翟中华、孟翔宇
深度学习原理与 PyTorch 实战	张伟振
ARKit 原生开发入门精粹——RealityKit ＋ Swift ＋ SwiftUI	汪祥春
Altium Designer 20 PCB 设计实战(视频微课版)	白军杰
Cadence 高速 PCB 设计——基于手机高阶板的案例分析与实现	李卫国、张彬、林超文
SolidWorks 2020 快速入门与深入实战	邵为龙
UG NX 1926 快速入门与深入实战	邵为龙
西门子 S7－200 SMART PLC 编程及应用(视频微课版)	徐宁、赵丽君
三菱 FX3U PLC 编程及应用(视频微课版)	吴文灵
全栈 UI 自动化测试实战	胡胜强、单镜石、李睿
软件测试与面试通识	于晶、张丹
深入理解微电子电路设计——电子元器件原理及应用(原书第 5 版)	[美] 理查德・C. 耶格(Richard C. Jaeger)，[美]特拉维斯・N. 布莱洛克(Travis N. Blalock)著；宋廷强　译
深入理解微电子电路设计——数字电子技术及应用(原书第 5 版)	[美]理查德・C. 耶格(Richard C. Jaeger) [美]特拉维斯・N. 布莱洛克(Travis N. Blalock) 著；宋廷强　译
深入理解微电子电路设计——模拟电子技术及应用(原书第 5 版)	[美]理查德・C. 耶格(Richard C. Jaeger) [美]特拉维斯・N. 布莱洛克(Travis N. Blalock) 著；宋廷强　译